Lecture Notes in Computer Science   8629

Commenced Publication in 1973
Founding and Former Series Editors:
Gerhard Goos, Juris Hartmanis, and Jan van Leeuwen

More information about this series at http://www.springer.com/series/7411

Miguel Garcia Pineda · Jaime Lloret
Symeon Papavassiliou · Stefan Ruehrup
Carlos Becker Westphall (Eds.)

Ad-hoc Networks
and Wireless

ADHOC-NOW 2014 International Workshops
ETSD, MARSS, MWaoN, SecAN, SSPA,
and WiSARN
Benidorm, Spain, June 22–27, 2014
Revised Selected Papers

 Springer

Editors
Miguel Garcia Pineda
Universitat de Valencia
Valencia
Spain

Jaime Lloret
Universitat Politècnica de Valencia
Valencia
Spain

Symeon Papavassiliou
National Technical University of Athens
Athens
Greece

Stefan Ruehrup
FTW – Telecommunication Center
Wien
Austria

Carlos Becker Westphall
Federal University of Santo Catarina
Florianopolis
Brazil

ISSN 0302-9743 ISSN 1611-3349 (electronic)
Lecture Notes in Computer Science
ISBN 978-3-662-46337-6 ISBN 978-3-662-46338-3 (eBook)
DOI 10.1007/978-3-662-46338-3

Library of Congress Control Number: 2015931375

Springer Heidelberg New York Dordrecht London

Printed on acid-free paper

Springer-Verlag GmbH Berlin Heidelberg is part of Springer Science+Business Media
(www.springer.com)

Preface

In 2014, the 13th International Conference on Ad-hoc Networks and Wireless (ADHOC-NOW) was accompanied by a workshop program consisting of six workshops, covering various topics in connection with ADHOC-NOW:

- The 2nd International Workshop on Emerging Technologies for Smart Devices (ETSD 2014) focused on 5G technologies, smart devices, and Internet of Things.
- The 2nd International Workshop on Marine Sensors and Systems (MARSS 2014) focused on emerging aspects pertaining to the new mechanisms for underwater sensor networks and applications.
- Multimedia Wireless ad hoc Networks 2014 (MWaoN 2014) focused on 4G multimedia services, efficient multicast communications, routing techniques, and cross-layer solutions as well as performance evaluations of multimedia services over wireless ad hoc and sensor networks.
- Security in Ad Hoc Networks (SecAN 2014) focused on topics of security in ad hoc networks: authentication and reliability, behavior and performance analysis, defense and detection of non-legitimate activities, active response and tolerance schemes, distributed security solutions, modeling, and validation.
- The 2nd Smart Sensor Protocols and Algorithms (SSPA 2014) focused on the design, development, analysis, or optimization of smart sensor protocols or algorithms, including artificial intelligence techniques for network management, network monitoring, quality of service enhancement, performance optimization.
- The 8th International Workshop on Wireless Sensor, Actuator and Robot Networks (WiSARN 2014) focused on the design, specification, and implementation of architectures, algorithms, and protocols for current and future applications of wireless sensor and actor networks, which are the confluence point where the traditional fields of wireless sensor networks (WSNs), robot networks, and control theory meet.

The commonality of these workshops is that they all cover aspects of ad hoc networking and wireless connectivity, which is the larger scope of the ADHOC-NOW conference. However, each workshop targets a specific aspect and/or field of application, which gives the participants a certain focus and the possibility of closer interaction. The ADHOC-NOW workshops attracted overall 59 submissions. A total of 25 papers were accepted for presentation after individual reviews by the respective Workshop Chairs and Program Committee members. The selected papers were compiled and distributed electronically to the workshop participants and presented in the workshops track of the ADHOC-NOW conference held in Benidorm, Spain, during June 22–27, 2014.

After the presentation in the workshops, the authors were given the opportunity to revise and update their papers in order to incorporate comments and suggestions from the discussions in the workshops. This volume contains the final (post-)proceedings of all ADHOC-NOW workshop papers and complements the proceedings of the ADHOC-NOW main conference, which also appeared in this series (LNCS 8487).

We would like to thank the Workshop Chairs for their ideas and their work to set up an interesting workshop program, and the Program Committee for their reviewing efforts. Last but not least, we thank the Springer team for their continuing support during all phases from submission to publication.

July/August 2014

Miguel Garcia Pineda
Jaime Lloret
Symeon Papavassiliou
Stefan Ruehrup
Carlos Becker Westphall

Organization

ADHOC-NOW Workshops General Committee

General Chairs

Jaime Lloret   Universitat Politècnica de València, Spain
Ivan Stojmenovic   University of Ottawa, Canada

Workshop Chairs

Symeon Papavassiliou   National Technical University of Athens, Greece
Carlos Becker Westphall   Federal University of Santa Catarina, Brazil

Submission Chair

Miguel Garcia Pineda   Universitat de València, Spain

Proceedings Chair

Stefan Ruehrup   FTW – Telecommunications Research
     Center Vienna, Austria

Workshop Chairs

ETSD Chairs

Angelos Antonopoulos   CTTC, Barcelona, Spain
Shahid Mumtaz   Instituto de Telecomunicações Portugal
Joel Rodrigues   Instituto de Telecomunicações, University of Beira
     Interior, Portugal
Christos Verikoukis   CTTC, Barcelona, Spain

MARSS Chairs

Miguel Ardid   Universitat Politècnica de València, Spain
Kevin Frank   SINTEF, Norway
Fernando de la Gándara   Instituto Español de Oceanografía, Spain
Jaime Lloret   Universitat Politècnica de València, Spain

MWaoN Chairs

Elsa Maria Macias Lopez   University of Las Palmas de Gran Canaria, Spain
Alvaro Suarez   University of Las Palmas de Gran Canaria, Spain

SecAN Chairs

José Camacho-Páez	University of Granada, Spain
Pedro García-Teodoro	University of Granada, Spain

SSPA Chairs

Jaime Lloret	Universitat Politècnica de València, Spain
Miguel Garcia Pineda	Universitat de València, Spain
Kayhan Zrar Ghafoor	Koya University, Iraq
Ali Safa	Universiti Teknologi Malaysia, Malaysia

WiSARN Chairs

Enrico Natalizio	Université de Technologie de Compiègne, France
Ioannis Paschalidis	Boston University, USA
Ivan Stojmenovic	University of Ottawa, Canada
Danilo Tardioli	Centro Universitario de la Defensa, Zaragoza, Spain

Panel and Tutorial Chairs

Sandra Sendra	Universitat Politècnica de València, Spain
Rashid Khokhar	Charles Sturt University, Australia
Victor Espinosa	Universitat Politècnica de València, Spain

Publicity Chairs

Rafael Falcon	Larus Technologies, Canada
Scott Fowler	Linköping University, Sweden
Valeria Loscrì	Inria Lille, Nord Europe, France
Eduardo Montijano	Centro Universitario de la Defensa, Zaragoza, Spain
Sandra Sendra	Universitat Politècnica de València, Spain

Web Chairs

Alejandro Cánovas Solbes	Universitat Politècnica València, Spain
João Dias	Instituto de Telecomunicações, University of Beira Interior, Portugal
Milos Stojmenovic	Singidunum University, Belgrade, Serbia
Dimitrios Zorbas	Inria Lille, Nord Europe, France

Technical Program Committee

Saman Abdullah	Koya University, Iraq
Javier Aguiar	University of Valladolid, Spain
Móonica Aguilar Igartua	Polytechnic University of Catalonia, Spain
Ramón Alcarria	Universidad Politécnica de Madrid, Spain
Luis Almeida	University of Porto, Portugal

Miguel Ardid	IGIC, Universitat Politècnica de València, Spain
Mohammed Arif Amin	Higher Colleges of Technology, United Arab Emirates
Joan Arnedo Moreno	Universitat Oberta de Catalunya, Spain
Nils Aschenbruck	University of Osnabrück, Germany
M. Bala Krishna	University School of Information and Communication Technology, Guru Gobind Singh Indraprastha University, New Delhi, India
Carlos Baladron	University of Valladolid, Spain
Novella Bartolini	University of Rome "La Sapienza", Italy
Firooz Bashashi Saghezchi	Instituto de Telecomunicações, Portugal
Elijah Blessing Rajsingh	Karunya University, India
Fernando Boavida	University of Coimbra, Portugal
Manuel Bou-Cabo	Universitat Politècnica de València, Spain
Azzedine Boukerche	University of Ottawa, Canada
Diana Bri Molinero	Universitat Politècnica de València, Spain
Alejandro Cánovas Solbes	Universitat Politècnica de València, Spain
Carlos Calafate	Universitat Politècnica de València, Spain
Francisco Camarena	Universitat Politècnica de València, Spain
Maria-Dolores Cano	Universidad Politécnica de Cartagena, Spain
Juan-Carlos Cano	Universitat Politècnica de València, Spain
Xianghui Cao	Illinois Institute of Technology, USA
Patricia Chaves	Escuela Politécnica del Litoral, Ecuador
Jiming Chen	Zhejiang University, China
Hassan Chizari	Universiti Teknologi Malaysia, Malaysia
Dumitru Dan Burdescu	University of Craiova, Romania
Gianni Di Caro	IDSIA Lugano, Switzerland
Giorgio Di Natale	University of Montpellier, France
Roberto Di Pietro	Roma Tre University, Italy
Jesús E. Díaz Verdejo	Universidad de Granada, Spain
Christos Douligeris	University of Piraeus, Greece
Milan Erdelj	Inria Lille, Nord Europe, France
Victor Espinosa	Universitat Politècnica de València, Spain
Romano Fantacci	University of Florence, Italy
Isabelle Fantoni	Université de Technologie de Compiègne, France
Farid Farahmand	Sonoma State University, USA
Apostolos Fournaris	Technological Educational Institute of Western Greece, Greece
Miguel Garcia Pineda	Universitat de València, Spain
Kayhan Ghafoor	Koya University, Iraq
Alberto J. Gonzalez	NicePeopleAtWork, Spain
Fabrizio Granelli	University of Trento, Italy
Stefano Guarino	Roma Tre University, Italy
Juan Carlos Guerri Cebollada	Universitat Politècnica València, Spain
Zygmunt J. Haas	Cornell University, USA
Shibo He	Arizona State University, USA

Jianping He	Zhejiang University, China
Dieter Hogrefe	University of Göttingen, Germany
Jonathan How	Massachusetts Institute of Technology, USA
Fuzhuo Huang	Google, USA
José Miguel Jiménez Herranz	Universitat Politècnica de València, Spain
Elli Kartsakli	Polytechnic University of Catalonia, Spain
Al-Sakib Khan Pathan	International Islamic University Malaysia, Malaysia
S. Khan	Kohat University of Science and Technology, Pakistan
Abdelmajid Khelil	Huawei European Research Center, Germany
Rashid Khokhar	Charles Sturt University, Australia
Jesse Kielthy	Waterford Institute of Technology, Ireland
George Kollias	Iquadrat, Spain
Javier López Muñoz	University of Málaga, Spain
Michail G. Lagoudakis	Technical University of Crete, Greece
Aristeidis Lalos	Polytechnic University of Catalonia, Spain
Giuseppina Larosa	Istituto Nazionale di Fisica Nucleare, Italy
Egons Lavendelis	Riga Technical University, Latvia
David Lizcano Casas	Open University of Madrid, Spain
Jaime Lloret	Universitat Politècnica de València, Spain
Jonathan Loo	Middlesex University, UK
Pascal Lorenz	University of Haute Alsace, France
Henrik Lundqvist	Huawei Technologies, Sweden
Jose M. Jimenez	Universitat Politècnica de València, Spain
Gabriel Maciá Fernández	Universidad de Granada, Spain
Reza Malekian	University of Pretoria, South Africa
Zoubir Mammeri	Paul Sabatier University, France
Sabato Manfredi	University of Naples "Federico II", Italy
Jose Manuel Moya	Universidad Politécnica de Madrid, Spain
Yuxin Mao	Zhejiang GongShang University, China
Ignacio Marín García	Escuela Politécnica del Litoral, Ecuador
Elsa Maria Macias Lopez	University of Las Palmas de Gran Canaria, Spain
Gregorio Martinez	University of Murcia, Spain
Chris McDonald	The University of Western Australia, Australia
Prodromos-Vasileios Mekikis	Polytechnic University of Catalonia, Spain
Tommaso Melodia	State University of New York, USA
Agapi Mesodiakaki	Polytechnic University of Catalonia, Spain
Nathalie Mitton	Inria Lille, Nord Europe, France
Alejandro Mosteo	Centro Universitario de la Defensa, Zaragoza, Spain
Octavio Nieto-Taladriz	Universidad Politécnica de Madrid, Spain
António Nogueira	University of Aveiro/Instituto de Telecomunicações, Portugal
Konstantinos Ntontin	CTTC, Barcelona, Spain

Jorge Ortín	Centro Universitario de la Defensa, Zaragoza, Spain
Lorena Parra	Universitat Politècnica de València, Spain
Juan R. Díaz	Universitat Politècnica de València, Spain
Tito Raúl Vargas Hernández	University of Santo Tomás, Colombia
Milena Radenkovick	Wollaton Road University of Nottingham, UK
Diego Real	Universitat de València, Spain
Tapani Ristaniemi	University of Jyväskylä, Finland
Joel Rodrigues	Instituto de Telecomunicações, University of Beira Interior, Portugal
Stefan Ruehrup	Telecommunications Research Center Vienna FTW, Austria
Anders Rundgren	PrimeKey, Sweden
Jorge Sá Silva	University of Coimbra, Portugal
Ali Safa Sadiq	Universiti Teknologi Malaysia, Malaysia
Kazi Saidul Huq	Instituto de Telecomunicações, Portugal
Angela Schoellig	University of Toronto, Canada
Ahmet Sekercioglu	Monash University, Australia
Sandra Sendra	Universitat Politècnica de València, Spain
Lei Shu	Guangdong University of Petrochemical Technology, China
Simone Silvestri	Pennsylvania State University, USA
Olivier Simonin	INSA de Lyon, France
Mohammad Siraj	King Saud University, Saudi Arabia
Nicolas Sklavos	Technological Educational Institute of Western Greece, Greece
Alvaro Suarez	University of Las Palmas de Gran Canaria, Spain
Fabrice Theoleyre	CNRS, France
Panagiotis Trakas	Open University of Catalonia, Spain
Vito Trianni	ICST-CNR, Italy
Georgia Tseliou	Open University of Catalonia, Spain
Mehmet Ufuk Çaglayan	Boğaziçi University, Turkey
Binod Vaidya	University of Ottawa, Canada
Fabrice Valois	INSA de Lyon, France
John Vardakas	Iquadrat, Spain
Spyros Vassilaras	Athens Information Technology, Greece
Wei Wei	South China University of Technology, China
Carlos Becker Westphall	Federal University of Santa Catarina, Brazil
Stephen D. Wolthusen	Gjøvik University College, Norway
Shuhui Yang	Purdue University, USA
Evsen Yanmaz	Klagenfurt University, Austria
Sherali Zeadally	University of Kentucky, USA
Liang Zhou	Nanjing University of Posts and Telecommunications, China
Juan de Dios Zornoza	Universitat de València, Spain

Contents

2nd International Workshop on Emerging Technologies for Smart Devices, ETSD 2014

Multimedia Content Delivery Between Mobile Cloud and Mobile Devices

Goran Jakimovski[1(✉)], Aleksandar Karadimce[2], and Danco Davcev[2]

[1] Faculty of Electrical Engineering and Information Technology,
Skopje, Macedonia
goranj@feit.ukim.edu.mk
[2] Faculty of Computer Science and Engineering, Skopje, Macedonia
akaradimce@ieee.org, danco.davcev@finki.ukim.mk

Abstract. The usage of mobile devices in everyday life poses new challenges for processing, adaptation and rendering of multimedia content, which can't be accomplished due to mobile device limitations (battery lifetime, storage limitation, processing capacity and etc.). The proposed Mobile Cloud Content Delivery (MCCD) framework shows how multimedia delivery applications and services can efficiently exploit the computing power of mobile cloud resources to achieve high efficient delivery of multimedia content. The experimental case study shows how local multimedia content from the mobile device can be offloaded and processed in the mobile cloud and the transformed multimedia content can be delivered back to the user's mobile device. The case study further shows that using different communication protocols (EDGE, 3G and LTE) to offload media content to MCCD can significantly influence the turnaround computational time.

Keywords: 3D scene · Cloud computing · Multimedia content delivery · Mobile computing

1 Introduction

Considering the challenges of mobile devices, like limited processor power, available memory and vast energy consumption, we conclude that delivery of multimedia content to mobile devices is still a big research opportunity. In order to improve user's perception of multimedia content delivery on mobile devices, we execute some of the applications on more powerful external computing servers. Cloud computing, according to the NIST (National Institute of Standards and Technology, USA), is a model that provides convenient, on-demand network access to a shared pool of configurable resources, like servers, storage, applications and services is the appropriate resource that will improve user experiences [1]. Therefore computing intensive applications, like content-based video analysis or 3D modeling running on a mobile device, could be offloaded onto a remote cloud infrastructure [2].

The integration of cloud computing and mobile devices has led to defining mobile cloud computing, as an infrastructure where both the data storage and the data processing happen outside of the mobile device [2, 3]. Applications based on mobile cloud

© Springer-Verlag Berlin Heidelberg 2015
M. Garcia Pineda et al. (Eds.): ADHOC-NOW Workshops 2014, LNCS 8629, pp. 3–11, 2015.
DOI: 10.1007/978-3-662-46338-3_1

computing are utilizing the computing and storage resources from the mobile cloud, thereby enabling much richer media experiences than what current native applications can offer [4]. The multimedia cloud computing frameworks are leveraging cloud computing in order to provide multimedia transformations that require a large amount of computing and are difficult to perform on the mobile device.

This paper presents our proposed Mobile Cloud Content Delivery (MCCD) framework, which enables adaptive distribution of multimedia content between the mobile cloud and mobile devices. The framework provides efficient exchange of multimedia content from the mobile cloud to the mobile device according to user demands, and other way back respectively. We have created two validation scenarios, one where the user sends two images and appropriate request that should be processed by the mobile cloud, and another in which the user sends video to be processed by the mobile cloud. After receiving the images (video) and the request, the mobile cloud performs proper adaptation and processing of the images (video) and responds to the mobile device by sending the processed multimedia content. In the first scenario it sends a single 3D image (model) and in the second one, it sends back the processed video.

The rest of the paper is structured accordingly to these steps. In Sect. 2, first we present the existing multimedia content delivery systems that support mobile cloud computing scenarios. Then, in Sect. 3, we briefly describe our proposed Mobile Cloud Content Delivery (MCCD) framework. Next, in Sect. 4 we present some of the conducted experimental case studies that illustrate our framework. We finish with conclusions and future directions of research in Sect. 5.

2 Related Work

The mobile multimedia services over cloud computing primarily solves the problem of how to increase efficiency in multimedia file playback using the cloud computing concept [5]. One of the first conceptual frameworks for mobile cloud services is the Context-Aware Cloud Services that enables tasks, such as, capturing context, tailoring services for context and running the adapted services [6]. As the context is changed the application invokes another service to adopt to the changed context [6]. With the introduction of the multimedia cloud computing architecture [7] the focus is changed on how to provide QoS provisioning for multimedia applications and services.

The mobile applications based on mobile cloud computing are utilizing the computing and storage resources from the mobile cloud, and enable delivery of more high quality multimedia content. These applications are known as Cloud Mobile Media (CMM) applications [4]. Authors Shaoxuan Wang and Sujit Dey in [4], have proposed a rendering adaptation technique, which can dynamically fluctuate the complexity of graphic rendering depending on the network and cloud computing constraints. This technique has an impact on the bit rate of the rendered video that needs to be streamed back from the mobile cloud to the mobile device, and the computation load on the CMG servers [4].

Algorithms that render 2D images into a 3D scene require images that are partially overlapping in order to automatically detect the depth. The algorithm creates a vector of points in the 2D images, each point representing the X and Y coordinate and the depth of the image. The algorithm evaluates each point from each image by comparing it to each other point of the other images, thus creating the 3D scene. This means that having a larger Array results in a better 3D scene but by severely increasing the computational time. Furthermore, algorithms that process video content have to extract frame by frame from the video, process the image (frame) accordingly and bind the frames back to a video stream. The video content can be supplied from a file or directly from the camera, making the computations even more extensive. The execution of computationally intensive applications like content-based video analysis or 3D modeling, as the main drawbacks of the existing mobile devices, have been addressed with the introduction of the Mobile Augmentation Cloud Services (MACS) [8].

After careful analysis of deficiencies in the existing frameworks, we have proposed the Mobile Cloud Content Delivery (MCCD) framework, which enables adaptive distribution of multimedia content between the mobile cloud and mobile devices. The framework provides efficient exchange of multimedia content from the mobile cloud to the mobile device according to the user demands, and other way back respectively.

3   Mobile Cloud Content Delivery (MCCD) Framework

The mobile cloud computing framework provides offloading of different kinds of multimedia transformations to the mobile cloud, many of which require a large amount of computing power and are difficult to perform on the mobile devices. The multimedia processing in the mobile cloud imposes a huge new challenge that needs to be addressed, such as storage and sharing multimedia content, adaptation and delivery, and rendering and retrieval, can optimally utilize cloud computing resources [7]. Considering the fundamental concept of multimedia cloud computing used in the MEC (Media-Edge Cloud), where media contents and processing are pushed to the verge of the cloud based computing and is based on the user's context or profile, has the main benefit to reduce latency [7]. Additionally, the framework should be context-aware, meaning that the most appropriate service is selected at the stage of adapting services described in the computing model of Context-Aware Cloud Services [6].

The proposed Mobile Cloud Content Delivery (MCCD) framework enables adaptive distribution of multimedia content between the mobile cloud and mobile devices. The MCCD framework provides efficient exchange of multimedia content from the mobile cloud to the mobile device according to user demands, and in same time it considers the context-aware conditions and mobile device characteristics. Mobile devices are connected to the cloud by the Internet connection using different kinds of protocols (WiFi, WiMax, GPRS, EDGE, 3G, LTE and etc.), as shown on Fig. 1 of the MCCD framework. Using these types of existing connection protocols, the mobile device is able to interface the mobile cloud and access the Cloud Server. This server provides scheduling for the execution of different tasks that will be executed on the mobile cloud. Following these communication protocols, users of mobile devices can generate images, video or send/receive different kinds of multimedia content. Using offloading servers can be very beneficial as it can run powerful multimedia processing algorithms that can produce more accurate and better results. For example, a cloud server can use insight3D for generating high definition 3d images from an array of 2D images. Furthermore, a cloud system can render images and process video streams faster than any mobile device can.

The first validation scenario is a middleware Android mobile application that does the heavy lifting of adaptive application partitioning, resource monitoring and computation offloads [8]. For the second scenario we are using a tablet PC that offloads the video-processing to the MCCD. We propose using a tablet in order to test the MCCD with various mobile devices.

Depending of the user's needs, this multimedia content can be easily adapted or transformed in the MCCD framework. The first step in this process is sending the request for processing from the users mobile device to the mobile cloud using one of the Internet protocols (see Fig. 1). Next, the Cloud Server of MCCD framework, which works as a cloud broker, is saving the user's request in the execution queue. Depending of the request type, the Cloud Server can require additional backend data and cloud services to complete the requested task.

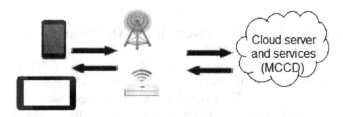

Fig. 1. Mobile Cloud Content Delivery (MCCD) framework

After the process of transformation on the multimedia finishes, the process of multimedia content adaptation starts, so that the transformation is returned according to the user's request. The final process in the MCCD framework is the delivery and distribution of adapted multimedia content to the user's mobile device, following the same connection channel backwards. In order to validate the MCCD framework we have used the following cloud generated 3D scene of an image presented with the experimental case study in Sect. 4. Additionally, we have used OpenCV algorithms to do the video processing on the MCCD.

4 Experimental Case Study

4.1 Generating 3D Image in the Cloud

Based on the proposed MCCD framework, we have done experimental case study that uses two images from the user's mobile device into a cloud based mobile application. This application uses research-grade algorithms to generate an OpenGL rendering of a 3D scene, computed from user's images [9]. The application wraps the two images and sends appropriate transformation request that should be processed by the Cloud Server, which is part of the mobile cloud. After processing the two images onto the mobile cloud, the completed single 3D scene is delivered back to the user's mobile device. The 3D Camera mobile application is available only for Android OS and has been tested on HTC Desire X mobile device, which has 4.0 inch display (480 × 800 pixels), with 5 MP camera (2592 × 1944 pixels), Android OS (v4.0), CPU Dual-core 1 GHz and 768 MB RAM. The case study starts by selecting two images from the user's phone gallery or by taking new pictures (Fig. 2).

Fig. 2. Process of collecting two images with 3D camera mobile application on HTC mobile device

After adding the two images into the mobile application and using the Internet connection, the images are uploaded for processing to the cloud server. This server is computing the depth of the scene automatically, which takes too much time on the mobile device (about 56 s).

Results from computing scene depth, which were executed on the cloud server, are returned back to the user's mobile device. The next step is local conversion of the single 2D picture into a 3D scene on the mobile device itself. This process of conversion takes about 2:05 min and it is done offline.

Considering that the last step of the image processing is done locally on the mobile device, it consumes much of HTC's processing power. During this case scenario we were able to measure, using the Task Manager on HTC mobile device, a 129 MB RAM memory consumption, see Fig. 3. The process of generating 3D scene becomes more complex as the number of polygons and entities in the 3D scene increases, so does the

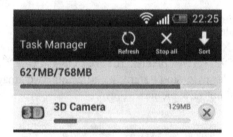

Fig. 3. Task manager on HTC mobile device

complexity for efficient management of the scene [10]. Rendering algorithm is going affecting each object in the scene graph and renders them each frame until it creates the 3D scene [10].

With the completed image processing, using the 3D Camera mobile application, we were able to see the 3D scene generated from the two images. In order to be able to see the 3D effect on the mobile device, we have rotated the screen in left and right side and we created two images, as shown in Fig. 4. Here we can clearly see the 3D effect, with the modified depth on the object that is achieved by using the mobile application.

The 3D Camera Android application used in this paper uses fewer points and user input for depth perception and tested against two images. Using these parameters, the Android system requires around 2 min doing offline calculations and rendering the 3D scene, which is with poor quality and prone to inconsistencies due to user-defined depth perception. Using a server, the input for depth is automatically calculated and the rendering is done in about 1 min, cutting the time in half. Furthermore, using a server can be highly beneficial in 3D rendering, where more complex and better algorithms can be used (such as insight3d), which cannot be executed on an Android system.

Fig. 4. Created 3D scene generated on HTC mobile device

Experimental case study has shown that local multimedia content from the mobile device can be offloaded and processed in the mobile cloud and the transformed multimedia content is delivered back to the user's mobile device. This experiment is first from series of case studies that we plan to conduct in the proposed Mobile Cloud Content Delivery (MCCD) framework.

Using Mobile Cloud to render 3D images is becoming more and more evident as the number of users that require server offloading increases. With a fast connection to the Mobile Cloud, a user can upload an average of 4 MB of pictures for less than 10 s, render the 3D object (with high quality) in about 20 s and receive the 3D scene in another 10 s, which makes the turnaround time approximately 40 s per user. Increasing the number of users can deteriorate the turnaround time, but Cloud systems are scalable and can grow in resources to meet the needs of the users. On the other hand, rendering pictures offline, with a total size of 4 MB, results in poor quality 3D scene that takes roughly about 4 min. Furthermore, rendering offline keeps the Android system busy, which drains the battery and makes the system useless for the entire rendering process. Table 1 shows how the communication protocol influences the 3D rendering time when the processing is offloaded onto cloud computing system. We are testing these three protocols as they are widely used in mobile communication. Even though, in recent time LTE is becoming main target of mobile device users, still 3G is the leading protocol used as older models of mobile devices have no support for LTE. EDGE is tested as it can become the only available protocol in rural areas or areas with low service signal.

Different communication protocols yield different data transfer rates ranging from EDGE having the lowest data transfer rate (around 384 kbps) and LTE having the highest data transfer rate (around 50 Mbps). Having different data transfer rates means that the multimedia content will be transferred in less time using the LTE compared with the EDGE, which would produce faster turnaround execution time [11].

Table 1. Offloading image processing using different protocols

Communication protocol	Execution time
EDGE	1:32 min
3G	1:06 min
LTE	0:48 min

4.2 Offloading Video Processing

The second validation scenario includes offloading video processing to a remote server using a tablet PC as a mobile device. Using OpenCV algorithms for image processing, first we are processing the video offline on the tablet. Furthermore, we are offloading the image processing to the MCCD using the same communication protocols from the first scenario (WiFi, EDGE, 3G and LTE). The OpenCV algorithm removes frame by frame and does image processing on each frame, adding additional info to each frame. After all the frames are processed, the algorithm combines the frames back to a video stream. The offline video processing is done in about 2:23 min, whereas the

Table 2. Offloading video processing content using different protocols

Communication protocol	Execution time
Offline	1:47 min
WiFi	1:29 min
EDGE	1:32 min
3G	1:06 min
LTE	0:48 min

MCCD processing is done in about 2:06 min. The MCCD waits for requests from a mobile device to execute the processing. It provides two interfaces, one for the image processing and another for video processing. Each interface creates another process upon request, making as much processes (threads) as requests. Each request uses as much resources as it requires from the MCCD resource manager. Results are shown in Table 2.

5 Conclusions and Future Work

Limitations of mobile devices like limited computing capacity, limited power and available memory are main drawbacks for large amount of processing and computing to be performed on the mobile device. One solution to improve user's perception of multimedia content delivery on mobile device is done by executing some of the applications on more powerful external computing servers. The proposed Mobile Cloud Content Delivery (MCCD) framework enables adaptive distribution of multi-media content between the mobile cloud and mobile devices. The presented framework should provide efficient exchange of multimedia content from the mobile cloud to the mobile device according to the user demands, and other way back respectively.

Furthermore, by using communication protocols (like LTE) that provide high transfer rates of large data (such as multimedia content), applications in different domains (like m-health, m-commerce or m-learning) can benefit from the faster execution time, lower battery consumption and powerful multimedia computational algorithms that are device independent.

References

1. Mell, P., Gance, T.: The NIST Definition of Cloud Computing (Final). NIST Special Publication 800-145, Sept 2011. http://csrc.nist.gov/publications/nistpubs/800-145/SP800-145.pdf
2. Kovachev, D., Klamma, R.: Framework for computation offloading in mobile cloud computing. Int. J. Interact. Multimedia Artif. Intell. IJIMAI 1(7), 6–15 (2012). doi:10.9781/ijimai.2012.171
3. Mobile cloud computing forum, July 2013. http://www.mobilecloudcomputingforum.com/
4. Wang, S., Dey, S.: Adaptive mobile cloud computing to enable rich mobile multimedia applications. IEEE Trans. Multimedia 15(4), 870–883 (2013)

5. Lai, C.F., Vasilakos, A.V.: Mobile multimedia services over cloud computing. E-Letter **5**(6), 39–42 (2012). Multimedia Communications Technical Committee, IEEE Communications Society

6. La, H.J., Kim, S.D.: A conceptual framework for provisioning context-aware mobile cloud services. In: IEEE CLOUD, pp. 466–473, IEEE. ISBN: 978-1-4244-8207-8

7. Wenwu, Z., Chong, L., Jianfeng, W., Shipeng, L.: Multimedia cloud computing. IEEE Signal Process. Mag. **28**(3), 59–69 (2011). doi:10.1109/MSP.2011.940269. ISSN: 1053- 5888

8. Kovachev, D., Yu, T., Klamma, R.: Adaptive computation offloading from mobile devices into the cloud. In: Proceedings of the 2012 IEEE 10th International Symposium on Parallel and Distributed Processing with Applications (ISPA 2012), pp. 784–791. IEEE Computer Society, Washington, DC (2012). doi:http://dx.doi.org/10.1109/ISPA.2012.115

9. 3D Camera, July 2013. http://www.androidpit.com/en/android/market/apps/app/comens.threedeecamera.lite/3D-Camera

10. Sazzad., K., Emdad, A., Lutful, K., Rokonuzzaman, M.: Scene graph management for OpenGL based 3D graphics engine. In: Proceedings of 6th International Conference on Computer and Information Technology (ICCIT 2003), vol. I, pp. 395–400, Dhaka, Bangladesh, 19–21 December 2003

11. Fein, D.: LTE Broadband and public safety. Division of emergency management – Nevada, November 2011

Delayed Key Exchange for Constrained Smart Devices

Joona Kannisto[1][(✉)], Seppo Heikkinen[1], Kristian Slavov[2], and Jarmo Harju[1]

[1] Tampere University of Technology, Tampere, Finland
{joona.kannisto,seppo.heikkinen,jarmo.harju}@tut.fi
[2] Ericsson Research, Jorvas, Finland
kristian.slavov@ericsson.com

Abstract. In the Internet of Things some nodes, especially sensors, can be constrained and sleepy, i.e., they spend extended periods of time in an inaccessible sleep state. Therefore, the services they offer may have to be accessed through gateways. Typically this requires that the gateway is trusted to store and transmit the data. However, if the gateway cannot be trusted, the data needs to be protected end-to-end. One way of achieving end-to-end security is to perform a key exchange, and secure the subsequent messages using the derived shared secrets. However, when the constrained nodes are sleepy this key exchange may have to be done in a delayed fashion. We present a novel way of utilizing the gateway in key exchange, without the possibility of it influencing or compromising the exchanged keys. The paper investigates the applicability of existing protocols for this purpose. Furthermore, due to a possible need for protocol translations, application layer use of the exchanged keys is examined.

1 Introduction

The Internet of Things (IoT) concept means that each and every device or thing can be accessed through network [1]. This enables new services and opportunities, but also generates a number of security requirements and needs for security related services. For instance, the amount of network connected devices in IoT can range depending on the use case, from as low as tens of nodes to even thousands of nodes, and the deployment environments and scenarios for IoT are versatile. Common examples include industrial systems, home automation, and intelligent traffic related services. Accordingly, the deployments can have justifications based on cost savings, safety, and even entertainment. Therefore, to make the deployment feasible for even the low margin scenarios, the nodes are often envisioned to be energy efficient and low-cost. Because of this requirement of energy efficiency and low cost, some of these devices are constrained and must spend considerable time in inactive sleep state. For example, 8-bit microcontrollers, which would be awake only for a few seconds each minute, could be considered as baseline examples of such devices.

However, end-users, as well as applications in networked devices often require that the services are available on-demand and that they respond with reasonable

M. Garcia Pineda et al. (Eds.): ADHOC-NOW Workshops 2014, LNCS 8629, pp. 12–26, 2015.
DOI: 10.1007/978-3-662-46338-3_2

delay. With sleeping constrained nodes this can be accomplished by utilizing a small number of more resourceful gateway devices that support the constrained nodes and make the services accessible to the end-users and other devices. This can mean, for example, protocol translations for the more capable end-user devices, so that they can access services in a standard manner.

While gateways enable energy efficiency for the smaller nodes, the use of a gateway can be problematic whenever the gateway is not trusted by both communication parties. Indeed, we might not trust the gateway in all aspects, and the integrity and confidentiality of the messages may have to be preserved end to end by using cryptography.

New protocols have been designed that suite constrained devices better, such as, Constrained Application Protocol (CoAP). Yet, since some of the clients using the services might support only legacy protocols, the gateway may have to do protocol translations as well [2,3]. However, it is not possible to do protocol translations if tunneling protocols are used [2]. Therefore, it is assumed that end-to-end security is handled at the application layer data objects, we explore how the keys from traditional key exchanges could be utilized in securing the data objects.

The following list summarizes the requirements and the problem statement for the presented IoT key exchange scenario:

- Constrained devices are sleeping most of the time, and are not available for direct communication without the help of a gateway device.
- The gateway may not be in the same trust domain, and trusting it should not be necessary.
- The gateway may need to do protocol translations from HTTP to CoAP.
- Cross-layer approaches offer advantages for constrained devices.

In the paper we evaluate existing key exchange protocols, and how they could be adapted to better fit into these requirements.

2   Related Work

The Internet of Things vision is for the Internet to expand into our every day objects. This requires some degree of interoperability with legacy solutions. Indeed, deployment of new protocols has traditionally been slow, and new solutions tend to be built on top of existing ones. Perhaps the most common key exchange protocols within the Internet context are Internet Key Exchange (IKE) [4] and Transport Layer Security (TLS) [5], even though the latter provides a more comprehensive framework for protecting the actual communication as well. Datagram TLS is an adaptation of TLS which makes it more suitable for constrained devices.

Constrained Application Protocol (CoAP) [6] is envisaged to be the way of providing RESTful, i.e. HTTP compatible services, within IoT. CoAP uses User Datagram Protocol (UDP) for transport instead of TCP. Therefore, Datagram TLS (DTLS) [7,8], which is a modification of TLS, is recommended for use with

CoAP [6]. Also IPsec, which uses IKEv2 [4] is mentioned as an option. End-users might not be able to support CoAP, for example, due to firewall policies and lack of application support. Therefore, protocol translations from HTTP to CoAP would need to be done by an intermediary device [3]. This poses a challenge for proposals, where a tunneling protocol, such as IPsec or DTLS is used to provide end-to-end security below CoAP [2, p. 18].

Key exchange could be done at the application layer as well. For example, DTLS handshake could be transmitted as application layer messages [9, p. 14]. In addition, WS-Trust [10] specifies an application layer key exchange, yet, it may not be suitable for constrained devices due the verbosity of the messages [11]. More compact format JSON Web Encryption (JWE) [12] defines public key and symmetric key based key distribution mechanisms for application layer, which are used by Sethi et al. [13] for constrained devices. However, the JWE specification excludes key exchange [12, p. 20].

Bianchi et. al. [14] suggests a key exchange protocol for Wireless Sensor Networks (WSN) that is loosely based on TLS. This protocol tries to minimize the transfer overhead but does not utilize the infrastructure. In addition, modifications to DTLS for constrained devices have been suggested [9], but infrastructure support for key exchange is a less investigated topic.

Proxy based key exchanges, such as Needham–Schroeder protocol [15] are well established, yet we assume a situation where the proxy is not trusted. In addition, using a gateway to mirror sensor readings is a well established practice in the IoT-world [16]. However, the security in these proposals is either not end-to-end, or could be based on key distribution rather than key exchange [13]. Indeed, we consider key exchanges, where both parties contribute to the resulting key material. Proposals where a middle man acts in a supportive role without a direct contribution to the resulting key material are more in this paper's scope [17,18]. Yet, we aim to accomplish this in a way that is transparent to the initiator of the exchange.

3   Overview of the Delayed Key Exchange Proposal

3.1   Assumptions and Security Properties

We base this proposal on a scenario where there is a *user*, a *gateway*, and a constrained device, which we will refer to as a *sensor* later on in this document. The user initiates the key exchange with the sensor. We expect that the user conforms to existing protocol specifications. The user's intention might be, for example, to fetch periodic readings of the sensor data. The gateway, which could also be called a reverse proxy, acts as a middle man between the sensor and the user. Its role is to enable the key exchange in an efficient way overcoming the sensor's constraints. There might be constraints in the access network of the sensor as well, such as narrow bandwidth or long delays. Also, the same gateway may serve multiple possibly unrelated sensors, each with independent security associations.

While the control of the participating nodes and the infrastructure might be distributed, some expectations can be made about the behavior of the nodes.

For example, the sensor could rely on the gateway to transmit its messages reliably, while the user should instead retransmit its requests as needed. In addition, neither one needs to trust the gateway to store or process the message contents unencrypted. The gateway assumes that the sensor and the user behave correctly but does not need to rely on that in any critical way.

We consider the initiating party i.e. the user as the prime actor, and focus mainly on authenticating the responder. Authenticated key exchange protocols usually support mutual authentication. For example, Pre Shared Key (PSK) based key exchanges authenticate both parties quite naturally. Moreover, they are usually the best choice for constrained environments whenever there is a possibility to setup symmetric secrets. Public key based mutual authentication is supported by the key exchange protocols, as well. However, end-users do not usually have verifiable public keys, and processing certificate chains can be a difficult task for a constrained device. Therefore, it is expected that application layer authentication credentials are used. Additionally, a viable approach to access control is to authenticate the user and decide on the access control policy by a more resourceful third party.

We base the used threat model on the Dolev-Yao -model [19]. Therefore, we expect the attacker to be able to read and insert messages to the channel but not break any strong cryptographic functions.

3.2   Pre-configuration and Delegation

As the sensor is sleeping most of the time, it is expected that the sensor uses the gateway to act as its representative and fetches information about requests periodically from there. This requires a pre-configuration phase, which means agreement on the messages that can be cached by the gateway. Pre-configuration steps differ from general service discovery scenarios by the inclusion of the gateway. Basically, the user will have to know that certain gateway is responsible for a resource that would normally be tied only to, for example, sensor's identity or locator. Moreover, it may be possible that the users will access the services without knowledge about the actual node providing the needed resource [20].

Authentication for the configuration between the gateway and the sensor could be handled either manually or in an opportunistic fashion. Opportunistic measures can be desirable, when better scalability is desired. In other words, in the initial negotiation the parties would learn their identities, which they would later trust. Such crypto-identities are also suggested in CoAP security architecture Internet draft [21].

If the gateway is not acting as a completely transparent device, the sensor should also provide authorisation, which will show that the gateway is authorised to act on its behalf. Moreover, there might even be need to signal more complex trust relationships. In other words, the sensor may have to have means to specify the actions that the gateway can be trusted to do on the sensor's behalf. Other pre-configurable items can include puzzle, or similar mechanisms for ensuring that the user is "honest" in his attempt to communicate. There might also be a need to share information needed for nonces, which can be used against replay attacks, especially if the gateway cannot be trusted to generate them.

3.3   Delayed Key Exchange

We assume a situation where the sensor wakes up on timed intervals, or as a result of some event. After waking up it sends any pending messages, and fetches new incoming messages from its designated gateway, processes those messages, sends the necessary replies, and goes back to sleep. Indeed, the sensor may rely on the gateway being available, and even delay going to sleep to in order to receive its response. Yet, it should not delay sleep waiting for external resources or clients.

The flow of events on a general level is depicted in Fig. 1. The user sends an initialisation message, which indicates his willingness to establish a secure communication to a service in question. Typically the gateway will issue some sort of a challenge, which is basically used to mitigate Denial of Service (DoS) concerns. Note that these parameters are not secret per se, i.e., the adversary will be able to learn them anyway and trying to change them basically results in DoS condition, which could be done anyway if the attacker is controlling the channel.

The user responds to the challenge (solve puzzle, re-send cookie, etc.) and sends the next message (*Response* in Fig. 1). Depending on the use case, the user might also need some authorisation statements to show that he really is allowed to access the service in question (i.e., gateway performs access control). Note that instead of sending this type of authorisation in the first message, it is better to do it here, because this way the gateway does not need to engage in any heavy processing after the first message. Access control could be based on user identities and predefined policies as well.

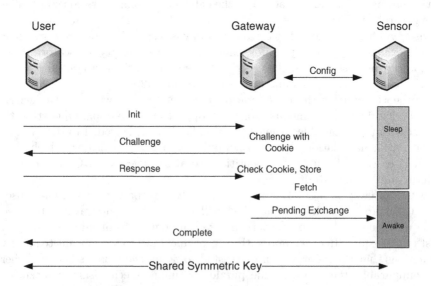

Fig. 1. Delayed key exchange on a general level

From the performance perspective, it is better if also the gateway is able to check any nonces or puzzles, so one needs to ensure that the gateway is in possession of proper information to do this. This might be a requirement if the gateway is responsible for storing the received valid messages, so that they can be later fetched by the sensor.

In case of multiple simultaneous requests for key exchange, the gateway has to implement suitable queuing, and possibly prioritisation, mechanisms. This may also protect the sensor from Denial of Service (DoS) attacks [22]. Depending on the application, it may be necessary for the gateway to send some sort of provisional acknowledgement to the user, so that the user can assume that the message has been received. Yet, the user may choose not to rely on these in a critical way, and timeout arrangements at the user side might be needed anyway.

In the fetching phase the sensor wakes up at its designated time and is then responsible for acquiring the legitimate pending handshake messages, which are waiting to be processed. It is worth noting that the fetching of messages may require the sensor to be awake a longer period of time as it needs to wait for the response to its fetch request.

Once the sensor has processed the messages, it will use the key exchange protocol's key agreement mechanism to generate the common keying material, based on the information it has received. After that it sends its response to the user, which will include sensor's contribution to the keying material if the gateway has not been able to transmit it in earlier messages. After this message, the user is also able to create the same keying material and establish the necessary common secrets. Once the common secret has been created, the user can transmit application level requests, for example, by using HTTP, and can protect the data object within. Also, the gateway may translate the HTTP to CoAP, but it needs to retain the data objects as they are.

3.4   Utilizing the Keys to Protect Application Layer Data

As the sleeping sensor is likely to utilize protocols that do not require sessions, for example CoAP, the gateway may have to be responsible for protocol translations. Therefore, one of the most interesting usages for the key material from the delayed exchange is securing data objects. An example approach presented here uses JavaScript Object Notation (JSON) to represent the data objects, which are to be secured. Signatures and encryption in JSON context are considered in JSON Web Signature [23], and JSON Web Encryption [12] draft specifications and we borrow concepts from there.

Listing 2.1. Protected data object header, comments and linebreaks for clarity

```
{
 "typ":"JWT",
  "alg":"HS256",
  //SPI or TLS session identifier
  "kid": 12010000001,
  "jti": "132312312"  // unique identifier
}
```

The Listing 2.1 illustrates what a protected data object header might look like. As the algorithms and keys are already negotiated we only need to provide a suitable index to the correct security association. The negotiated security association is indicated with a security parameter index (SPI) value. The signature could of course host a public key based signature, but we take advantage of our key exchange procedure so that we are able to use the negotiated keying material to calculate the message authentication code.

We also want to provide a unique identifier *jti* to the protected data objects and it could be used as a sequence number if one wants to avoid potential replays of messages. Note that in case of limited devices, having an increasing number might be preferable option because it is not feasible to keep track of all the received id numbers.

Another use case for the protected data objects is the possibility to encrypt the message. Such example is given in Listing 2.2. The header format follows the same principles as the earlier signed version. The initialization vector and the encrypted payload are given separately, as per JWE [12].

Listing 2.2. Encrypted message header

```
{
  "typ":"JWE",
  "alg":"dir"
  "enc":"A128GCM"
  // index the security associtation
  "kid": 1201000001,
  "jti": "121213"
}
```

Whenever data objects are protected with symmetric keys they hold value only to the parties who have the corresponding keys. Therefore, a gateway should serve data protected with different keys as separate resources. key wrapping and content encryption keys are used [12].

4   Applicability of Existing Key Exhance Protocols

Rather than completely devising a new key exchange mechanism, we have decided to investigate the feasibility of applying existing protocols.

4.1   DTLS Delayed Key Exchange and Its Security Analysis

This section investigates the applicability of Datagram Transport Layer Security (DTLS) [8] as the base protocol for the delayed key exchange.

DTLS adds a DoS protection mechanism to regular TLS, which makes sure that the connecting addresses are return routable. This exchange is optional, and for sleepy sensors it can be expected that the gateway does this stateless cookie exchange on behalf of the sensor. Because ServerHello message in DTLS contains the server nonce, the gateway cannot proceed in the handshake any further.

Yet, if the DTLS implementation of the sensor is modified slightly, a supporting gateway is able to reduce the key exchange messages to the sensor to two flights. Hartke and Bergmann [9] suggest that retransmitting ClientHello with a statelessly verifiable ServerHello to the responder enables it to delay establishing state until it receives the ClientKeyExchange message. We assume that the gateway is able to respond with a valid ServerHello, and so it can transmit all these messages to the sensor at once. This is shown in Fig. 2.

The user and the gateway complete the first handshake flights. After waking up and sending a fetch message (Fig. 2), the sensor receives the pending handshake messages. After it extracts the premastersecret from the ClientKeyExchange message, it sends ChangeCipherSpec message as well as a Finished message, which uses the derived mastersecret and authenticates the handshake. It should be noted that the messages may need to arrive or be processed in the right order, as implementations, for example Californium [24], expect to be able to derive the master secret when processing the ClientKeyExhange message.

The returning ServerHello is marked as optional in Fig. 2 as the sensor is able to construct it. Indeed, the sensor should have the previously chosen nonce stored,

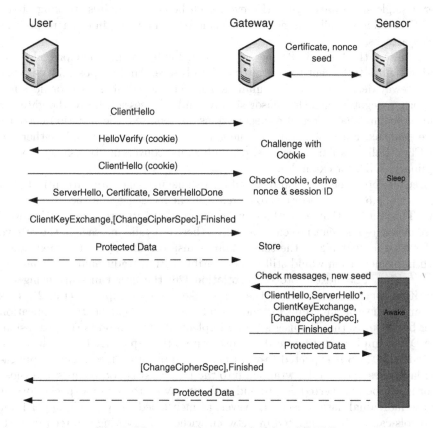

Fig. 2. Message exchange with delayed DTLS

and the Cipher Suite selection is deterministic. However, this restricts the amount of delayed handshakes per sleep cycle to one. As the gateway is not expected to perform DoS, the ClientHello message with a correct cookie is sufficient proof of the return routability of the initiator address. It should be noted that the DTLS return routability cookie does not protect against local DoS, and the gateway or any other node in the wireless network is able to send spoofed initiator packets with correct cookies.

From a security standpoint, the only change to the basic TLS handshake is in the responder's timing, as the ServerHello is transmitted before the ClientHello. This timing change cannot break the security of the key exchange protocol under the indistinguishability assumption [25]. It also does not weaken the sensor's contribution to the resulting symmetric key, if a stored ServerHello is used, or when the used pseudo random function is secure if the gateway derives the nonce.

Indeed, even though the gateway cannot be trusted to choose the TLS server nonce, it could reasonably use a pseudo random function to derive it from a renewed nonce using the client information as a seed. This is why the message in the Fig. 2 is marked as nonce seed, and not as ServerHello, which is the simplest option. Deriving the ServerHello at the gateway enables multiple handshakes over a single sleep-wake cycle. However, such behavior requires tracking of the used ServerHellos while the nonce is considered to be valid, in order to prevent replay attacks.

As noted, the ServerHello message in Fig. 2 is optional. Therefore, also the session ID has to be sent to the gateway in advance. Another possibility is that the gateway derives it in a deterministic fashion. Indeed, if the sensor does not support re-negotiations, the sensor simply sends the session ID to the gateway before the handshake. Yet, if re-negotiations are required, the situation becomes more complex. For re-negotiations the gateway and the sensor could either use the ClientHello session ID as it is, or apply some converging transformation to it (binary AND, for example).

Because both the sent and received handshake messages are authenticated in (D)TLS [5, p.61], the sensor has to recreate all the messages that the gateway has sent. This verifies that everything went according to its preferences. However, these messages need not be sent. Indeed, as there is a valid message in the receive buffer for the next flight, the sensor may transition directly to the next state. Such implementation would still be compatible with a traditional handshake.

By using TLS next protocol negotiation [26] the user can start using the security association even before the handshake is fully completed (Fig. 2). This minimizes the waiting time for the actual communication on the application layer but requires that the user's DTLS implementation supports this extension. The DTLS implementation needs to also expose the capability for applications.

Indeed, DTLS is expected to work close to application layer. Thus, application layer messages can be wrapped into the protected DTLS records and those records can be transferred end-to-end without having to devise a special format for the individual data objects. However, as mentioned the possible upper layer protocols cannot be translated in between without decrypting the records first, i.e., one cannot first run data objects on top of HTTP and then expect to switch to CoAP, unless one were to make more exotic protocol stack choice, such as data objects over DTLS over HTTP/CoAP.

Using DTLS derived keys with data objects as described in Sect. 3.4, requires an interface to the negotiated keys. However, there already is work concerning exporting keys from TLS and DTLS [27] to be used in the application layer, for example, DTLS-SRTP [28] uses this mechanism for its key derivation.

4.2 Case IKEv2

In this section we investigate the possibility of employing Internet Key exchange version 2 (IKEv2) [4] to implement the suggested key exchange procedure. While the "full" IKEv2 has been used with IPsec security suite for protecting IP traffic, we also consider the "light" version of IKEv2 [29], which might be more suited for limited environments. The minimal IKEv2 supports only limited amount of exchanges and optional features have been mainly left out [29], even though the draft specification does not consider so much suitable algorithmic options. However, it assumes that the device initiating the communication is a limited one, while our architecture has the opposite use case. One should also note that IKEv2 runs on top of UDP.

The pre-configuration phase expects that the sensor informs the gateway about its public key and Diffie–Hellman (DH) parameters. A very minimal sensor implementation might just have one DH group, but a more complex one could use several ones. The gateway just has to make sure that the sensor knows which one has been used in the messages it stores.

IKEv2 negotiation in its basic form consists of two different exchanges: IKE_SA_INIT and IKE_AUTH. They are shown in Fig. 3 as Ix, Rx pairs. The first actual message contains the typical IKE header with a chosen Security Parameter Index (SPI) and the suggested ciphers, which the user is able to use for the establishment of the IKE security association. The message also contains the Diffie–Hellman parameters and a random nonce. In Fig. 3, we have added the cookie mechanism to mitigate DoS concerns, even though IKE minimal [29] leaves that out. The message is re-sent with the cookie attached, in order to prove the return routability of the initiator's address.

After the cookie is validated, the gateway sends the chosen cipher, another random nonce, and the Diffie–Hellman parameter of the sensor. The gateway needs to select a suitable SPI_s on behalf of the sensor and use a $Nonce_s$ that the sensor has provided to it. The used values need to be communicated to the sensor later, if they are not fixed. The same security considerations for the nonce are needed as in DTLS. In addition, an optional CERTREQ could be used to inform about the trust anchors the sensor is willing to trust.

The next pair of messages (I2 and R2) can be protected using the negotiated cipher and the keying material generated with the help of the nonces and the common DH secret. The initiator's message contains its identity, which can be, e.g., IP address, but a more interesting possibility is to base it on a public key. That can be used with AUTH parameter to prove the ownership of the

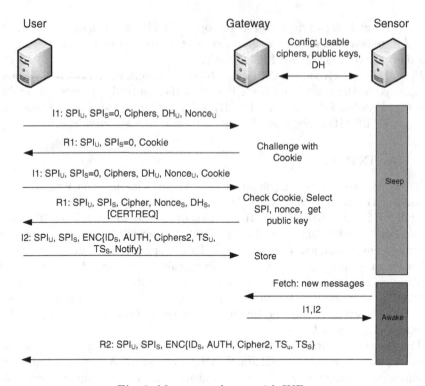

Fig. 3. Message exchange with IKE

key by signing data (basically the received message with some additional data). Alternatively, one could authenticate based on a shared secret.

The message also contains proposal for algorithms, which are to be used for the child security association, i.e., the one that is going to protect data payloads. The Notify parameter in Fig. 3 is used as in [29] to indicate that this is the first association to be created, thus effectively deleting any previous ones between these participants.

Next the I2 message is stored by the gateway to wait for retrieval. When the sensor wakes up and sends a fetch message, the gateway sends both I1 and I2 messages to it. Even though, the sensor might not worry about downgrade attacks, for example, because of limited crypto options, the I1 message is still necessary in IKE, as it is needed for the symmetric secret used in the IKE_AUTH message.

As stated before, the sensor has to support using some values that were used in the gateway's IKE_SA_INIT-reply (message R1), like the aforementioned SPI_s and $Nonce_s$. The sensor will then send the IKE_AUTH message (R2 in Fig. 3) using the values that the gateway has used previously. This message states the selected cipher and the traffic selectors. After that, protected application level messages can be sent.

5   Discussion

It should be noted that while key exchanges are done in the traditional Internet to protect communication flows, it may not always be preferable for constrained nodes. This is true especially for message authentication done by a constrained device. For instance, storing multiple symmetric keys and maintaining sessions can be a burden for constrained devices if the associations are many. Moreover, public key cryptography has other advantages for message authentication. For example, non-repudiation is present when public key based digital signatures are used. Additionally, the signatures can be verified by anyone who has the public key, including the gateway in the presented scenario. This is in contrast to symmetric secrets where, in order to authenticate the same payload for multiple receivers the sensor may have to calculate a separate message authentication code for every recipient.

However, exchanged keys are a natural choice for confidentiality protection. For instance, the amount of stored key material is less for the symmetric keys, as symmetric keys are smaller for the same security level. While public keys could perhaps be stored somewhere else in an authenticated directory and the symmetric keys could be wrapped, some form of key exchange is needed in cases where the recipient lacks an authenticated public key or the liveliness of the other party is a concern.

The biggest benefits from the delayed key exchange materialize when the key exchange can be made asynchronous, and amount of transmitted data in the sensor's access network can be minimized. Certificates and certificate chains are lengthy messages, so schemes using certificates for authentication, such as DTLS, are more efficient if the gateway or the infrastructure is able to transmit such messages.

It is also important to minimize the sensor's waiting time for external responses, as denial of sleep can be a concern. Because a normal handshake lasts multiple flights, a sleepy sensor might need multiple sleep cycles to complete it. Moreover, direct key exchange might require the sensor to be awake at the same time the user is accessing it. For the first message this requires either synchronisation and knowledge about the sensor's wake up times, or constant polling. If the gateway would store only the first message and pass through the others, it would have to be aware of the wake up times or the sensor would have to poll the subsequent messages.

The presented key exchanges would both work in the delayed key exchange scenario. These protocols have traditionally been used for different purposes and direct comparison may not make that much sense. However, from the further development point of view, one can consider which is the more promising target. Overall, if we consider the current adoption, and integrating with existing clients, the best choice might be DTLS. For example, IETF CoRE Working Group mainly focuses on DTLS [6] for CoAP. Moreover, DTLS already has some support for using negotiated keys in applications. In addition, the TLS next protocol negotiation enables the client to send its first payload even before the sensor has woken up.

CoAP specification defines an application profile for DTLS, which includes two ciphersuites: PSK based ciphersuite[1], and a ciphersuite based on Ephemeral Elliptic Curve Diffie–Hellman key agreement[2] [6]. The ephemeral public key ciphersuite signs the initiator's nonce in an early phase of the handshake, and is not as such compatible with the presented procedure. However, the same primitives can be combined to produce a non-ephemeral version of the ciphersuite, which is compatible with our proposal. Also, the suggested PSK based ciphersuite should work in the presented scenario, particularly if the gateway is able to give identity hints, if necessary.

The security assumptions in the presented key exchange protocols are not broken by the suggested modifications as these protocols are not sensitive to timing changes. There could be benefit in offloading more functions to the gateway, for instance, the integrity calculation in DTLS is heavy. This would avoid the need for the sensor to receive and store all the handshake messages. However, an adversary would then be able to, for example, change the ServerHello message, and control the chosen ciphersuites. Also, the aforementioned replay attacks would be possible if the gateway would be allowed to choose the nonce, for example.

6   Conclusions

In this paper we have proposed a delayed key exchange method, which would take into account the sleeping nature of sensors and use an intermediary gateway to cache key exchange messages for later retrieval. Even though the gateway is used, the suggested method would still be able to provide end-to-end security. In our case, the keys are intended to be used to secure application layer messages, that is, data objects carried over HTTP and CoAP.

We have presented how the suggested procedure would work with two different existing key exchange protocols. Both investigated protocols can be used to do a key exchange in a delayed fashion. Moreover, the procedure is opaque for a client initiating the key exchange. However, some configuration and modification can be needed in the constrained node and the gateway element.

Part of our proposal is the data object security, for which we suggest using JSON based security tokens. Those tokens get their security material from the key exchange, which has been run prior to actual data exchange. JSON provides suitably light weight option and enjoys already widespread adoption in the web world, thus facilitating the integration of the Internet and Internet of Things.

Acknowledgements. The research was conducted in the Internet of Things program of DIGILE (Finnish Strategic Centre for Science, Technology and Innovation in the field of ICT), funded by Tekes.

---

[1] TLS_PSK_WITH_AES_128_CCM_8.
[2] TLS_ECDHE_ECDSA_WITH_AES_128_CCM_8.

References

1. Giusto, D., Lera, A., Morabito, G., Atzori, L.: The Internet of Things. Springer, New York (2010)
2. Garcia-Morchon, O., Keoh, S., Kumar, S., Hummen, R., Struik, R.: Security Considerations in the IP-based Internet of Things. Internet-Draft draft-garcia-core-security-04, Internet Engineering Task Force, March 2012, Work in progress
3. Castellani, A., Loreto, S., Rahman, A., Fossati, T., Dijk, E.: Best Practices for HTTP-CoAP Mapping Implementation. Internet-Draft draft-castellani-core-http-mapping-05, Internet Engineering Task Force, July 2012, Work in progress
4. Kaufman, C., Hoffman, P., Nir, Y., Eronen, P.: Internet Key Exchange Protocol Version 2 (IKEv2). RFC 5996 (Proposed Standard), September 2010, Updated by RFC 5998
5. Dierks, T., Rescorla, E.: The Transport Layer Security (TLS) Protocol Version 1.2. RFC 5246 (Proposed Standard), August 2008, Updated by RFCs 5746, 5878, 6176
6. Shelby, Z., Hartke, K., Bormann, C., Frank, B.: Constrained Application Protocol (CoAP). Internet-Draft draft-ietf-core-coap-11, Internet Engineering Task Force, July 2012, Work in progress
7. Rescorla, E., Modadugu, N.: Datagram Transport Layer Security. RFC 4347 (Proposed Standard), April 2006, Obsoleted by RFC 6347, updated by RFC 5746
8. Rescorla, E., Modadugu, N.: Datagram Transport Layer Security Version 1.2. RFC 6347 (Proposed Standard), January 2012
9. Hartke, K., Bergmann, O.: Datagram Transport Layer Security in Constrained Environments. Internet-Draft draft-hartke-core-codtls-02, Internet Engineering Task Force, July 2012, Work in progress
10. Nadalin, A., Goodner, M., Gudgin, M., Barbir, A., Granqvist, H.: Oasis ws-trust 1.4. Specification Version 1 (2008)
11. Shelby, Z.: Embedded web services. IEEE Wirel. Commun. 17(6), 52–57 (2010)
12. Jones, M., Rescorla, E., Hildebrand, J.: JSON Web Encryption (JWE). Internet-Draft draft-ietf-jose-json-web-encryption-05, Internet Engineering Task Force, July 2012, Work in progress
13. Sethi, M., Arkko, J., Keranen, A.: End-to-end security for sleepy smart object networks. In: 2012 IEEE 37th Conference on Local Computer Networks Workshops (LCN Workshops), pp. 964–972. IEEE (2012)
14. Bianchi, G., Capossele, A.T., Mei, A., Petrioli, C.: Flexible key exchange negotiation for wireless sensor networks. In: Proceedings of the Fifth ACM International Workshop on Wireless Network Testbeds, Experimental Evaluation and Characterization, WiNTECH '10, pp. 55–62. ACM, New York (2010)
15. Needham, R., Schroeder, M.: Using encryption for authentication in large networks of computers. Commun. ACM 21(12), 993–999 (1978)
16. Vial, M.: CoRE Mirror Server. Internet-Draft draft-vial-core-mirror-proxy-01, Internet Engineering Task Force, July 2012, Work in progress
17. Kadyk, D., Fishman, N., Seinfeld, M., Kramer, M.: Negotiating secure connections through a proxy server, 7 February 2006, US Patent 6,996,841
18. Ylitalo, J., Melén, J., Nikander, P., Torvinen, V.: Re-thinking security in IP based micro-mobility. In: Zhang, K., Zheng, Y. (eds.) ISC 2004. LNCS, vol. 3225, pp. 318–329. Springer, Heidelberg (2004)
19. Dolev, D., Yao, A.: On the security of public key protocols. IEEE Trans. Inf. Theor. 29(2), 198–208 (1983)

20. Nikander, P., Arkko, J., Ohlman, B.: Host identity indirection infrastructure (hi3). In: Proceedings of the 2nd Swedish National Computer Networking Workshop SNCNW 04, 1–4 (2004)

21. Arkko, J., Kernen, A.: CoAP Security Architecture. Internet-Draft draft-arkko-core-security, Internet Engineering Task Force, July 2011, Expired

22. Ylitalo, J., Salmela, P., Tschofenig, H.: Spinat: Integrating ipsec into overlay routing. In: First International Conference on Security and Privacy for Emerging Areas in Communications Networks, SecureComm 2005, pp. 315–326. IEEE (2005)

23. Jones, M., Bradley, J., Sakimura, N.: JSON Web Signature (JWS). Internet-Draft draft-ietf-jose-json-web-signature-05, Internet Engineering Task Force, July 2012, Work in progress

24. Jucker, S.: Securing the constrained application protocol (2012)

25. Canetti, R., Krawczyk, H.: Analysis of key-exchange protocols and their use for building secure channels. In: Pfitzmann, B. (ed.) EUROCRYPT 2001. LNCS, vol. 2045, pp. 453–474. Springer, Heidelberg (2001)

26. Langley, A.: Transport Layer Security (TLS) Next Protocol Negotiation Extension. Internet-Draft draft-agl-tls-nextprotoneg-04, Internet Engineering Task Force, May 2012, Work in progress

27. Rescorla, E.: Keying Material Exporters for Transport Layer Security (TLS). RFC 5705 (Proposed Standard), March 2010

28. McGrew, D., Rescorla, E.: Datagram Transport Layer Security (DTLS) Extension to Establish Keys for the Secure Real-time Transport Protocol (SRTP). RFC 5764 (Proposed Standard), May 2010

29. Kivinen, T.: Minimal IKEv2. Internet-Draft draft-kivinen-ipsecme-ikev2-minimal-00, Internet Engineering Task Force, February 2011, Expired

Concept of IoT 2.0 Platform

Jordi Mongay Batalla[✉], Mariusz Gajewski, and Konrad Sienkiewicz

National Institute of Telecommunications, ul. Szachowa 1, 04-894 Warsaw, Poland
{j.mongay,m.gajewski,k.sienkiewicz}@itl.waw.pl

Abstract. In the near future, Internet of Things (IoT) should integrate an extremely large amount of heterogeneous entities. However this process runs slower than expected due to the difficulty to develop services related to heterogeneous things. Environment for easy creation of IoT services (based on open interfaces) is one of the solutions for speeding up the IoT massive usage. In order to deploy such an environment, we propose to follow the track of the so-called Telco Application Programming Interface (known as Telco 2.0) initiatives. In this paper we show the concept of IoT 2.0 platform (based on Telco 2.0) and describe functional architecture of the proposed solution.

1 Introduction

Internet of Things (IoT) refers to a global network infrastructure linking a huge amount of everyday things, where physical and virtual objects may communicate without human interaction. In IoT, objects are active participants of network ecosystem: they can recognize changes in their surroundings, share information about those changes and perform appropriate actions in an autonomous way. This paper proposes an implementation strategy for IoT services on the track of the so-called Telco Application Programming Interface (API) -known also as Telco 2.0- initiatives [1,2] in order to maximize the IoT usage[1].

Telco 2.0 aims at exploiting the services of (mobile) telecom operators (sms, mms, etc.). Its long tail business model involves letting the market innovate, by allowing third party developers to implement new personalized services. It allows telcos to receive instant and direct feedback from developers and end users. On technology wise, open APIs shorten service innovation cycle and help to create competitive service products. It is worth mentioning that, the number of API calls on the AT&T network has grown from 300 million USD per month in 2009 to more than 4.5 billion USD by the end of 2011 thanks to the introduction of open APIs [3].

On the track of Telco 2.0 we propose platform for programming of composed services based on open IoT APIs, which we call IoT 2.0. The proposed IoT 2.0

---

[1] This strategy is being used for the development of an extended IoT environment (called IoT 2.0) on the framework of POLLUX II IDSECOM project. Note that the proposed solution is not still fully implemented and no results are exposed in this paper, but we think that the idea and functional architecture are worth enough to be presented in a positioning paper.

© Springer-Verlag Berlin Heidelberg 2015
M. Garcia Pineda et al. (Eds.): ADHOC-NOW Workshops 2014, LNCS 8629, pp. 27–34, 2015.
DOI: 10.1007/978-3-662-46338-3_3

environment allows the creation of composed services on the basis of elementary IoT services exposed by sensors/actuators/tag readers, wherein these objects are connected to Internet directly or via gateways. The elementary services provided by the IoT object vendors are firstly unified for spread use and exposed in the platform. At last, composed services are made available by their creators for the use and/or further adaptation by other consumers. This approach aims to extend the popularity and interest of IoT by making easier the collaboration in programming new services and applications as well as introducing new business models as it occurred in the case of Telecom APIs exposed by telcos.

2   Perspectives for IoT

The solutions for IoT implementation focus on uniform naming [4] and addressing, and common protocols for ubiquitous smart objects. It concerns objects reachable both directly by IP network or connected using dedicated protocols specific for given area of use (e.g., in case of home automation the most common standards are X-10, Z-Wave and Insteon). This approach makes feasible the creation of common platforms which aim to provide tools for management, data analysis and adequate reaction, and makes use of numerous IoT middleware solutions providing interoperability. This is a crucial step towards the exposition of IoT services. Many currently running international projects deal with exposition of objects/services for use at the middleware layer. Generally, these efforts center on exposing virtual representations of physical objects in order to develop a common communication platform. In this way, SENSEI project [5] introduces the concept of resource which corresponds to a physical entity in the real world. Moreover, the resource is related to a software process (called Resource End Point), which represents the physical resource in the Resource layer and implements a set of Resource Access Interfaces. For this purpose, SENSEI proposes each resource to provide one or several standard and non-standard access interfaces, using different protocols. For example, one resource could provide a GET operation for retrieving data (i.e., temperature, pressure, etc.) using REST (Representational State Transfer) protocol while another resource could provide an operation called Read_Temperature using full Web Services.

The solution proposed in IoT-A project [6] assumes that the objects expose vendor API to the unified communication platform. In consequence, the virtualization of devices is required (so-called virtual entities) and an information model should be developed in order to describe virtual entities. This concept is independent of specific technologies and use-cases, and IoT-A does not investigate the specific ways for implementing it (only it offers some general methods and tools). As we describe below, our proposal assumes an uniform approach of exposed API functions, which is in line with the idea of resources virtualization from IoT-A.

Another pending challenge addressed by some few initiatives is the broader usage of IoT platforms by encouraging advanced users/programmers to implement attractive services. These services could be next subscribed by consumers and exploited according to their needs. The COMPOSE project [7] uses cloud-based infrastructure featuring Platform as a Service (PaaS) for hosting back-end applications and, on the other hand, it introduces an IoT Marketplace for offering IoT applications. The scope of this project is to deliver a Software Development Kit (SDK) for service high-level development as well as runtime environment for service configuration and execution. The creation of composed services is assumed to be performed by advanced programmers which are experts on the delivered SDK. Whereas, our approach aims at delivering a set of tools which supports application creation/adaptation process performed by the general public (including non-advanced programmers). Moreover, our approach tends to reach wide spread of IoT use thanks to the native sharing of simple and composed services due to the introduction of open source-like model, avoiding dedicated platforms.

Opengate [8] is a commercial Machine-to-Machine (M2M) platform provided and hosted by Amplia that allows the user to manage M2M communications without having to deploy its own infrastructure. Opengate develops a set of tools for creating and running M2M/IoT services. It addresses mainly both the IoT services and resources layer and the device connectivity layer, implementing also some security functions. A similar approach is taken in Xively [9] and Nimbits [10] projects, which, in turn, focus on support of smart objects (e.g., modules for Arduino, Raspberry Pi, etc.). Both platforms provide Web access to registered smart objects as well as set of analytical tools for retrieved data. In these cases, the creation of advanced application environment (i.e., Integrated Development Environment, IDE) is not supported.

There are two more projects where service creation function is strongly addressed: ClickScript [11] and homeBLOX [12]. Both platforms engage home automation devices. ClickScript allows users to visually create Web mash-ups by connecting building blocks of resources (Web sites) and operations, whereas homeBLOX extends a process engine with the capabilities to communicate with heterogeneous smart devices, to integrate virtual devices and to support different home automation protocols. It is equipped with a graph-based user interface which abstracts from the complexity of process specification. Both solutions are specific and not open to general use and sharing of IoT resources.

3   IoT 2.0

The proposed IoT 2.0 platform is described in three subsections. In the first one we present the general idea, the second one includes the description of functional elements and the last one highlights the benefits of this approach.

3.1   General Idea

We propose an universal open programmable platform that uses different APIs located at the borders of the sensor networks and are accessible in an open way. The proposed environment allows the creation of composed services on the basis of existing and future elementary IoT services exposed by sensors/actuators/tag readers, wherein these objects are connected to Internet directly or via gateways [13]. The elementary services provided by the vendors are firstly unified for spread use and exposed in the platform. Figure 1 presents overall idea of the platform where the elementary services provided by the vendors are firstly unified for spread use (black lines in Fig. 1) and exposed in the platform. At last, composed services are made available by their creators (green lines in Fig. 1) for the further use and/or adaptation by other consumers (red line in Fig. 1).

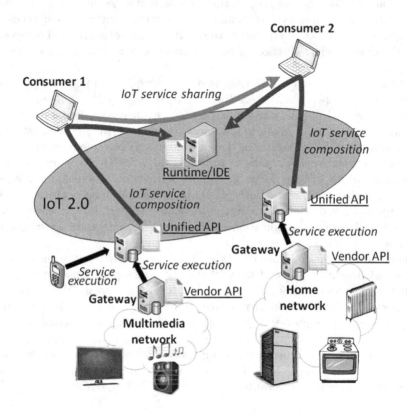

Fig. 1. IoT 2.0 overall idea (Color figure online)

3.2 Functional Elements of the Platform

The proposed IoT 2.0 platform consists of several main functional elements which are:

- unified Application Programming Interface (API),
- semantic interoperability,
- Integrated Development Environment(IDE),
- runtime environment,
- application sharing platform.

Unified API provides functions to create, deploy and manage services and applications based on IoT elementary services (made available by the vendors of the things) for users/ developers. This approach is crucial since the APIs from different vendors are not unified with other vendors(e.g., vendors of different fridges in home networks). In this way the proposed unified API provides common interface to services offered by objects produced by different vendors. The functions exposed by the unified API should be independent from physical specific capabilities of IoT objects so that service creators are not forced to consider device constraints. That is the reason why the implementation of the unified API should take into account different types of smart objects and networks of smart objects (e.g., RFID tags, 6LowPAN).

The core of the idea is the reasoning, that only a uniform description of the objects and functionalities of the sensors can lead to a wide use of the platform and, as a consequence, of the sensors in business and everyday activities. Therefore, **semantic technologies** are key enablers for IoT because they resolve semantic interoperability and integration, as well as facilitate reasoning, classification and automation. Moreover semantic methods should assure a uniform namespace for variety of smart objects. In result, it should be possible to search object/service across different sensor networks according to specific criteria. This process may engage advanced methods based on hermeneutic profile (which takes into accounts individual, personalized ontological model) of a user [14].

Used ontology is divided into upper ontology and domain-specific ontologies, as described in [15]. The upper ontology is a high-level ontology which captures general context knowledge about the physical world. In our solution we propose Semantic Sensor Network (SSN) ontology [16] for upper ontology. SSN is commonly used in IoT initiatives (e.g., IoT-A, SENSEI, Ebbits). The domain-specific ontologies define the details of general concepts and their properties in each domain (e.g., home) and cover particular area of usage. The ontology encompasses the specific features for IoT objects and, especially in our case, services. For example, the validity time of service [13] indicates the significance of the service and its value is configurable and strongly depends on the place and the context of object usage. Moreover it should cover as many potential areas of IoT use as possible. Therefore, IoT 2.0 platform provides mechanisms to easy import existing ontologies (designed for specific purposes). It is important because nowadays there are many specific ontologies. As an example, the Semantic Web for Earth and Environmental Terminology (SWEET) [17] includes more than 200 ontologies; the capacity of importing them will shorten IoT services implementation. On the other hand, the IoT 2.0 platform makes possible to create any new domain ontology which exactly fits to IoT services developer needs.

The **IDE** is another functional element of the platform. It is used to accelerate development of composed services and applications by the users/developers (not familiar with programming languages). The IDE consists of a set of tools used to develop cross platform applications which make use of variety of sensors. This is performed in an integrated development environment which can be comprised of graphical front-ends for code editing, compilation, documentation, source versioning, change management, debugging and profiling. Moreover, IDE is suited to the platform where it is run (e.g., it takes into account constraints of devices with reduced accessibility). The IDE functionalities include among others:

- managing service creation process including testing and debugging tools;
- easy drag&drop designing for service composition, with support for different modelling languages.

Composed services designed in the IDE can be then tested and finally run according to service developer needs by means of the runtime environment.

The **runtime environment** functional element encompasses functions responsible for running created services. We assume that IoT users may execute their services on standalone machines or virtual instances (provided by clouds). In order the platform to be scalable, we use reference protocols, which are widely considered as scalable. Web Services is used for communication between involved devices, where addressing IoT services is performed by the URI of the HTTP protocol.

Composed IoT services can be shared with other users by means of the **application sharing platform** (web portal), which makes consumers able to download the code of foreign applications, as well as to test and adapt it. Moreover portal provides detailed documentation and the informative blog, whereas the developers community discusses the latest innovations in forums. The application sharing platform also includes catalogue of available IoT services and searching tool. Dependencies between functional elements described above are shown in Fig. 2.

There are three types of actions between functional elements: describes, uses and exposes. The IDE uses semantic technologies, which describes elementary

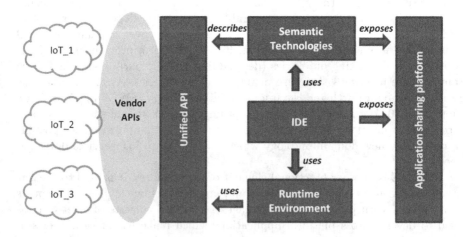

Fig. 2. IoT 2.0 platform

IoT services and objects available through unified API, to create IoT services. Created services are executed in runtime environment, which uses the unified API to communicate with objects. On the other hand, application sharing platform exposes IoT services definitions (elementary and composed) to developers.

3.3   Benefits of IoT 2.0

Our approach tends to reach wide spread of IoT use thanks to the native sharing of simple and composed services due to the introduction of open source-like model, avoiding dedicated platforms. Proposed API, IDE and runtime environment create a powerful ecosystem, which, in turn, can be used like a "Sensor network as a Service" platform. This idea encompasses the business model that assumes that general consumers are able to create and modify existing services, and then share them with other consumers in open source-like manner.

The proposed solution goes beyond the current idea of an unified and established IoT communication and proposes to use the sensor islands in an anarchical way, as current spread approaches as telcos show (case AT&T, see [3]).

4   Summary

This paper shows the concept of IoT 2.0 platform and describes its functional architecture. This platform allows the creation of composed services on the basis of existing and future elementary IoT services exposed by sensors/actuators/tag readers, wherein these objects are connected to Internet directly or via gateways.

This approach takes into account current works on IoT research as well as successful business models inherited from telecommunication market. It aims to extend the popularity and interest of IoT by making easier the collaboration in programming new services and applications as well as to introduce new business models as it occurred in the case of Telecom APIs exposed by telcos.

We argue that a good description of the services (based on proposed ontology) together to their easy management by the IDE will provide to a fast development of the IoT environment. The new IoT services will gain in complexity while maintaining their handiness and finally it will result in a vast spread of IoT usage.

Further work will center on the definition of unified API as the core functionality of the IoT 2.0 platform. Moreover, it should be analysed the data model and select an adequate set of tools supporting semantic interoperability.

Acknowledgments. This work was undertaken under the POLLUX II IDSECOM project. We would also like to thank our project partners who have implicitly contributed to the ideas presented here.

References

1. Davies, J., Duke, A., Mehandjiev, N., lvaro Rey, G., Stini, S.: SOA4 All Project, D8.5 Telco 2.0 Recommendations 2010

2. Chudnovskyy, O., Weinhold, F., Gebhardt, H., Gaedke, M.: Integration of telco services into enterprise mashup applications. In: Harth, A., Koch, N. (eds.) ICWE 2011. LNCS, vol. 7059, pp. 37–48. Springer, Heidelberg (2012)
3. Voxeo Labs Tropo whitepaper: Make the Shift From Telco Power to Telco Powered with the Tropo API 2013
4. Mongay Batalla, J., Krawiec, P., Gajewski, M., Sienkiewicz, K.: ID layer for internet of things based on name-oriented networkig. J. Telecommun. Inf. Technol. (2), 40–48 (2013)
5. EU FP7 SENSEI Project Consortium: Final SENSEI Architecture Framework. SENSEI Project Deliverable D3.6 (2011)
6. EU FP7 Internet of Things Architecture IoT-A Consortium. Project Deliverable D1.5 Final Architectural Reference Model for IoT (2013)
7. Doukas, C., Antonelli, F.: COMPOSE: Building smart & context-aware mobile applications utilizing IoT technologies. In: 5th IEEE Global Information Infrastructure & Networking Symposium, October 2013
8. Watanabe, K., Otani, M., Tadaki, S., Watanabe, Y.: Opengate on the cloud. In: 26th International Conference on Advanced Information Networking and Applications Workshops (2012)
9. Xively Public Cloud for the Internet of Things. https://xively.com/
10. Nimbits. http://www.nimbits.com/
11. Guinard, D., Trifa, V., Mattern, F., Wilde, E.: From the internet of things to the web of things: Resource oriented architecture and best practices. In: Uckelmann, D., Harrison, M., Michahelles, F. (eds.) Architecting the Internet of Things, pp. 97–129. Springer, Berlin (2011)
12. Walch, M., Rietzler, M., Greim, J., Schaub, F., Wiedersheim, B., Weber, M.: Home-BLOX: making home automation usable. In: International Joint Conference on Pervasive and Ubiquitous Computing (2013)
13. Mongay Batalla, J., Krawiec, P.: Conception of ID layer performance at the network level for internet of things. Pers. Ubiquitous Comput. **18**(2), 465–480 (2014). Springer
14. Chudzian, C., Klimasara, E., Paterek, A., Sobieszek, J., Wierzbicki, A.P.: Personalized search using knowledge collected and made accessible s model of ontological profile of the user and group in PrOnto system. In: SYNAT Workshop, Warsaw Poland, 1 July 2011
15. Guarino, N.: Ormal ontology and information systems. In: First International Conference on Formal Ontologies in Information Systems (1998)
16. Compton, M., et al.: The SSN ontology of the W3C semantic sensor network incubator group. J. Web. Semant. **17**, 25–32 (2012). ISSN 1873-7749
17. SWEET ontology. http://sweet.jpl.nasa.gov/ontology/

A Cooperative End to End Key Management Scheme for E-health Applications in the Context of Internet of Things

Mohammed Riyadh Abdmeziem[1]([✉]) and Djamel Tandjaoui[2]

[1] LSI, USTHB: University of Sciences and Technology Houari Boumedienne,
BP 32, El Alia Bab Ezzouar, Algiers, Algeria
`rabdmeziem@usthb.dz`
[2] CERIST: Center for Research on Scientific and Technical Information,
03, Rue Des Freres Aissou, Ben Aknoun, Algiers, Algeria
`dtandjaoui@mail.cerist.dz`

Abstract. In the context of Internet of Things where real world objects will automatically be part of the Internet, e-health applications have emerged as a promising approach to provide unobtrusive support for elderly and frail people. However, due to the limited resources available and privacy concerns, security issues constitute a major obstacle to their deployment. Among these issues, key distribution for heterogeneous nodes is problematic due to the inconsistencies in their cryptographic primitives. This paper introduces a new key management scheme that aims to establish session keys for highly resource-constrained nodes ensuring security protection through strong encryption and authentication means. Our protocol is based on collaboration by offloading heavy asymmetric cryptographic operations to a set of third parties. The generated shared secret is then used to derive further credentials. Security analysis demonstrates that our protocol provides strong security features while the scarcity of resources is taken into consideration.

Keywords: Internet of things · E-health · Wireless Body Area Networks (WBAN) · Confidentiality · Key management · Cooperation

1 Introduction

Internet of things (IoT) has recently received research attention. Through this concept, it is possible to connect everyday sensors and devices to each other and to the Internet. According to [1], the main concept behind IoT is the pervasive presence around us of various wireless technologies such as Radio-Frequency IDentification (RFID) tags, sensors, actuators or mobile phones in which computing and communication systems are seamlessly embedded. Based on unique addressing schemes, these objects interact with each other and cooperate to reach common goals.

M. Garcia Pineda et al. (Eds.): ADHOC-NOW Workshops 2014, LNCS 8629, pp. 35–46, 2015.
DOI: 10.1007/978-3-662-46338-3_4

Technology advances along with popular demand will foster the widespread deployment of IoTs services, it would radically transform our corporations, communities and personal spheres. From the perspective of a private user, IoTs introduction will play a leading role in several services. E-health is one of the most interesting applications as it will provide medical monitoring to millions of elderly and disabled patients while preserving their autonomy and comfort anywhere. Using sensors planted in or around the body, physiological data are gathered and transmitted to qualified personnel that can intervene in case of an emergency. Nevertheless, e-health applications are unlikely to fulfil a widespread diffusion until they provide strong security foundations. Making sure that only authorized entities can access and modify data is particularly relevant in e-health applications where data are very sensitive and any unwanted modification could lead to dramatic events. Securing communications in e-health systems necessarily passes through key management protocols. They are in charge of delivering security credentials to the involved entities. However, classical solutions are hindered due to the scarcity of resources available, either energy power or computation capabilities.

In this paper, we propose a new lightweight key management scheme based on collaboration to establish a secured communication channel between a highly resource constrained node and a remote entity (server). The secured channel allows the constrained node to transmit captured data while ensuring confidentiality and authentication. Our solution is based on the offloading of highly consuming cryptographic primitives to third entities (not necessarily trusted). Constrained nodes obtain assistance from more powerful entities in order to securely establish a shared secret with any remote entity.

The structure of this paper is organized as follows. In Sect. 2, e-health applications in the context of IoT are briefly introduced along with the main security threats that might hinder their deployment. Thereafter, we provide an overview on the state of the art of the proposed security approaches in Sect. 3. In Sect. 4, we present in detail our novel cooperative key management scheme. We continue in Sect. 5 with an analysis of our protocol in terms of security requirements. Section 6 concludes the paper and gives future directions.

2 E-health Applications in the Context of Internet of Things

Internet of Things deployment will open doors to a huge number of applications that would deeply improve our daily life. Among IoT applications, e-health systems are gaining more and more attention [1]. An e-health system is a radio-frequency-based wireless networking technology that provides ubiquitous networking functionalities. It is based on the interconnection of tiny nodes enhanced with sensing and/or actuating capabilities planted in, on, or around a human body. E-health systems are context-aware, personal, dynamic and anticipative. IoT meets all these characteristics providing a suited environment for e-health applications. An extensive research on using IoT paradigm in e-health has recently been reported [2]. In fact, population ageing and the increasing survival rates from disabling accidents and illnesses will lead to a more important part of the population that requires a continuous health care and monitoring [3]. E-health applications could spare the patient from long stays in hospitals, which is especially sought in emerging countries that lack medical infrastructures and well-trained personnel. Additionally, the continuous monitoring anticipates emergency situations allowing rapid and effective intervention of health teams in case of emergency. Moreover, early stage diagnostics could also be achieved remotely [4]. In sum, e-health systems, in the context of IoT, constitute a cost effective and unobtrusive solution without interrupting the patient's everyday activities. Nevertheless, e-health applications deployment could be hindered if privacy challenges are not addressed efficiently.

Studies in [5–8] have underlined that e-health applications might be more vulnerable to attacks compared to other IoT applications as the generated data are very sensitive. The health related data are always private in nature; any breach in the confidentiality of personal captured data would seriously repulse patients from adopting e-health solutions. For instance, many people would not like their health personal information, such as early stage of pregnancy or details of certain medical conditions, be divulged to the public domain [9]. The eavesdropped communications could be used in many illegal purposes. Moreover, any modification in the captured data could lead to disastrous consequences as it could engender wrong medical prescriptions or delay an emergency intervention.

Classical countermeasures are not suited to the constrained environment of IoT due to several factors such as power and computation scarcity, weak reliability of wireless links or the scalability issue. Thus, a considerable effort has been undertaken by the research community to provide viable solutions to secure IoT applications. The next section provides an overview on the state of the art of the proposed security approaches and positions our contribution regarding the literature.

3 Related Work

The research community has focused its attention on proposing security protocols that take into consideration the constrained environment of IoT. In our discussion of related work, we distinguish two research directions: (i) specific solutions for e-health systems, (ii) tailoring of security protocols for the IP-based IoT.

Several specific solutions for e-health systems have been proposed in the literature. TinySec is part of the official TinyOS release, it aims to achieve link-layer encryption and authentication of data in biomedical sensor [10]. The protocol is based on a single key shared among nodes which constitutes its main weakness as node capture would give access to the entire network. Otherwise, hardware solutions are proposed to deal with the scarcity of resources [11,12]. Nevertheless, these approaches present some drawbacks as they do not offer AES (Advanced Encryption Standard) decryption (only base stations can decrypt the transmitted data), they are highly platform-dependent and not all the nodes are equipped with hardware encryption capabilities. A different approach is based on biometric techniques [13,14]. These techniques use the human body to manage the key establishment process based on physiological values (e.g., electrocardiogram). One of their main drawbacks is the recoverability that is not complete at nodes over the network.

A different but complementary research direction has seen several approaches that aim to *tailor security protocols for the IP-based IoT*. The main focus of these works is to make standard based security protocols more suitable for the constrained environment of IoT. Specifically, several compression schemes for the IP-based IoT have been proposed. In [15,16], the compression of IPV6 headers, extension headers, along with UDP (User Datagram Protocol) headers have been standardized through 6LoWPAN (IPv6 over Low power Wireless Personal Area Networks). Authors in [17,18] have presented 6LowPAN compressions for IPsec payload headers: AH (Authentication Header) and ESP (Encapsulating Security Payload). In [19], authors have proposed a tailoring to Mikey-Ticket protocol for e-health applications in the context of IoT. Furthermore, an IKE (Internet Key Exchange) compression scheme [20] has also been proposed in order to provide a lightweight automatic way to establish security associations for IPsec.

Apart from packet compression schemes, further design improvement approaches have been introduced to tailor security protocols to the IoT. Authors, in [21], have proposed complementary lightweight extensions to HIP DEX (Host Identity Protocol Diet Exchange) that could be generalized to DTLS (Datagram Transport Layer Security) and IKE. Furthermore, delegation procedures of protocol's primitives have also been proposed in the literature. They aim to offload the computational load to third entities. Authors in [22] have introduced collaboration for HIP. The basic idea is to take advantage of more powerful nodes in the neighborhood of a constrained node to carry heavy computations in a distributed way. Likewise, an IKE session establishment delegation to gateways has been proposed in [23]. Furthermore, authors in [24], have introduced a delegation procedure that enables a client to delegate its certificate validation to a trusted server. While the proposed delegation approaches reduce the computational load at the constrained nodes, they break the end to end principle by requiring a third trusted party. Our novel cooperative key management scheme overcomes this limitation by providing an end to end secured channel between constrained nodes and remote entities. The end to end principle is highly sought in e-health applications as captured data are highly sensitive. In fact, we do believe that securing IoT applications will be achieved through the tailoring of current security protocols to IoT environment rather than developing specific solutions to each application scenario as it is safer to build on tested and trusted security protocols.

4   The Proposed Scheme

In this section, we present our lightweight end to end key management scheme. The proposed solution ensures key exchange with minimal resource consumption.

Firstly, we present the network model and a set of assumptions. Afterwards, we provide a broad overview of our approach along with a summary of the notations used throughout the paper. Finally, we describe in detail the different phases of our protocol.

4.1 Network Model

We consider in our network model four main components: the mobile and contextual sensors (constrained nodes), the third parties, the remote server and the certification authority. (See Fig. 1).

- *Mobile and contextual sensors:* the sensors are planted in, on or around a human body to enable health-related data to be collected (e.g. blood pressure, blood glucose level, temperature level, etc.).
- *Third Party:* the third parties represent a key component in our protocol. A third party could be any entity that is able to perform high consuming computations on behalf of the sensor nodes. The resource constrained sensors rely on them by offloading the high consuming cryptographic primitives in a cooperative way.
- *Remote server:* the remote server receives the gathered data for further processing. A remote server could be used by caregiver services in order to take appropriate decisions according to patient's data.
- *Certification authority:* the certification authority is required to guarantee authentication between the third parties and the remote server by delivering authenticated certificates.

The network is thus heterogeneous combining nodes with various capabilities both in terms of computing power and energy resources. We distinguish two categories of entities:

- Highly resource constrained nodes (mobile and contextual sensors), unable to perform public key cryptographic operations.
- Nodes with high energy, computing power and storage capabilities (the third parties and the remote server).

We assume an end to end secured communication between sensor nodes and the remote server due to the high sensitivity of gathered data. Hence, a key exchange protocol is required between the two entities to secure their communications. The protocol has to deal with resource capabilities of the involved entities along with the fact that no prior knowledge has been established between them.

4.2 Assumptions

For the implementation of our protocol we assume that:

- Sensor nodes are able to perform symmetric encryption.
- Third parties are able to perform asymmetric cryptographic operations (either public or private).

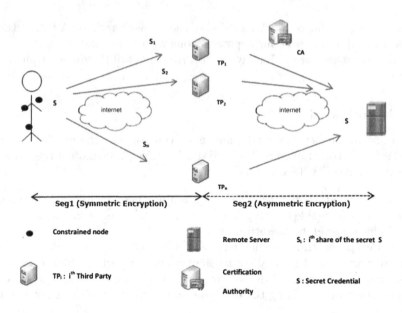

Fig. 1. Network model

- Third parties are not necessarily trusted.
- The remote server is powerful enough to support asymmetric encryption.
- The certification authority is a trusted entity. It delivers authenticated cryptographic credentials to the third parties and to the remote server.
- Each sensor node is able to keep a list of remote third parties pre-established during the initialization phase.
- Each sensor node shares pairwise keys with each third party. These keys may be generated during the initialization phase.
- Both third entities and the remote server own a pair of public/private key.

4.3   Overview of the Proposed Scheme

We provide a broad overview of our protocol before diving into a formal description in the next section.

Once a resource constrained node is willing to establish a shared secret with a remote server, it initiates our protocol which goes through successive phases. We propose an offloading of the heavy computational asymmetric operations through the third entities in a cooperative way. The shared secret generated by the constrained node is split and transmitted to the third parties. The encryption of each part is based on symmetric algorithms (less resource consuming than asymmetric one) using pre-shared keys. MAC (Message Code Authentication) messages are used to ensure authentication.

Each third party secures the delivery of the received secret part to the remote server. The encryption is based on asymmetric algorithms using the remote server's public key. Authentication is provided using digital signatures. Upon successful authentication and decryption of the different parts, the remote server reassembles the shared secret, which will be used to derive further keying materials. In our protocol, we make sure that each third party proves to the remote server that it is actually a legitimate entity, authorized by the constrained node to act on its behalf. The following section describes in detail each phase of our protocol.

Table 1. Terminology table

Notation	Description
CN	Constrained node (the sensor)
UN	Unconstrained node (the remote server)
TP_i	Third party
CA	Certification authority
N_x	Nonce generated by node X
$K_{x,y}$	Shared pairwise key between X and Y
K_x	Public key of node X
K_x^{-1}	Private key of node X
$[data]_K$	Data encrypted with the key K
$SIGN_X$	X's digital signature
S	Secret credential used to generate further keying material when required

4.4 Formal Description

After an initialization phase where each constrained node is pre-loaded with a set of third parties identities along with pre-shared keys, our protocol proceeds with successive phases. Table 1 summarizes the notations used to present the exchanged messages and Fig. 2 illustrates the succeeding phases of our protocol.

- *Phase 1: Initial exchange.* Node CN initiates the exchange by sending a CN_hello message (A) to UN. The message informs UN about the security policies (e.g. encryption algorithm, HMAC algorithm, key life time,... etc.) and the cooperative key establishment process it supports. If UN agrees, it selects one of the proposed security policies and responds with a UN_hello message. Nonces are included in the exchanged messages to prevent replay attacks.
- *Phase 2: Securing connection between entities.* This phase follows the successful connection between CN and UN. It aims to establish a secure channel either between CN and TP_i or between TP_i and UN.

    In message C, CN informs the third parties about UN identity. The message includes a Message Authentication Code (MAC) and is encrypted using

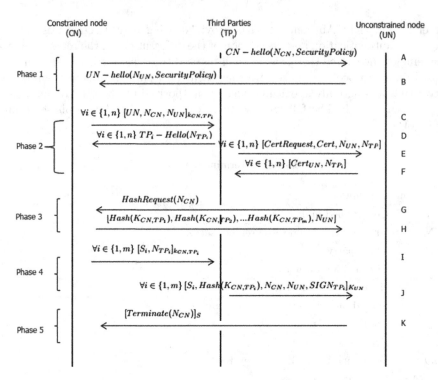

Fig. 2. Illustration of the different phases and message exchanges of our scheme

K_{CN,TP_i}. The third parties express their willingness to be part of the key exchange protocol through message D. It is worth noticing that not all the asked third parties respond with message D due to possible resource exhaustion or any other reason. Hence, we consider that only m TP_i ($m <= n$) respond with message D expressing their willingness to take part in the key exchange process.

In message E, each TP_i provides UN with its certificate containing its public key (delivered by CA) and requests UN for its own certificate. UN verifies that the third party has supplied a valid public key. It, then, responds with message F that contains the requested certificate. We highlight that all messages contain nonces against replay attacks.

- *Phase 3: Proving third parties's representativeness of CN to UN.* This phase aims to prove the representativeness of CN by the third parties to UN. The authentication is achieved using the pre-shared keys between CN and TP_i. In message G, UN requests the pairwise keys shared with the TP_i. CN applies a hash function on each key to keep it confidential and send it to UN through message H. The authentication will occur later after receiving message J from the third entities containing the key's hashes.
- *Phase 4: Secret generation and delivery.* Upon successful preparation of the involved entities, CN generates a premaster secret S used later to generate

further keying materials at both CN and UN sides. Because wireless connection is the main media in the IoT context, CN applies an error redundancy scheme to the original secret S. The aim is to enable UN retrieving the secret without requiring the reception of all the packets, in case where some of them where altered during the transmission process. In our solution, we have chosen the widely used Reed-Solomon code [25].

The secret is split into m parts S_1, S_2, ... S_m. Each part is sent to the appropriate TP_i in message I. The communication is secured using the symmetric key K_{CN,TP_i}. Upon receiving message I, each TP_i uses UN's public key to encrypt message J that contains the secret part S_i, K_{CN,TP_i}'s hash and TP_i's signature that covers all the fields of the message. After its decryption, UN verifies the authenticity of each message using TP_i's public key. If the messages are authenticated, UN verifies TP_i's representativeness of CN. The verification is done by the comparison of the hashes received in message H and those received in message J. If the hashes match, the TP_i act on behalf of CN as they pretend.

UN, then, reconstructs the secret S after having received enough packets. The secret is used to derive further key credentials along with the exchanged nonces during previous messages. Both messages I and J contains nonces to avoid replay attacks.

- *Phase 5: Termination phase.* This phase concludes the exchanges through message K by proving to CN the knowledge of the secret S.

The pre-master key S, along with the exchanged nonces are used by UN and CN to derive a master key. The derivation process is ensured by a hash function agreed upon during the first phase. Both parties are then able to derive state connection keys for encryption and authentication of the exchanged data. A secure end to end channel is hence created between highly constrained sensors and remote unconstrained servers.

5   Security Analysis

We provide an analysis of security features provided by our scheme based on the proprieties presented in [26]. We have added an analysis concerning integrity and confidentiality as we consider them being critical in an e-health system. For the following discussion, we consider our communication channel split into two segments: (Seg1) from CN to the TP_i and (Seg2) from the TP_i to UN (See Fig. 1)

- *Confidentiality:* the exchanged data between the different involved entities in our protocol are kept confidential. For Seg1, symmetric encryption is used based on the pre-shared keys set during an initialization phase. We recommend the use of the AES-CCM mode that defines AES-CBC for MAC generation with AES-CTR for encryption [27]. Nowadays, more and more tiny sensors include AES hardware coprocessor which would help to decrease the overhead. Regarding Seg2, communications are secured using Public Key Encryption (e.g. RSA algorithm). The CA is charged to deliver the required certificates to the involved entities. Our protocol can be run periodically to update the established keys in order to strengthen confidentiality and prevent long term attacks.

- *Authentication and integrity:* through the use of MACs in Seg1 and digital signatures in Seg2, our protocol makes sure that the exchanged data are genuine. The aim is to ensure that the data have not being altered and have been sent from legitimate nodes. Our scheme also ensures that the involved TP_i prove their authenticity to UN. This is done through the comparison of the secret shared between CN and the TP_i. (we refer to Sect. 4.4 for more details). Nonces (e.g. time-stamps, random values, ... etc) are included in the exchanged messaged to avoid any replay attacks.
- *Distribution:* the distribution of security credentials in Seg1 is performed by an off-line dealer during an initialization phase. However, in Seg2, through the use of Public Key Encryption, the involved entities establish a secure channel in an online mode. Thus, upon key's distribution in Seg1, our protocol can be run without any external intervention allowing updates to be proceed in an automatic way.
- *Overhead:* computation overhead is relatively low. Through the different messages of our protocol, constrained nodes are only involved in symmetric encryption primitives, which are much less resource consuming than asymmetric ones. All asymmetric operations are offloaded to the third parties that are much more powerful. Limiting computation requests for the constrained nodes decreases their power consumption and thus increases their battery life-time.
- *Resilience:* the resilience of our scheme is high. To compromise the exchanged secret S, an attacker has to take control of all third parties as S is split into several parts. Thus, the third parties are not required to be trusted. Unless all TP_i are compromised, it is nearly impossible to recover the secret.
- *Extensibility and scalability:* our network model allows new sensors to be integrated (e.g. we can imagine a physician prescribing the implantation of a new sensor for medical reasons). The new sensor has to pass through an initialization phase. The sensor will receive a set of TP_i identities to rely on, along with pairwise keys shared with each of them. This phase is performed by the network administrator. No operation is required concerning the TP_i or the remote servers that will be involved later in the protocol. Upon successful initialization phase, the new sensor can establish an end to end secure channel with any remote entity.
- *Storage:* nowadays, smart objects provide vast amounts of storage space due to the recent advances in flash memory technology [28]. In our protocol, we rely on this space to make the constrained nodes store the TP_i's identities list, along with the corresponding shared keys. We also consider that the number of TP_i will not exceed a certain threshold defined by the network administrator. Storage space will therefore not hinder our scheme's deployment.

6 Conclusion and Future Work

In this paper, we have proposed a new key management scheme for e-health applications in the context of Internet of Things. We have based our solution on the offloading of heavy cryptographic primitives to third parties in an end to end way. Our goal is to allow highly resource-constrained nodes to establish a shared end to end secret with any remote server. This is achieved through simple message exchanges with third parties. These messages are much less energy consuming than the use of asymmetric cryptographic primitives. Security analysis has shown that our solution provides strong security protection while resources scarcity is taken into consideration. In fact, the third parties support the heavy part of the computational load performing asymmetric cryptographic tasks instead of the constrained nodes. The scheme is then suitable to be applied in e-health applications deployed in a resource-constrained environment. As future work, we aim to implement and experiment our protocol in real testbeds to provide quantitative analysis as well as a comparative study with existing schemes. We also project to develop a lightweight trust model to allow constrained nodes automatically select effective third parties.

References

1. Atzori, L., Iera, A., Morabito, G.: The internet of things: a survey. Comput. Netw. **54**, 2787–2805 (2010)
2. Istepanian, R., Jara, A., Sungoor, A., Philips, N.: Internet of things for m-health applications (IOMT). In: AMA-IEEE Medical Technology Conference on Individualized Healthcare, Washington DC (2010)
3. Dohr, A., Modre-Opsrian, R., Drobics, M., Hayn, D., Schreier, G.: The internet of things for ambient assisted living. In: Information Technology: New Generations (ITNG), pp. 804–809, April 2010
4. Patel, M., Wang, J.: Applications, challenges, and prospective in emerging body area networking technologies. Wirel. Commun. **17**, 80–88 (2010)
5. Li, M., Lou, W., Ren, K.: Data security and privacy in wireless body area networks. IEEE Wirel. Commun. **17**(1), 51–58 (2010). doi:10.1109/MWC.2010.5416350
6. Javadi, S.S., Razzaque, M.A.: Security and privacy in wireless body area networks for health care applications. In: Khan, S., Khan Pathan, A.-S. (eds.) Wireless Networks and Security. SCT, pp. 165–187. Springer, Heidelberg (2013)
7. Lim, S., Oh, T., Choi, Y., Lakshman, T.: Security issues on wireless body area network for remote healthcare monitoring. In: IEEE International Conference on Sensor Networks, Ubiquitous, and Trustworthy Computing (SUTC), pp. 327–332, February 2010
8. Ng, H.S., Sim, M., Tan, C.: Security issues of wireless sensor networks in healthcare applications. BT Technol. J. **24**(2), 138–144 (2006)
9. Ameen, M.A., Liu, J., Kwak, K.: Security and privacy issues in wireless sensor networks for healthcare applications. J. Med. Syst. **36**, 93–101 (2012)
10. Karlof, C., Sastry, N., Wagner, D.: Tinysec: a link layer security architecture for wireless sensor networks. In: Second ACM Conference on Embedded Networked Sensor Systems, November 2004

11. Healy, M., Newe, T., Lewis, E.: Analysis of hardware encryption versus software encryption on wireless sensor network motes. In: Mukhopadhyay, S.C., Gupta, G.S. (eds.) Smart Sensors and Sensing Technology. LNEE, vol. 20, pp. 3–14. Springer, Heidelberg (2008)

12. Meingast, S., Roosta, T., Lewis, E.: Security and privacy issues with health-care information technology. In: Proceedings of the 28th Annual International Conference of the IEEE Engineering in Medicine and Biology Society, pp. 5453–5458 (2006)

13. Cherukuri, S., Venkatasubramanian, K., Gupta, S.: Biosec: a biometric based approach for securing communication in wireless networks of biosensors implanted in the human body. In: Proceedings of International Conference on Parallel Processing Workshops, October 2003

14. Poon, C., Zhang, Y.T., Bao, S.D.: A novel biometrics method to secure wireless body area sensor networks for telemedicine and m-health. IEEE Commun. Mag. 4, 73–81 (2006)

15. Montenegro, G., Kushalnagar, N., Hui, J., Culler, D.: Transmission of IPv6 packets over IEEE 802.15.4 networks. RFC 4944, IETF (2007)

16. Hui, J., Thubert, P.: Compression format for IPv6 datagrams over IEEE 802.15.4-based networks. RFC 6282, IETF (2011)

17. Granjal, J., Monteiro, E., Silva, J.S.: Enabling network-layer security on IPv6 wireless sensor networks. In: Proceedings of IEEE GLOBECOM (2010)

18. Raza, S., Duquennoy, S., Chung, T., Yazar, D., Voigt, T., Roedig, U.: Securing communication in 6LoWPAN with compressed IPsec. In: Proceedings of IEEE DCOSS (2011)

19. Abdmeziem, R., Tandjaoui, D.: Tailoring mikey-ticket to e-health applications in the context of internet of things. In: Proceedings of International Conference on Advanced Networking, Distributed Systems and Applications (INDS'2014) (2014)

20. Raza, S., Voigt, T., Jutvik, V.: Lightweight IKEv2: a key management solution for both compressed IPsec and IEEE 802.15.4 security. In: IETF/IAB Workshop on Smart Object Security (2012)

21. Hummen, R., Wirtz, H., Ziegeldorf, J.H., Hiller, J., Wehrle, K.: Tailoring end-to-end IP security protocols to the internet of things. In: Proceedings of IEEE ICNP (2013)

22. Saied, Y.B., Olivereau, A.: D-hip: a distributed key exchange scheme for hip-based internet of things. In: Proceedings of IEEE WoWMoM (2012)

23. Bonetto, R., Bui, N., Lakkundi, V., Olivereau, A., Serbanati, A., Rossi, M.: Secure communication for smart iot objects: protocol stacks, use cases and practical examples. In: Proceedings of IEEE WoWMoM (2012)

24. Freeman, T., Housley, R., Malpani, A., Cooper, D., Polk, W.: Server-based certificate validation protocol (SCVP). RFC 5055, IETF (2007)

25. Reed, S., Solomon, G.: Polynomial codes over certain finite fields. J. Soc. Ind. Appl. Math. 8, 300–304 (1960)

26. Roman, R., Alcaraz, C., Lopez, J., Sklavos, N.: Key management systems for sensor networks in the context of internet of things. Comput. Electr. Eng. 37, 147–159 (2011)

27. Dworkin, M.: Recommendation for block cipher modes of operation: the CCM mode for authentication and confidentiality. SP-800-38c, NIST, US Department of Commerce (2007)

28. Tsiftes, N., Dunkels, A.: A database in every sensor. In: Proceedings of the 9th ACM Conference on Embedded Networked Sensor Systems (2011)

2nd International Workshop on Marine Sensors and Systems, MARSS 2014

The Time Calibration System of KM3NeT: The Laser Beacon and the Nanobeacon

Diego Real[(⊠)] and David Calvo

IFIC. Instituto de Física Corpuscular, CSIC-Universidad de Valencia,
C/Catedrático José Beltrán, 2, 46980 Paterna, Spain
{real,dacaldia}@ific.uv.es

Abstract. The KM3NeT collaboration has started the construction of a deep
sea neutrino telescope in the Mediterranean with an instrumented volume of
several cubic kilometers. The objective of the KM3NeT telescope is to observe
cosmic neutrinos. For this, the detector will consist of a tri-dimensional array of
optical modules, each one composed of a pressure resistant glass sphere housing
31 small area photomultipliers. An important element of the KM3NeT detector
is the system for the relative time calibration between optical modules with a
precision of about 1 ns. The system comprises two independent devices: a
nanobeacon inside each optical module for calibration of optical modules in the
same vertical detection unit and a laser beacon for the calibration of optical
modules of vertical units. After a general introduction of the KM3NeT project, a
detailed description of the KM3NeT time calibration devices is presented.

Keywords: Neutrino telescope · Time calibration · Laser beacon ·
Nanobeacon

1 Introduction

KM3NeT [1] is a deep sea cabled research infrastructure to be deployed at the bottom
of the Mediterranean Sea. The infrastructure will host a very large volume neutrino
telescope distributed over three locations at depths of several kilometers. The main
objective of the telescope is to detect extraterrestrial neutrinos with energies above
50 GeV to investigate the origin of the highest energy cosmic rays. The detection
technique of the telescope is based on the detection of Cherenkov photons induced by
the passage of relativistic charged particles through the sea water. If a neutrino interacts
in the sea water in or in the vicinity of the detector or in the rock beneath it, it can
produce a subatomic particle called *muon*, which travels through the detector at a speed
exceeding that of light in water. Such an electrically charged particle generates
Cherenkov radiation, visible as faint blue light that will be recorded by the highly
sensitive light detectors of the telescope. The arrival times of the light collected by
optical detectors disposed in a three dimensional array are used to reconstruct the *muon*

D. Real and D. Calvo: On behalf of the KM3NeT collaboration.

M. Garcia Pineda et al. (Eds.): ADHOC-NOW Workshops 2014, LNCS 8629, pp. 49–56, 2015.
DOI: 10.1007/978-3-662-46338-3_5

trajectory and consequently the direction of the neutrino, as these are strongly corre-
lated. The instrumented volume of the KM3NeT detector will consist of a tri-dimensional array of *optical modules*, which are the core elements of a neutrino telescope.
Vertical *detection units* are moored on the seabed, each comprising 18 optical modules
distributed over a height of about 700 m. Horizontally, the detection units are separated
by about 100 m. In the case of KM3NeT, the Digital Optical Module (DOM) is a
pressure resistant glass sphere housing 31 small photomultipliers to measure the
Cherenkov light and transform it into electronic signals. The electronics signals are
read-out the front-end electronics inside the DOM, and the relevant information is sent
to shore.

The angular resolution of the reconstructed *muon* track depends highly on the
accurate measurement of the arrival time of Cherenkov photons reaching the photomultipliers in the optical modules. The quality of time and position calibration of the
detector is therefore of utmost importance to achieve a good angular resolution. Deep
sea neutrino telescopes have intrinsic and unavoidable limitations in the time precision
due to the chromatic dispersion and scattering of light in sea water ($\sigma \sim 2$ ns for a
travelling distance of 50 m), and the combined effect of the photomultiplier transit time
spread and electronics ($\sigma \sim 1.5$ ns). Taking into account these intrinsic limitations, the
required precision of a time calibration system to measure the relative time between
optical modules should be $\sigma \leq 1$ ns. Pulse light emitters (beacons) [2] are successfully
used in the ANTARES neutrino telescope [3] (experiment precursor of KM3NeT) to
measure *in situ* relative time offsets between optical modules [4]. LEDs and lasers
located throughout the detector [5, 6] produce short duration and powerful light pulses
which are detected by the photomultipliers, allowing the measurement of the time delay
between the arrival of the photon at the photocathode and the time stamp by the front-
end electronics. In KM3NeT, the time calibration procedure has been decoupled in two
different techniques:

- Intra-DU Calibration: LED beacons, called *nanobeacon*s in KM3NeT, will be used
  to calibrate DOMs in the *same* detection unit (DU). The system determines the time
  offset of the DOMs of the same DU. Each DOM is equipped with a nanobeacon
  with its LED pointing upwards.
- Inter-DU Calibration: Laser beacons will be used to calibrate DOMs at *different*
  detection units. The sideward light emitted from laser beacons allocated at the
  bottom of the KM3NeT detector in well-chosen positions will allow illuminating
  the first DOMs of several neighboring detection units.

2 The KM3NeT Laser Beacon

The KM3NeT Laser Beacon is a device that emits light by means of a diode pumped
Q-switched Nd-YAG laser head (In the Fig. 1 is shown an open laser beacon). The
laser head, together with all the associated electronics to control the device is housed
inside a cylindrical titanium container high pressure and seawater corrosion resistant.
A voltage-controlled optical attenuator consisting of a liquid crystal retarder and a
linear polarizing beam-splitter cube is used in order to remotely control the light

intensity emitted by the laser. A glass rod, mounted in an opening in the top-cap of the container, permits the laser beam to go outside. A flat disk diffuser that spreads the light beam outside following a Lambertian distribution is mounted in the inner side of the glass rod. Hence, the light leaves the cylinder through its vertical wall where biofouling is negligible; the upper part of the glass rod is painted black to avoid the light go outside. The dimensions of the glass rod and the top-cap have been calculated carefully to maximize the light going through the walls of the glass rod. The bottom end-cap of the titanium container holds a penetrator for the power supply and the external communications. Currently, two prototype KM3NeT Laser Beacons have been produced and are successfully operational since April 2013: one integrated in the instrumentation line of ANTARES and another one integrated in a prototype tower of NEMO Phase II project.

Fig. 1. An open KM3NeT Laser Beacon; visible are the glass rod, with its top painted black, the penetrator, the inner mechanics, the laser head and all the power and control electronics.

2.1   The Laser Head

The main component of the KM3NeT Laser Beacon is a diode pumped Q-switched Nd-YAG laser which produces short pulses with time duration of ~ 400 ps (FWHM) and a total energy of about 3.5 µJ (data given by the manufacturer, 4.15–4.25 µJ measured in our lab as shown in Fig. 2). The laser head is the model STG-03E-1S0 from Teem-photonics which emits light with a wavelength of 532 nm. Other laser heads with higher energy per pulse are being evaluated, in particular one with energy per pulse of 25 µJ.

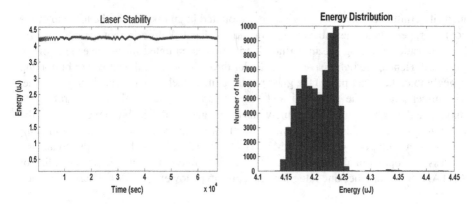

Fig. 2. Left:Measurement of the time-stability of the pulse-energy. Right: Histogram of the pulse-energy.

2.2   The Anti-biofouling System

The anti-biofouling system is composed of a diffuser disk and a quartz cylinder rod inserted in the upper titanium cap (details of the anti-biofouling system on Fig. 3). The laser beam points to the optical disk that spreads light out following a cosine distribution. The diffuser disk is glued to the lower surface of the quartz cylinder rod using a transparent epoxy resin.

Disk Diffuser. The disk diffuser has a Lambertian theoretical distribution. The laser head is slightly deviated with respect to the vertical direction in order to prevent damages from its own reflected beam. Measurements in the laboratory have shown that this small tilt has no effect on the outgoing light distribution and that the distribution follows the theoretical one.

Cylinder Quartz Rod. The cylinder quartz rod is made of pressure-resistant quartz. The dimensions of the quartz cylinder have been calculated to optimize the maximum and minimum angle of the outgoing light. With the current design it is possible to illuminate DOMs located at a horizontal distance of 200 m and 50 m above the seabed.

2.3   The Voltage-Controlled Optical Attenuator

The amount of light emitted by the laser head is fixed. In order to overcome this limitation, a voltage-controlled optical attenuator is located in the beam path. Liquid crystal variable retarders consist of a thin crystal liquid layer placed into a small cavity made of parallel fused silica. The anisotropy of the liquid crystal molecules causes its birefringence. When a voltage is applied the molecules align parallel to the electric field. The higher the voltage, the lower the birefringence and the delay of the optical phase are. This allows the electrical tuning of a linearly polarized beam. Since the light from the laser is linearly polarized, the attenuation can be achieved by the combination of a liquid crystal variable retarder and a linear polarizer. The liquid crystal head is the retarder model LVR-100-532 from Meadowlark Optics. The polarizer used is a broadband polarizing cube beam-splitter from Newport, model 05FC16 PB.3, together with his holder, model CH-0.5. The beam-splitter cube consists of two right-angle prisms where the hypotenuse of one of the prisms is coated with a multilayer dielectric polarizing beam-splitter. The incoming beam is divided into two orthogonal, linearly polarized components in such a way that p-polarized light is transmitted while s-polarized light is reflected. The used model is optimized to work in the 420–680 nm range. The reason to use a beam-splitting polarizing cube is that it presents a higher damage threshold to laser exposure than that of a standard linear polarizer and that it can fulfill the expected life time of KM3NeT.

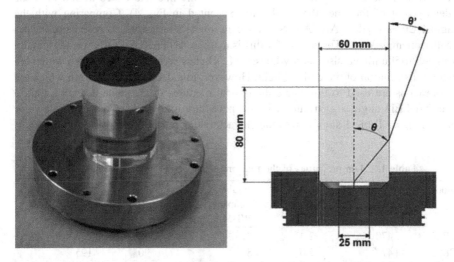

Fig. 3. Detail of the Laser Beacon anti-biofouling system (left) and a technical drawing of the system (right).

3 The KM3NeT Nanobeacon

The main goal of the KM3NeT Nanobeacon is to perform *intra-DU* calibration, i.e. calibration between the DOMs of the same detection unit. The nanobeacon is integrated in the DOM and consists of a small electronics board that controls an LED pointing upwards. The mechanical integration inside the DOM avoids the need for a mechanical container for the device which reduces substantially the cost of this calibration device. The LED emits an ultra-short light pulse ($\sim$2–3 ns of rise time) when the appropriate command is received. The main component of the nanobeacon electronics is the pulser that provides the electrical signal to enable the LED to flash. The nanobeacon light intensity and rate of emission can be changed. In order to do this the pulser is controlled remotely via two I2C control signals. Geometrical considerations show that a 15° opening angle is sufficient to illuminate DOMs located on the same detection unit allowing potential misalignments smaller than 10°. At present, 11 prototype KM3NeT Nanobeacons have been already been produced; eight of them have been are successfully operational integrated in a prototype tower of the NEMO Phase II project; three more have been integrated in the DOMs in a prototype KM3NeT detection unit that is awaiting deployment.

3.1   The LEDs

Several LED models have been tested in the laboratory. Based on comparative studies of amplitude and rise-time of the emitted pulses and angular distribution of light, four models were preselected as suitable for use in the KM3NeT Nanobeacon device. Following the recovery of ANTARES line 12, these new models were incorporated in the lowest of the LED Optical Beacon (LOB) of the line and tested *in situ* after the redeployment of the line (the results are presented in Fig. 4). Comparing with the original LEDs used in ANTARES, these new models are more powerful but have a smaller opening angle. The reason for this is that the ANTARES LOBs were originally designed to illuminate also nearby lines, so LED caps were machined off to widen the angular distribution of the emitted light. However, in a decoupled system in which the nanobeacons are used to illuminate the optical modules in the same detection unit, a modified LED angular distribution is not necessary and uncleaved LEDs allow for longer ranges. Table 1 summarizes the characteristics of the four pre-selected models.

Table 1. Main properties of the four preselected LED models for KM3NeT

Model	Wavelength (mm)	Rise Time (ns)	Angular Occupancy, FWHM (°)	Intensity (pJ)	Range for 0.1 p.e./flash (m) (see Fig. 4)
CB26	470	2.4	23	150	230
CB30	472	2.0	28	90	195
NSPB500	470	3.2	20	170	250
AB87	470	2.4	51	130	235

3.2   Electronics

The KM3NeT Nanobeacon electronics consists of two components, the pulser and the control electronics. Currently, the control electronics has been integrated in the central logic board and power board of the DOM. The pulser circuit is based on an original design from Kapustinsky [7] that has been modified for KM3NeT. The trigger is provided by a 1.5 V negative square pulse of around 150 ns superimposed on a negative DC bias that can be varied from 0 to 24 V. The DC component charges the capacitor and the rising edge of the differentiated 1.5 V pulse switches on the pair of transistors, triggering the fast discharge of the 100 pF capacitor through the low impedance path that includes the LED. The parallel inductor develops charge in opposition to the discharging capacitor reducing its time constant. The level of the DC voltage determines the amount of current through the LED and thus the intensity of the emitted pulse. The foreseen typical trigger frequency will be between 5 kHz up to

20 kHz. The nanobeacon control board is in charge of providing the pulser with two control signals, one to set up the light intensity emitted by the LED and the other one to set up the flashing frequency. It is also possible to select how the trigger signal is provided, either internally (self-triggering) or fixed by an external signal. The electronics consists of two main blocks: the trigger, which can provide a variable signal to the Nanobeacon pulser to fix its flashing frequency, and the booster, that provides the power needed to fix the intensity of the flash emitted by the LED. The trigger signal is controlled with a variable DC voltage that is provided by the Digital Analog Converter (DAC) block. This voltage is set up via I2C. The output provided by the trigger is a squared signal changing from 0 to 3 V, whose frequency can vary from 5 to 20 kHz. The power supply to set the light intensity of the pulser is provided by the booster.

The booster is provided directly from the power supply and the output voltage is controlled with a variable potentiometer also via I2C. The output varies from 3 V to 24 V. The trigger configuration, external or internal, is chosen using an I2C controlled DAC.

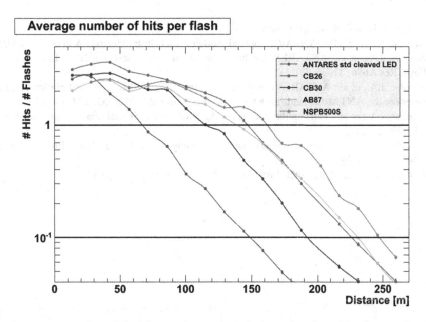

Fig. 4. Average number of hits (photo – electrons) per flash as a function of the distance. For the four preselected LED model, measurements were carried out in the ANTARES detector.

4 Summary and Conclusions

The KM3NeT Laser Beacon and KM3NeT Nanobeacon time calibration devices have been presented. Two prototype Laser Beacons have already been deployed and are successfully operational: one integrated in the ANTARES instrumentation line (3.5 μJ) and another one in a prototype tower of the NEMO-phase II project (3.5 μJ). For the first phase of construction of KM3NeT, a more powerful head with 25 μJ per pulse is under evaluation. Four LED precandidates for the KM3NeT Nanobeacon have been evaluated in ANTARES. Eleven prototype KM3NeT Nanobeacons have already been deployed and are successfully operational since April 2013: eight in a prototype tower of the NEMO-phase II project; another one in a prototype KM3NeT DOM integrated in the ANTARES instrumentation line. Currently, three prototype KM3NeT Nanobeacons have been integrated in DOMs in a prototype KM3NeT detection awaiting deployment.

Acknowledgments. The authors acknowledge the financial support of the Spanish Ministerio de Ciencia e Innovación (MICINN), grants FPA2009-13983-C02-01, FPA2012-37528-C02-01, ACI2009-1020, Consolider MultiDark CSD2009-00064, RYC-2012-10604, European Community's Sixth Framework Programme under contract n° 011937 and the Seventh Framework Programme under grant agreement n° 212525 and of the Generalitat Valenciana, Prometeo/2009/026.

References

1. KM3NeT Technical Design Report
2. Amram, P., et al.: The ANTARES optical module. Nucl. Instr. Methods **A484**, 369 (2002)
3. Aguilar, J.A., et al.: ANTARES: the first undersea neutrino telescope. Nucl. Inst. Methods Phys. Res **A656**, 11–38 (2011)
4. Aguilar, J.A., et al.: ANTARES collaboration, "study of large hemispherical photomultiplier tubes for the ANTARES neutrino telescope". Nucl. Instr. Meth. **A555**, 132–141 (2005)
5. Aguilar, J.A., et al.: Time calibration of the ANTARES neutrino telescope. Astropart. Phys. **34**, 539 (2011)
6. Toscano, S., et al.: Time calibration and positioning for KM3NeT. Nucl. Instrum. Meth. **A602**, 183 (2009)
7. Kapustinsky, J.S., et al.: Nucl. Instr. Meth. Phys. Res **A241**, 612. (1985)

Adaptive Data Collection in Sparse Underwater Sensor Networks Using Mobile Elements

M.J. Jalaja[✉] and Lillykutty Jacob

Department of Electronics and Communication Engineering,
National Institute of Technology Calicut, Kozhikode, India
jalaja@rit.ac.in, lilly@nitc.ac.in

Abstract. Underwater Wireless Sensor Network (UWSN) is a group of
sensors and underwater vehicles, networked via acoustic links to perform
collaborative tasks. Due to hostile environment, resource constraints and
the peculiarities of the underlying physical layer technology, UWSNs tend
to be sparse or partitioned, and energy-efficient data collection in a sparse
UWSN is a challenging problem. We consider mobility-assisted routing as
a technique for enabling connectivity and improving the energy efficiency
of sparse UWSN, considering it as a Delay/Disruption Tolerant Network
(DTN) or Intermittently Connected Network (ICN). The DTN framework shows superior performance in terms of energy efficiency and packet
delivery ratio, at the cost of increased message latency. We investigate the
effectiveness of a *polling model* to analyze the delay performance and propose a *dynamic* optimization technique to minimize latency adaptively,
thereby supporting delay-sensitive applications also. The effectiveness of
the proposed technique in modelling the dynamically changing environment and minimizing the data collection latency is validated using NS-2
based simulation.

Keywords: Underwater Wireless Sensor Network · Delay Tolerant
Network · Mobile sink · Polling · Dynamic optimization · Scheduling
Preference Index

1 Introduction

Underwater Wireless Sensor Networks (UWSNs) have emerged as powerful sys-
tems for providing autonomous support for several potential applications [1,2].
Acoustic communication is the underlying physical layer technology used in
UWSNs, though research is in progress on the use of radio frequency waves [3].
Underwater sensor nodes are more expensive and energy-consuming than terrestrial sensor nodes and it is not feasible to deploy them in large quantities.
Also, due to sparse deployment, harsh environment, node mobility and resource
limitations, a contemporaneous end-to-end path may not exist between any two
nodes. These factors result in intermittent connectivity and at any given time,
when no path exists between source and destination, network partition is said
to occur. Hence sparse UWSNs need to be treated as Intermittently Connected

© Springer-Verlag Berlin Heidelberg 2015
M. Garcia Pineda et al. (Eds.): ADHOC-NOW Workshops 2014, LNCS 8629, pp. 57–65, 2015.
DOI: 10.1007/978-3-662-46338-3_6

Networks (ICN) or Delay/Disruption Tolerant Networks (DTN) [4]. While traditional routing protocols require an end-to-end contemporaneous path for data transfer, such a path may never exist in a DTN and DTN protocols make use of *contact* or forwarding opportunity for data transfer. The primary objective of DTN routing is to achieve eventual delivery of data, rather than reducing data collection latency. Due to energy constraints, conventional multipath DTN approaches can not be used in UWSNs and hence, it is better to consider mobile sink or mobile collectors for providing connectivity and for facilitating energy-efficient data collection in sparse UWSNs.

In mobility-assisted data collection, node mobility is exploited to fill the connectivity gaps in the network and to improve energy efficiency of sensor nodes. Compared to direct transmission and ad hoc multi hop network, mobility-assisted routing has three main advantages: it (i) takes care of sparse and disconnected networks, (ii) eliminates the relaying overhead, and (iii) reduces the transmit power requirement. However, due to the limited travel speed of the mobile elements, the data collection latency will be quite large. Such large latency may be acceptable in certain environmental sensing applications which are not time-critical and which give more weightage to energy efficiency and network lifetime than message latency. Since the delay performance of simple mobility-assisted data collection scheme is not comparable with that of ad hoc network, techniques for reducing data collection latency and thus supporting delay-sensitive applications in the former one is an interesting research problem.

Mobility of the data collector can be random, predictable, or controlled; and the last one is preferred (wherever possible) to provide latency bounds. Flexibility in path selection and speed control of individual mobile elements and scheduling of multiple mobile elements can be exploited to optimize the delay performance. While optimal path selection is a well addressed problem, speed control is not much useful in UWSNs since the travel speed can not be increased beyond a limit, say $20\,m/s$ due to practical reasons. Also, increasing the number of mobile elements will increase the cost considerably. Hence, we propose a scheme that simultaneously addresses the path selection of a single mobile sink (MS) and the assignment of sensors to it, such that the data collection delay is minimized.

We start with a basic DTN framework for energy efficient on-demand data collection in sparse underwater sensor networks using a mobile sink; and then augment it with a technique to optimise its data collection performance dynamically. Analytical results for energy saving, packet delivery ratio, message latency, and sensor buffer occupancy are presented. The analytical results are validated using our own simulation model developed in Aqua-Sim [5], an NS-2 [6] based network simulator. The rest of the paper is structured as follows. A brief review of the related work is given in Sect. 2. The system model is presented in Sect. 3. The expressions used for analytical results are developed in Sect. 4. Section 5 discusses the results and the paper is concluded in Sect. 6.

2   Related Work

A number of routing protocols have been developed for UWSNs, such as VBF, HH-VBF, FBR, DBR, ICRP, DUCS and so fourth [1,2,5,10], where it is implicitly assumed that the network is connected and there exists a contemporaneous end-to-end path between any source and destination pair. This assumption need not be valid in a physical network. Recently, considerable effort has been devoted to developing architectures and routing algorithms for DTNs and routing in DTNs is investigated in [7]. In [8], an adaptive routing protocol has been proposed for UWSNs, considering it as a DTN. Jain et al. [9] have presented a three-tier architecture based on mobility to address the problem of energy efficient data collection in a terrestrial sensor network. Energy analysis of underwater sensor networks is done in [10]. In [11], an M/G/1 queueing model is used for mobility-assisted routing, proposed for reducing and balancing the energy consumption of sensor nodes. The use of controlled mobility for low energy embedded networks has been discussed in [12]. AUV-aided routing for UWSNs is discussed in [13,14]. The usage of message ferries in ad-hoc networks is considered in [15]. In this paper, we propose an analytical model based on polling to evaluate the delay performance of mobility-assisted data collection and augment it with dynamic optimization for supporting delay-sensitive applications.

3   System Model

We consider sparse underwater sensor networks with possibly disconnected components and with a mobile sink (MS) used for data collection. The static sensors, anchored to the bottom of the ocean, monitor the underwater surroundings, generate data, and store it in the sensor buffer. The sensors have limited power and memory and they can communicate to the MS using acoustic links only. The MS is an entity with large processing and storage capacity, renewable power, and the ability to communicate with static sensors and the surface gateway. Though the sensors can use direct or multi-hop paths for sending service requests to the gateway, their bulk data communications are limited to single-hop transmission to the nearby MS only, so as to reduce energy consumption. As the MS comes in close proximity to (i.e. within transmission range of) a static sensor, the sensor's data is transferred to the MS and buffered there for further processing.

Before the beginning of any data collection cycle, the MS broadcasts beacon messages in the network. As a response to this, the static sensors, having buffered data awaiting transmission, place service requests before the sink. In the basic model, the request contains the identification of the source node, and the data generation rate. The MS visits the static nodes according to an FCFS policy, collects the data and then proceeds to the next location. The sensor data is assumed to be delivered successfully, once it has been collected by the MS. The sensors are assumed to be equipped with sufficient buffer space, so that no data is lost due to buffer overflow.

In the enhanced model, the cycle time of the MS is minimized for sensors with time-critical data, by using optimum scheduling of visits, based on proximity to the sink, buffer occupancy, packet generation rate and service time requirement of individual sensors. At the beginning of each data collection cycle, the gateway prepares the optimum visit sequence and it is assigned to the MS. Thus, instead of visiting the nodes in a cyclic order or based on FCFS policy, the MS visits the nodes according to the optimum visit sequence that results in minimum time for completing the data collection cycle. The minimization of cycle time improves the support for delay-sensitive applications that may use mobility-assisted data collection for energy efficiency and improved connectivity.

4   Analytical Study

In this section, we develop the necessary analytical expressions, the numerical results of which are compared with the simulation results in Sect. 5. Since we focus on the delay performance and because of space constraints, other aspects like energy efficiency, network lifetime, and packet delivery ratio are only briefly discussed, not elaborated.

4.1   Energy Efficiency and Network Lifetime

For a given target signal-to-noise ratio SNR_{tgt} at receiver, available bandwidth $B(l)$, and noise power spectral density $N(f)$, the required transmit power $P_t(l)$ can be expressed as a function of the transmitter-receiver distance l [10]. If P_r is the receive power, L is the packet size in bits, M is the number of packets transferred from the source node to the destination and α is the bandwidth efficiency of modulation, the energy consumption for the single hop data transfer is given by Eq. 1 as

$$E_{hop}(l) = \frac{M(P_r + P_t(l))L}{\alpha B(l)} \tag{1}$$

Due to the very small value of l, hop energy consumption is much less in MS-based scheme. Also, the number of transmissions required for successful reception of a packet is 1, while the relaying of packets in ad-hoc multi hop communication leads to more and unequal number of transmissions [12]. Due to the reduced and balanced communication overhead of all the static sensors in MS-based scheme, it is more energy efficient and of enhanced lifetime compared to ad hoc network.

4.2   Data Collection Latency

A queueing theoretic approach is used to analyze and optimize the delay performance of MS-based data collection. The system is modelled as multiple queues accessed by a single server in cyclic order. In the basic *polling model*, a single server visits (or polls) the queues in a cyclic order and after completing a visit to queue i, it incurs a switch over period or *walk time* [16]. The time between the server's visit to the same queue in successive cycles is called *polling cycle time*. Mobile sink and the static sensor buffers in our model correspond to the single server and queues of the polling model, respectively. Travel time of the MS to move from one location to the next is modelled as the *walk time* and the time spent at each location to transfer data from the sensor buffer to the MS is modelled as the *service time*. Assuming Poisson arrival of packets at rate λ at each sensor buffer, the offered load is given by $\rho = N\lambda\overline{X}$ for N number of static sensors, where $\overline{X}$ is the mean message service time. For system stability, ρ should be less than 1.

Let $\overline{X^2}$ denote the second moment of the packet transfer time and the MS travel time between two consecutive locations be a random variable with mean and variance $\overline{W}$ and $\overline{W^2}$, respectively. Under the assumption of symmetric queues and *exhaustive* service, the mean waiting time of the packet in the sensor buffer before the MS approaches it for data transfer can be evaluated as [16]:

$$W_q = \frac{\overline{W^2}}{2\overline{W}} + \frac{N\lambda\overline{X^2} + \overline{W}(N - \rho)}{2(1 - \rho)} \tag{2}$$

The influence of factors like packet arrival rate, packet length, data transfer rate, MS velocity, number of static sensors, dimension of sensing area, etc.; on the average waiting time can be easily studied using Eq. 2. Also, the expected response time of a message, the average buffer size and the average number of messages in the system (in queue and in service) can be evaluated as $\overline{X} + W_q$, $W_q\lambda$ and $(\overline{X} + W_q)\lambda$, respectively.

A major goal of this modelling and analysis is to improve (optimize) the system performance by identifying the design parameters to prioritize the queues. The delay performance can be optimized by changing the order and/or frequency of visiting the sensors statically or dynamically: in *static* optimization, the visit sequence is decided prior to operation, while in *dynamic* optimization, it will change dynamically according to the system state during operation. In the *dynamic* control of server's visits, the server will modify its order of visits in response to the stochastic evolution of the system which is difficult to analyse. Hence we use a *semi dynamic* scheme [17], i.e., instead of polling *cycles*, the server perform Hamiltonian tour, in which every sensor is visited exactly once. But the visit sequence in each tour may differ from the previous one, depending on the dynamic state of the system at the *beginning* of each tour, so as to minimize the cycle time of the MS.

Suppose the initial system state (buffer occupancies of the sensors as reported by the sensors in the service request) be $Q = (q_1, q_2, ...q_N)$ and data generation rates be $(\lambda_1, \lambda_2, ...\lambda_N)$. To determine the optimum sequence of visiting the sensors, let us define the *Scheduling Preference Index* (SPI) of sensor node i as

$$SPI_i = \frac{\overline{W}_i + q_i\overline{X}_i}{\rho_i}; \tag{3}$$

where $\overline{W}_i$ and $\overline{X}_i$ are the *walk time* to and packet *service time* requirement at sensor node i and $\rho_i = \lambda_i\overline{X}_i$ is the traffic load at sensor node i. Following the approach used in [17], it is seen that, performing an MS visit tour that follows an increasing order of $SPIs$ of static sensors will minimize the data collection cycle duration and hence the mean waiting time. If $visit(i)$ is the node-i in the optimum visit sequence (a permutation of sensor nodes) computed based on the SPIs, the minimum cycle time is given by Eq. 4 as

$$CycleTime_{min} = \sum_{i=1}^{N} \frac{q_{visit(i)}\overline{X}_{visit(i)} + \overline{W}_{visit(i)}}{\prod_{r=i}^{N}(1 - \lambda_{visit(r)}\overline{X}_{visit(r)})} \tag{4}$$

4.3   Packet Delivery Ratio (PDR)

Since all the data generated at one sensor in one *cycle time* is transferred in one visit of the MS, assuming sufficiently large buffer space and no errors in communication, the PDR will be 1.

5   Simulation Study and Results

Extensive simulations have been carried out to validate our analytical results using the NS-2 based network simulator for underwater applications, Aqua-Sim. It is an event-driven, object-oriented simulator written in C++ with an OTCL interpreter as the front-end. We have augmented it with DTN framework, polling based (*exhaustive* service) data collection and dynamic optimization of MS scheduling using our own code in C++ and OTCL.

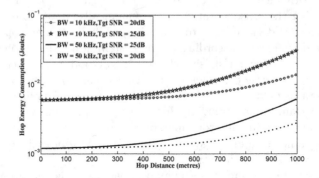

Fig. 1. Hop energy consumption for varying hop length and bandwidth

The variation of hop energy consumption (as given by Eq. 1) with hop length and channel bandwidth is shown in Fig. 1 and the variation of packet delivery ratio with node density is shown in Fig. 2. For multi-hop ad-hoc network, delivery ratio is very small for low node density. The result shows that MS-based scheme is the better option for sparse networks and the only option for disconnected networks.

Assuming symmetric queues, 10 sensor nodes uniformly distributed in an area of size $1000\,\text{m} \times 1000\,\text{m}$, data rate $10\,\text{kbps}$, packet size $50\,\text{bytes}$, and controlled motion of the MS, the mean waiting time of packets (as given by Eq. 2) using the

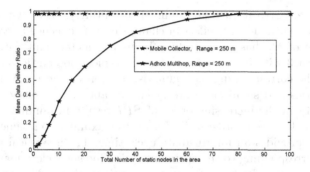

Fig. 2. Packet delivery ratio: ad hoc multi hop vs mobility-assisted

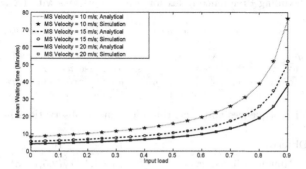

Fig. 3. Variation of mean waiting time with input load and MS velocity

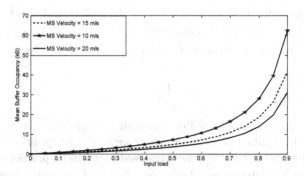

Fig. 4. Variation of mean buffer occupancy with input load and MS velocity

basic MS-based model is plotted in Fig. 3. As expected, the mean waiting time of a packet increases with the packet arrival rate and decreases with the speed of the MS. The buffer occupancy, as illustrated by Fig. 4 also shows exactly the same behaviour. This gives an indication about the sensor buffer requirement for a particular arrival rate and/or MS velocity, so as to avoid buffer overflow and thus packet loss.

With different arrival rates and number of waiting packets at varying number of static sensor nodes, the effect of change in visit sequence in each cycle on the cycle time of MS has been studied. Input parameters and the $SPIs$ computed using Eq. 3 with 5 sensor nodes for two typical observations are given in Table 1 and the corresponding optimum visit sequences and cycle times (based on both analysis and simulation) are given in Table 2. It is observed that the tour of the MS in the increasing order of SPI results in minimum cycle time and that in the decreasing order of SPI leads to maximum cycle time. Any other visit sequence provides an intermediate cycle time. The analytical and simulation results corresponding to the optimum cycle time show close matching, thus validating the proposed scheme for dynamic optimization of MS tour.

Table 1. Scheduling Preference Index for two observations with $(\overline{X} = 0.1467\,\text{s})$

Initial state (No. of pkts in buffer)					Arrival rate (Packets/s)					Scheduling Preference Index (SPI)				
q_1	q_2	q_3	q_4	q_5	λ_1	λ_2	λ_3	λ_4	λ_5	SPI_1	SPI_2	SPI_3	SPI_4	SPI_5
4	486	572	65	269	0.0012	0.0833	0.1	0.0083	0.042	180870	8320	7790	32640	11420
1014	1994	242	520	1484	0.167	0.333	0.04	0.083	0.25	7311	6610	10776	8724	6764

Table 2. Visit sequences and cycle times for sample observations in Table 1

Observation	I	II
Optimum visit sequence	(3,2,5,4,1)	(2,5,1,4,3)
Minimum cycle time (Analytical)	6.05 min	16.7 min
Minimum cycle time (Simulation)	6.15 min	17.05 min
Maximum visit sequence	(1,4,5,2,3)	(3,4,1,5,2)
Maximum sequence cycle time	7.52 min	20.37 min
Cycle time for visit seq.(2,1,3,5,4)	6.83 min	18.52 min

6   Conclusion

In this paper, we have investigated the suitability of a mobility-assisted data collection scheme for sparse underwater sensor network to support delay-sensitive traffic. A polling based analytical model is used to evaluate, predict and optimise its performance like latency and buffer occupancy. The basic scheme improves energy efficiency and delivery ratio at the cost of increased latency and hence it is suited for sparse or disconnected networks and in situations where network lifetime is more important than message delay. Techniques to support delay-sensitive applications in mobility-assisted scheme have been explored. The proposed optimized scheduling of MS has been found to be effective in adapting to the network conditions and reducing the cycle time of data collection, thus minimizing message latency. As a future work, we plan to develop energy-efficient and adaptive data collection schemes for large 3-dimensional networks with a variety of application requirements.

References

1. Akyildiz, I., Pompili, D., Melodia, T.: Underwater acoustic sensor networks: research challenges. Ad Hoc Netw. **3**, 257–279 (2005)
2. Garcia, M., Sendra, S., Atenas, M., Lloret, J.: Underwater wireless ad-hoc networks: a survey. In: Mobile Ad hoc Networks: Current Status and Future Trends. CRC Press/Taylor and Francis, Boca Raton (2011)
3. Lloret, J., Sendra, S., Ardid, M., Rodrigues, J.J.P.C.: Underwater wireless sensor communications in the 2.4 GHz ISM frequency band. Sensors **12**(4), 4237–4264 (2012)
4. Zhang, Z.: Routing in intermittently connected mobile ad hoc networks and delay tolerant networks: overview and challenges. IEEE Commun. Surv. Tuts. **8**(1), 24–37 (2006)
5. Xie, P., Zhou, Z., Peng, Z., Yan, H.: Aqua-sim: an NS-2 based simulator for underwater sensor networks, Underwater Sensor Networks Lab, University of Connecticut, OCEANS (2009)
6. ns-2 Network Simulator. http://www.isi.edu/nsnam/ns/
7. Jain, S., Fall, K., Patra, R.: Routing in a delay tolerant network. In: Proceedings of ACM SIGCOMM 2004, pp. 145–158 (2004)
8. Guo, Z., Colombi, G., Wang, B., Cui, J.H., Maggiorini, D., Rossi, G.P.: Adaptive routing in underwater delay/disruption tolerant sensor networks. In: Proceedings of the Fifth Annual Conference on Wireless on Demand Network Systems and Services, WONS (2008)
9. Jain, S., Shah, R., Brunnette, W., Borriello, G., Roy, S.: Exploiting mobility for energy efficient data collection in wireless sensor networks. Mob. Netw. Appl. **11**, 327–339 (2006)
10. Zorzi, M., Casari, P., Baldo, N., Harris III, A.F.: Energy-efficient routing schemes for underwater acoustic networks. IEEE J. Sel. Areas Commun. **26**(9), 1754–1766 (2008)
11. He, L., Pan, J., Zhuang, Y., et al.: Evaluating on-demand data collection with mobile elements in wireless sensor networks. In: Proceedings of the IEEE VTC (2010)
12. Somasundara, A.A., Kansal, A., Jea, D.D., Estrin, D., Srivastava, M.B.: Controllably mobile infrastructure for low energy embedded networks. IEEE Trans. Mob. Comput. **5**(8), 1–16 (2006)
13. Yoon, S., Azad, A.K., Kim, S.: AURP: an AUV-aided underwater routing protocol for underwater acoustic sensor networks. Sensors **12**, 1827–1845 (2012)
14. Hollinger, G.A., Choudhary, S., Qarabaqi, P., Mitra, U., Sukhatme, G.S., Stojanovic, M., Singh, H., Hover, F.: Underwater data collection using robotic sensor networks. IEEE J. Sel. Areas Commun. **30**(5), 899–911 (2012)
15. Kavitha, V., Altman, E.: Queueing in space : design of message ferry routes in static adhoc networks. In: Proceedings of the 21st International Teletraffic Congress (ITC 21), France, Paris (2009)
16. Takagi, H.: Queueing Analysis of Polling Models: An Update. Stochastic Analysis of Computer and Communication Systems, pp. 267–318. Elsevier Science Publishers B.V., Amsterdam/North Holland (1990)
17. Yechiali, U.: Optimal dynamic control of polling systems. In: Cohen, J.R., Pack, C.D. (eds.) Queueing, Performance and Control in ATM, vol. Elsevier Science Publishers B.V., pp. 205–217. North Holland, North Holland (1991)

Acoustic Signal Detection Through the Cross-Correlation Method in Experiments with Different Signal to Noise Ratio and Reverberation Conditions

S. Adrián-Martínez, M. Bou-Cabo, I. Felis, C.D. Llorens,
J.A. Martínez-Mora, M. Saldaña, and M. Ardid[✉]

Institut d'Investigació per a la Gestió Integrada de Zones Costaneres (IGIC),
Universitat Politècnica de València, Paranimf 1, 46730 Gandia, Spain
mardid@fis.upv.es

Abstract. The study and application of signal detection techniques based on cross-correlation method for acoustic transient signals in noisy and reverberant environments are presented. These techniques are shown to provide high signal to noise ratio, good signal discernment from very close echoes and accurate detection of signal arrival time. The proposed methodology has been tested on different signal to noise ratio and reverberation conditions using real data collected in several experiences related to acoustic systems in astroparticle detectors. This work focuses on the acoustic detection applied to tasks of positioning in underwater structures and calibration such those as ANTARES and KM3NeT deep-sea neutrino telescopes, as well as, in particle detection through acoustic events for the COUPP/PICO detectors. Moreover, a method for obtaining the real amplitude of the signal in time (voltage) by using cross correlation has been developed and tested and is described in this work.

Keywords: Acoustic signal detection · Cross-correlation method · Processing techniques · Positioning · Underwater neutrino telescopes · Astroparticle detectors

1 Introduction

Acoustic signal detection has become an object of interest due to its utility and applicability in fields such as particle detection, underwater communication, medical issues, etc. The group of Acoustics Applied to Astroparticle Detection from the Universitat Politècnica de València collaborates with the particle detectors ANTARES [1], KM3NeT [2] and COUPP/PICO [3]. Acoustic technologies and processing analyses are developed and studied for positioning, calibration and particle detection tasks of the detectors.

Acoustic emitters and receivers are used for the positioning systems of underwater neutrino telescopes ANTARES [4] and KM3NeT [5] in order to monitor the position of the optical detection modules of these telescopes. The position of optical sensors need to be monitored with 10 cm accuracy to be able to determine the trajectory of the muon

© Springer-Verlag Berlin Heidelberg 2015
M. Garcia Pineda et al. (Eds.): ADHOC-NOW Workshops 2014, LNCS 8629, pp. 66–79, 2015.
DOI: 10.1007/978-3-662-46338-3_7

produced after a neutrino interaction in the vicinity of the telescope from the Cherenkov light that it produces [6]. An important aspect of the acoustic positioning system is the time accuracy in the acoustic signal detection since the positions are determined from triangulation of the distances between emitters and receivers, which are determined from the travel time of the acoustic wave and the knowledge of the sound speed. The distances between emitters and receivers are of the order of 1 km. Therefore, the acoustic emitted signals suffer a considerable attenuation in the medium and arrive to the acoustic receivers with a low signal to noise ratio. The environmental noise may mask the signal making the detection and the accurate determination of its arrival time a difficult goal, especially for the larger future telescope KM3NeT with larger distances.

On the other hand, an acoustic test bench has been developed for understanding the acoustic processes occurred inside of the vessels of the COUPP Bubble Chamber detector when a particle interacts in the medium transferring a small amount of energy, but very localized, to the superheated media [7]. This interaction produces a bubble through the nucleation process. Under these circumstances the distance from the bubble to the vessel walls are very short (cm order) and a reverberant field generated by multiple reflections in the walls takes place. With these conditions, the distinction of the direct signal from reflection is quite difficult to achieve, being also quite complex to determine the time and amplitude of the acoustic signal produced.

The elaboration of protocols and post-processing techniques are necessary for the correct detection of the signals used in these tasks. Methods based on time and frequency analysis result insufficient in some cases. The first step consists of using the traditional technic of cross-correlation between the received signals and the emitted signals (expected) for localizing the source distance. In addition, the use of specific signals with wide band frequency or non-correlated such as sine sweep signals or Maximum Length Sequence (MLS) signal together with correlation methods increase the amplitude and the correlation peak narrows, this allows a better signal detection, improves the accuracy in the arrival time and the discernment of echoes.

In this work the detection of acoustic signals with a unique receiver under a reverberant field or a high noise environment is shown. The correlation method has been studied and applied for this purpose. Moreover, a method for obtaining the real amplitude of the signal (voltage) by using cross-correlation technique has been developed. Its validation has been done by comparing the results with the ones obtained by analytic methods in time and frequency domain, achieving a high reliability for the accurate detection of acoustic signals and the analysis of them. The results obtained in these tests in different environments using different kind of signals are shown.

In Sect. 2 the cross-correlation technique is described, as well as the method proposed for signal detection. The application of the method under different situations: high reverberation, low signal-to-noise ratio (S/N) or very low S/N, is presented in Sect. 3. Finally, the conclusions are summarized in Sect. 4.

2 The Cross-Correlation Method for Signal Detection

Cross-correlation (or cross-covariance) consists on the displaced dot product between two signals. It is often used to quantify the degree of similarity or interdependence between two signals [8]. In our case, since all measurements were recorded using a digital acquisition system, all signals under study have been worked in discrete time, so that the correlation between two signals x and y with the same N samples length is expressed by the following expression:

$$Corr\{x, y\}[n] = \sum_{m=1}^{N} x[m] \cdot y[m+n] \tag{1}$$

If we do $y = x$ we obtain the autocorrelation of the signal x.

Figure 1 shows the appearance of the signals used in these studies: tones, sweeps, and MLSs. On top, there are these ideal signals in the time domain, that is, the generated signals by the electric signal generator equipment. In the middle row, the spectrum of each signal can be seen, where the different bandwidths can be appreciated. At the bottom, the autocorrelations of each signal show that the higher bandwidth signals have a narrower correlation peak, so, in principle, they are easier to detect. To understand the importance and convenience of using these signals in each detector, the reader can look at articles [9, 10].

It is worth to note that, in the cases shown, the correlation peak amplitude ($V_{max,\ corr}$) is the same and equal to the number of samples of the signal in question (N). Therefore, it can be obtained the peak voltage of the signal (V_p) by the following expression:

$$V_p = \frac{2\, V_{max,corr}}{N} \tag{2}$$

Furthermore, this ratio does not vary with the amplitude of the signal and is less susceptible to the presence of noise.

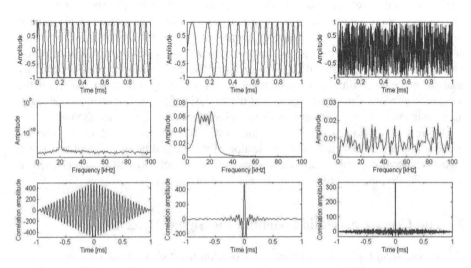

Fig. 1. Signals used for acoustic studies: tone, sweep and MLS.

However, the interest is the use of the method for the accurate detection of signals and the recorded signals will be influenced by reflections and noise that may vary the amplitude and profile of the direct signal detection.

Figure 2 shows the case of a tone, a sweep and MLS received signals with a distance of 112.5 m between emission and reception (E-R). On the top, the receiving signals in time domain after applying a high order band pass filter are shown (the original recorded signal in time is so noisy that the receiving signal is completely masked). On the bottom, it can be seen the cross-correlation of each signal (without prefiltering) where direct signal and reflections are easier and more effective to discern that working in the time or frequency domains, especially for high bandwidth signals (narrower auto-correlation peak).

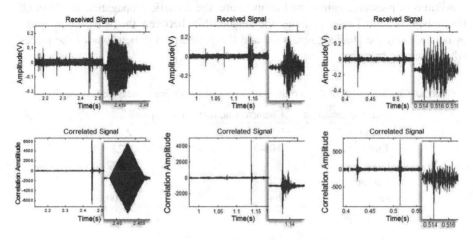

Fig. 2. Example of recorded signals at 112.5 m Emitter-Receiver distance in the harbor of Gandia.

Nevertheless, using this it is only possible to locate the signal but cannot know a priori the peak amplitude of the signal. This is because trying to tackle the problem from both time and frequency domains is completely crucial windowing temporarily the direct signal avoiding reflections to obtain a reliable value of its amplitude, which is not always possible.

Then, it would be important to obtain the corresponding relation between the maximum of the cross-correlation between received and emitted signal with the amplitude of the received signal avoiding reflections. This issue has been studied and has been found that if the amplitude of the signal sent ($V_{p,\,env}$), its number of samples (n_{env}) and the maximum correlation value ($V_{max,\,corr}$) corresponding to the detection of this signal are known, then it is possible to obtain the peak-amplitude voltage of the received signal applying the following expression:

$$V_{p,rec} = \frac{V_{max,corr}}{V_{p,env}} \frac{2}{n_{env}} \tag{3}$$

In the following sections the results of applying this equation to the results of the correlations obtained and compared with values obtained applying time and frequency domain methods are presented. In addition, the improvements obtained by using this technique in terms of detection accuracy in different acoustical environments are also shown.

3   Application

The different conditions in which the measurements of acoustic detection were performed are: inside a small vessel, in a tank of acoustic test, in a pool, in the harbor of Gandia, and in ANTARES deep sea neutrino telescope. Although under different conditions of pressure, salinity and temperature, the acoustic propagation media in all the tests is water. Table 1 shows the relationship between the wavelength range associated with the studied signals (λ) and the geometrical dimensions of the places where acoustic processes occur (l).

Table 1. Characteristics of the acoustic conditions of the different measurements and tests.

Measure condition	Characteristic distance l [m]	λ/l
Vessel	0.02	2.2
Tank	0.05–1	0.22
Pool	4	0.022
Harbor	120	0.0005
Sea	200	0.0003

With this, it follows that conditions with higher ratio λ/l means working in a reverberant field, with a higher complexity, while configurations with a smaller λ/l ratio means that there is a less reverberant field, but usually a lower S/N ratio. As discussed below, both extreme situations make difficult the process of acoustic detection.

The results obtained in these conditions, the acoustic systems used in transmission and reception, and the results in terms of improvement of signal detection and S/N using cross-correlation method are shown in the following sections.

3.1   High Reverberation Conditions: Vessel and Tank

When emitter and receiver are close and the dimensions of the enclosure where the acoustic processes occur are comparatively small, both signal and reverberation are high. This is the case of the configurations shown in Fig. 3 that corresponds to a part of the acoustic test bench for COUPP detector [11]. On the left, the two experimental setups are shown. The first one corresponds to acoustic propagation studies inside a vessel, and the second one was used to study the acoustic attenuation. On the right the transducers used are shown. The signal was emitted with the pre-amplified ITC 1042 transducer and received with the needle-like RESON TC 4038 transducer.

Fig. 3. Experimental setups (left) and transducers used (right).

Figure 4 shows an example of a 30 kHz tone of 5 cycles of duration emitted and recorded under these conditions and their cross-correlation. It can be seen that the maximum of the correlation corresponds with the reception time of the received signal.

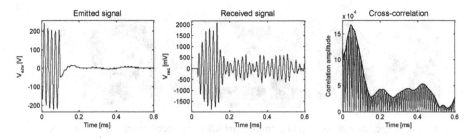

Fig. 4. Example of emitted signal, received signal, and cross-correlation.

Figure 5-left shows that for the tones studied between 10 kHz and 100 kHz the accuracy of this method is quite good, with an error smaller than 10 %. Considering the characteristic dimensions of the problem and 1500 m/s as sound propagation speed, this uncertainly is of the same order of magnitude of the experimental uncertainly (1 mm). As expected, the maximum deviation corresponds to lower frequencies, and it seems there is some frequency dependent fluctuations. This can be another argument in favor of using broadband signals for cross-correlation techniques.

The received amplitudes of the signals have been obtained using Eq. (3). The results are shown in Fig. 5-right compared to the results obtained with standard techniques in time and frequency domains. It can be observed that the results are very similar.

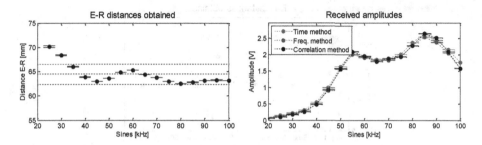

Fig. 5. Left: distances obtained between emitter and receiver by cross-correlation with tones between 20 kHz to 100 kHz. Right: Received amplitudes through the cross-correlation method using Eq. (3) and using time and frequency domain methods.

3.2 Low Signal to Noise Ratio Conditions: Pool

The following configuration used is an intermediate step between high signal to noise ratio (Sect. 3.1) and very low signal to noise (measurements in the harbor and in the ANTARES neutrino telescope, presented in the next Sects. 3.3 and 3.4, respectively). This is the case of measurements taken on a pool as shown in Fig. 6 (left). In this experimental setup, the transmitter consists of an array of three transducers FFR SX83 (middle) and an electronic board to generate and amplify the different acoustic signals. This system can operate in three different modes: emitting with a single element, with the three elements connected in series and the three elements connected in parallel [10]. Our measures were made with the transducers connected in parallel so, in this embodiment, higher transmission power is obtained. The reception was performed using a FFR SX30 (right).

Fig. 6. Experimental setup (left), emitter (middle) and receiver (right) transducers.

Using tones between 10 kHz and 60 kHz in these conditions, we have calculated the emitter-receiver distances from flight times, as described above. The results are shown in Fig. 7 and compared with those obtained directly in time-domain method. In this case, we can see that the deviation of the measurements relative to a mean value is 5 %, which corresponds to an uncertainty less than ±20 cm. However, if we discard some out-layer measure (sine of 40 kHz) the deviation of the values is reduced to 2.3 %, i.e., ±9 cm. We think that a reason for the relatively large variation between different measurements at different frequencies might be the interference between the three emitters of the array, which depends on the frequency. Again here, the use of broadband signals with the cross-correlation method may help to mitigate this problem since it will average the response of the different frequencies.

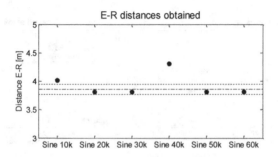

Fig. 7. Emitter-receiver distances obtained by cross-correlation method using tones between 10 kHz to 60 kHz (considering 1500 m/s as the sound propagation speed).

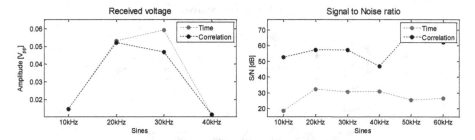

Fig. 8. Comparison of cross-correlation and time domain method to obtain the received voltage amplitude (left) and the S/N ratio (right).

The plots of Fig. 8 show the results obtained by comparing the voltages (left) and the S/N ratios (right) both in cross-correlation method and time-domain method (in this case, since the signals can be windowed properly, avoiding the presence of reflections, values obtained in time and frequency domains are coincident).

As before, using the Eq. (3) very similar results to the usual techniques are obtained. On the other hand, the S/N ratio increases considerably (at least 20 dB) for the set of signals used using correlation method. This improvement is crucial for a correct detection of the signals.

3.3   Very Low Signal to Noise Ratio Conditions: Harbor

The first kind of measures taken in very low signal to noise ratio condition has been done in the port of Gandia. The acoustic signals were emitted from one side of the harbor and received from the other side, as it can be seen in Fig. 9 (left), the distance between the emitter and the receiver was about 120 m. The emitter and receiver transducers used were the same as the ones used in pool measurements (Sect. 3.2).

Fig. 9. Measured conditions (left), emitter (middle) and receiver (right) transducers.

In this case, tones of 10 kHz and 30 kHz, a sweep between 5 to 50 kHz and a MLS signals were used. As it can be shown in Fig. 10, with these signals we have obtained the distances between emitter and receiver around 113 m with a precision of ±30 cm.

In these measures, the amplitudes obtained in time and cross-correlation present more deviation by using a 10 kHz tone, however these amplitudes are very close if we look at higher frequencies, as it can be shown in Fig. 11. Additionally, the signal to noise ratio increases more than 10 dB in the signals analyzed using the cross-correlation method, which helps significantly its detection in this noisy environment.

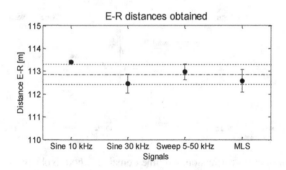

Fig. 10. Emitter-receiver distances obtained by cross-correlation method using tones of 10 kHz and 30 kHz, 5 to 50 kHz sine sweep and MLS signals.

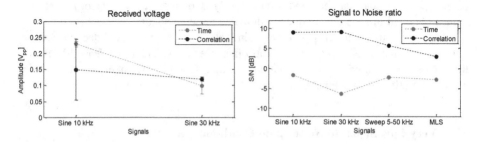

Fig. 11. Comparison of cross-correlation and time domain method to obtain the received voltage amplitude (left) and the S/N ratio (right).

3.4   Very Low Signal to Noise Ratio Conditions: Sea

The more complex environment in which this study has been performed is the acoustic measurements made in situ in deep-sea at the ANTARES site. In this case, the distance between emitter and receiver was about 180 m and the S/N ratio was quite low. Figure 12 shows on the left an artistic and schematic view of the telescope. The emitter was a FFR SX30 transducer, shown in the middle, with an electronic board designed specifically for this type of transducer to optimize and amplify the signal sent [10, 12], and the receiving hydrophone was a HTI-08 transducer, shown on the right [13].

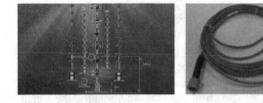

Fig. 12. View of the ANTARES neutrino telescope (left) and pictures of the emitter FFR SX30 (middle) and of the receiver HTI-08 (right) transducers.

Since in this ANTARES test synchronization between transmitter and receiver was not available, it is not possible to calculate absolute flight times. However, the received amplitude expression as well as the increase of the S/N ratio obtained by cross-correlation method can be evaluated here, as shown in Fig. 13. In this case, sine signals of 20, 30 and 40 kHz, 20 to 48 kHz and 28 to 44 kHz sweep signals and MLS (order 11) signals were used.

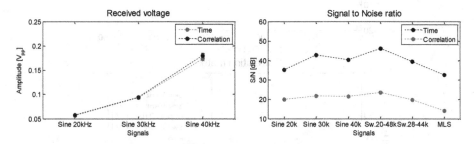

Fig. 13. Received amplitude (left) and S/N ratio (right) both in cross-correlation and time domain method.

It can be concluded from these measurements that using the cross-correlation method is possible to obtain the signal amplitude accurately and obtain an increase of 15 to 20 dB in the S/N ratio, with a consequent improvement in the acoustic detection.

Additionally, and with the aim of applying this technique for post-processing signals in the future KM3NeT neutrino telescope, simulations of propagation of signals measured in ANTARES over longer distances have been done. The received signals measured have been propagated to further distances (up to 2.16 km) in order to know the pressure levels reached and the amplitude signal received by the hydrophones and its correlation amplitude. The signals received are propagated applying the spherical divergence loss transmission and the sea water absorption coefficient per frequency. The propagation has been performed following the steps shown in the

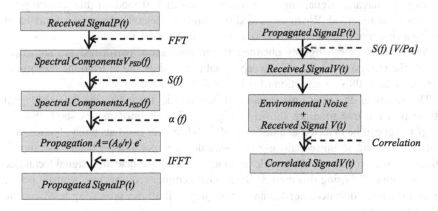

Fig. 14. Block diagrams of the received signal propagation

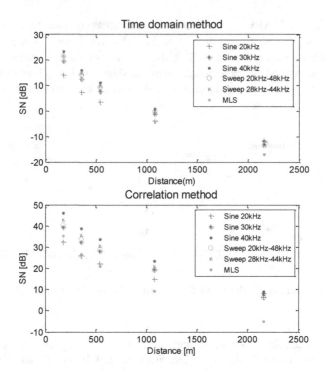

Fig. 15. S/N ratio obtained using time domain method (top) and cross-correlation method (bottom).

diagrams of Fig. 14. The diagram on the left shows the digital processes in order to know the expected amplitude per distance. The diagram on the right shows the processes to obtain the time detection by correlation method with the propagated signals added to real noise in different time position. Figure 15 shows, the improvement in the S/N ratio as a function of the distance using the different signals and methods.

Finally, in order to determine the reach of the system in terms of time detection accuracy, propagated signals have been introduced in 100 random (but known positions) of noise recorded. We have studied the ability of detecting them as a function of the distance (from 180 m to 2.16 km) using the correlation method.

An example of the figures obtained in the processing performed are shown in Fig. 16, the analysis with the sine sweep signal received at 180 m from the transmitter inserted to the 0.08 s of a section of 0.14 s noise is shown:

The values of deviation time (mean and standard deviation) for the detected signal with respect the true time are shown in Fig. 17. The obtained results show that the sweep signal can be detected at a distance of 2.16 km with good accuracy. Notice, that even if the signal to noise ratio is not favorable, it is possible to detect the signals emitted up to 2.16 km with reasonable accuracy by using the detection signal technique describe above. By using this method the acoustic emitter requires less acoustic power to reach the large distances needed and consequently it allows producing less acoustic pollution in the media.

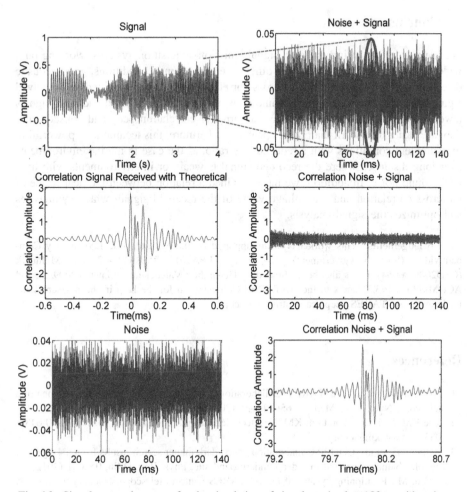

Fig. 16. Signal processing steps for the simulation of signal received at 180 m with noise.

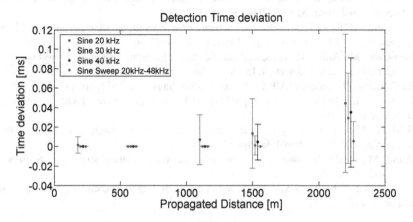

Fig. 17. Values of deviation time (mean and standard deviation), with respect the true time, for the detected propagated signal with noise. It is determined by the correlation detection method.

4 Conclusions

We have seen that, using different signal emission-acquisition systems, working on a wide range of distances and in very different environmental conditions, good acoustic detection through the technique of cross-correlation between the emitted and received signals can be obtained. This technique is more favorable for broadband signals (sweeps and MLS) because they have a narrower correlation peak and consequently they are easier to discern than others peaks. Furthermore, this technique is powerful in measurement conditions with a reduced S/N ratio, as the case in marine environments over long distances where the recorded signal is weak, or in environments with high background noise. In addition, we have obtained a relation between the peak value of the cross-correlation and the voltage value of the received signal, which synthesizes and optimizes the signal analysis.

Acknowledgements. This work has been supported by the Ministerio de Economía y Competitividad (Spanish Government), project ref. FPA2012-37528-C02-02 and Multidark (CSD2009-00064). It has also being funded by Generalitat Valenciana, Prometeo/2009/26, and ACOMP/2014/153. Thanks to the ANTARES Collaboration for the help in the measurements made in the ANTARES deep-sea neutrino telescope.

References

1. Ageron, M., et al. (ANTARES Collaboration): ANTARES: the first undersea neutrino telescope. Nucl. Instr. Meth. A **656,** 11–38 (2011)
2. The KM3NeT Collaboration: KM3NeT technical design report (2010). ISBN 978-90-6488-033-9. www.km3net.org
3. Behnke, E., et al. (COUPP Collaboration): First dark matter search results from a 4-kg CF3I bubble chamber operated in a deep underground site. Phys. Rev. D **86**, 052001 (2012)
4. Ardid, M.: Positioning system of the ANTARES neutrino telescope. Nucl. Instr. Meth. A **602**, 174–176 (2009)
5. Larosa, G., Ardid, M.: KM3NeT acoustic position calibration of the KM3NeT neutrino telescope. Nucl. Instr. Meth. A **718**, 502–503 (2013)
6. Ardid, M.: ANTARES: an underwater network of sensors for neutrino astronomy and deep-sea research. Ad Hoc Sensor Wirel. Netw. **8**, 21–34 (2009)
7. Bou-Cabo, M., Ardid, M., Felis, I.: Acoustic studies for alpha background rejection in dark matter bubble chamber detectors. In: Proceedings of the IV International Workshop in Low Radioactivity Techniques. AIP Conference Proceedings, vol. 1549, pp. 142–147 (2013)
8. Proakis, J.G., Manolakis, D.G.: Digital Signal Processing, 3rd edn. Prentice Hall, Upper Saddle River (1996)
9. Saldaña, M.: Acoustic system development for the underwater neutrino telescope positioning KM3NeT, Bienal de Física (2013)
10. Ardid, M., et al.: Acoustic transmitters for underwater neutrino telescopes. Sensors **12**, 4113–4132 (2012)
11. Felis, I., Bou-Cabo, M., Ardid, M.: Sistemas acústicos para la detección de Materia Oscura, Bienal de Física (2013)

12. Llorens, C.D., et al.: The sound emission board of the KM3NeT acoustic positioning system. J. Instrum. **7**, C01001 (2012)
13. Graf, K.: Experimental studies within ANTARES towards acoustic detection of ultra high energy neutrinos in the deep sea. Ph.D. thesis, U. Erlangen, FAU-PI1-DISS-08-001 (2008)

Multimedia Wireless Ad Hoc Networks 2014, MWaoN 2014

CARPM: Cross Layer Ant Based Routing Protocol for Wireless Multimedia Sensor Network

Mohammed Abazeed[1]($\boxtimes$), Kashif Saleem[2],
Suleiman Zubair[1], and Norsheila Fisal[1]

[1] Faculty of Electrical Engineering, University Technology Malaysia,
81310 Johor Bahru, Johor Darul Ta'zim, Malaysia
mohmbaz@gmail.com
[2] Center of Excellence in Information Assurance (CoEIA),
King Saud University (KSU), Riyadh, Kingdom of Saudi Arabia (KSA)

Abstract. Applying multimedia to Wireless Sensor Network (WSN) adds more challenges due to WSN resource constraints and the strict Quality of Service (QoS) requirements for multimedia transmission. Different multimedia applications may have different QoS requirements, so routing protocols designed for Wireless Multimedia Sensor Network (WMSN) should be conversant of these requirements and challenges in order to ensure the efficient use of resources to transfer multimedia packets in an utmost manner. The majority of solutions proposed for WMSN depends on traditional layered based mechanisms which are inefficient for multimedia transmission. In this paper we propose a cross-layer Ant based Routing Protocol for WMSN (CARPM) by using modified ant colony optimization (ACO) technique to enhance the routing efficiency. The proposed protocol uses an improved ACO to search for the best path that are satisfied with the multimedia traffic requirements. While making best decision the weightage is given to energy consumption, and queuing delay. The proposed cross layer scheme works between the routing, MAC, and physical layers. Since, the remaining power and timestamp metrics are exchanged from physical layer to network layer. Dynamics duty cycle assignment is proposed at MAC layer which changes according to traffic rate. The presented algorithm is simulated using NS2 and is proven to satisfy its goals through a series of simulations.

Keywords: Ant colony optimization · Multimedia · Network simulator 2 · Cross-layer · Dynamics duty cycle · Wireless sensor networks

1 Introduction

Recently in last few years Wireless Multimedia Sensor Networks have appeared and attracted the interest from typical sensors to multimedia sensors. The development toward wireless multimedia sensor network has been the result of progress in the Complementary metal–oxide–semiconductor (CMOS) technology which leads to development of single chip camera module that could be easily integrated to sensor nodes [1]. The multimedia sensor nodes are capable to catch video, images, audio as

© Springer-Verlag Berlin Heidelberg 2015
M. Garcia Pineda et al. (Eds.): ADHOC-NOW Workshops 2014, LNCS 8629, pp. 83–94, 2015.
DOI: 10.1007/978-3-662-46338-3_8

well as scalar sensor data, and then deliver the multimedia content over wireless network. Wireless Multimedia Sensor Networks (WMSNs) have more additional features and requirements than Wireless Sensor Network (WSN) such as high band-width demand, intolerable delay, acceptable jitter and low packet loss ratio; these characteristics add more resource constraints that involve energy consumption, memory size, bandwidth and processing capabilities because of the physically limited small size of sensors and the nature of multimedia application which produces huge data traffic [2]. The challenge is how to handle these constraints with limited resources. Many factors having manipulation on WMSN design, should compromise between them to get better performance. Routing protocols are important to Quality of Service (QoS) assurance for multimedia data because it is responsible for selecting the best path that meets the QoS metrics and energy efficiency, also the routing layer serves as intermediate to exchange performance parameters between application and medium access control (MAC) layer [1]. As well as the researches in the routing protocol stand behind the improvement of WSN, the same is correct for WMSN. When the number of nodes increases in the WSN, and so does the size, the routing in such situation becomes critical and challenging, to handle such problem biologically-inspired intelligent algorithms can be applied. Ants, bees and other social swarms are used as a model. Agents can be generated to solve complex routing problem in different networks. The most famous and successful swarm intelligence algorithm is named Ant Colony Optimization (ACO) [3]. ACO uses artificial ants to find a solution by moving in the problem area, simulating real ants where Pheromone is left in the selected path for the use of ants coming in future. ACO was applied effectively to the number of complex optimization problem like travelling sales man's problem. A survey on swarm intelligence routing protocols proposed for wireless sensor network can be found in [4, 5]. Another survey of the challenges and the state of the art of routing protocols in the wireless multimedia sensor network can also be found in [6] where the most common WMSN routing protocols are presented.

The proposed protocol depends on Biological Inspired Secure Autonomous Routing Protocol (BIOSARP) as a baseline routing protocol. Most of ACO based routing protocols follow the standard approach where forward ant and backward ant agents are utilized. The forward ant collects the paths information while the backward ant confirms the selected paths. This approach is heavy and not suitable for WMSN. In BIOSARP search ant and data ant agents are proposed. The search ant finds the optimal best neighbor node then the data ant carries information to next best node till it reaches the destination. This decreases the overhead and energy consumption while increase the delivery ratio which is proven through simulation and test bed implementation.

BIOSARP is not designed specifically for multimedia application and is untested in simulation with it. So we aim to enhance the performance of BIOSARP by increasing the delivery ratio and minimizing energy consumption while guarantee the delay requirements. We changed the metrics used to calculate the optimal node, we depended on queuing delay as a main metric since multimedia application generates huge data traffic, and the queuing delay is the most dynamic component of delay and normally dominates all other delay components. This is also a sign of bandwidth utilization and congestion level.

Our proposed routing protocol aims to satisfy the QoS requirements for efficient multimedia transmission while considering the resource constraint nature of WMSN. The proposed protocol is based on cross-layer design and improved ant colony algorithm. The QoS metrics that are used are energy and queuing delay. The cross-layer design works through routing, Mac, and physical layers. Dynamic duty cycle is involved at MAC layer aiming to save energy and enhance throughput. Simulation results acquire through a series of simulations conducted in NS-2, show that the proposed algorithm satisfies the goals.

Next section reviews the related literature and BIOSARP. Section 3 presents the approach. The implementation, results and comparison are demonstrated in Sect. 4. Section 5 states the conclusion and future work.

2 Related Work

The mentioned characteristics, challenges and requirements of WMSNs designing open many research area issues and future research directions to develop protocols, algorithms, architecture devices and test beds to maximize the network lifetime while satisfying the quality of service requirements of the different applications [7]. While a significant amount of research has been conducted on WSN routing problems, WSN multimedia data routing remains vastly unexplored. The different solutions exist for traditional wireless environments and the Interne, yet these solutions cannot be directly applied to WMSN. Consequently, there is a growing need for research efforts to address the challenges of WSN multimedia communications to help realize many currently available multimedia applications [8]. In the following discussion, we give a brief summary of the various routing protocols and techniques proposed for WMSN.

A Meta heuristic ant colony technique proposed in [9] the cost function is calculated based on energy consumption, link quality and link reliability. The link quality is defined as the bit error on the link, while the link reliability is defined as the percentage of the time link up. The transmission probability depends on the pheromone and heuristic values, which is based on the link cost. The pheromone value is updated globally when all ants finish constructing their paths. Luis copo et al. [8] combined hierarchical structure of the network with principle of ACO. Each node has four QoS metrics. These are: available memory, queue delay, packet loss rate and remaining energy. To become a cluster head the node should have a cluster ant and meet the cluster head requirements such as energy level. The proposed work introduces packet scheduling policy to give different priorities for different traffic classes. The protocol works through four phases and each phase has specific ant type, these ants types are: forward ant, data ant, backward ant and maintenance ant. ACO WMSN [10] achieves multimedia transmission through two stages, routing discovery and routing confirm. There are two ants, forward ant used to find paths and backward ant to update the pheromone value globally. The probability equation for data transmission depends on bandwidth, delay, packet loss rate and energy consumption. Other Routing protocol for visual sensor proposed by Adam Muetelaa et al. [11] is an improved version of EEABR [12] which is designed for WSN. The proposed work optimizes the memory usage by only keeping two records of last two visited nodes. Some modifications are introduced on EEABR to enhance energy consumption, and reduce flooding by giving the nodes near to destination more priority. The number of ant lunched by every node is limited to 5 ants. The routing decision is based on the same rule proposed in the ACO Meta heuristic. In [13] the routing discovery phase starts by calling forward ant to search paths between source and destination. The protocol considers de-lay, bandwidth and hop count. Bandwidth is calculated as a minimum bandwidth of all links along the path. The initial pheromone value is set to zero at the beginning then increased by 0.1 when detecting a neighbor through hello massages. If the link goes down its phero-mone value becomes zero. In case the load increases in optimal path then the path preference probability is automatically decreased and then alternate paths can be used. BIOSARP [14–16] is a routing protocol based on ant colony optimization proposed for WSN. The proposed work has improved performance as compared to other routing protocols shown in [17]. Two types of ant are used where the search ant explore new neighboring nodes and data ant move hop by hop on the base of best pheromone value for neighboring nodes until the destination. The optimal routing decision is taken depending on end-to-end delay, PRR and remaining battery power. The optimal decision provides real-time communication, load distribution to enhance WSN lifetime and better data throughput over WSN.

Although there are still many routing protocols based on ant colony optimization yet most of these protocols consider the energy efficiency as a main goal and do not address the QoS requirements of WMSN.

3   Approach

This paper reports the following main contributions. Firstly, it proposes ANT colony optimization based routing protocol that uses cross layer design between physical, mac and routing layer to enhance routing decision efficiency and computes optimal for-warding node based on remaining power and the queuing delay. By choosing the forwarding nodes with the minimum packet queuing delay, the multimedia data transfer is ensured and the congested nodes are avoided. Additionally, choosing nodes with the highest remaining power level ensure Variety selection of forwarding neighbor node and distribute transmission load among such nodes which prolong the network lifetime. Secondly, dynamic duty cycle assignment is proposed where the duty cycle changes according to the required time for transmission depending on the transmission rate.

3.1   Cross-Layer Design in the Proposed Protocol

The cross layer design is defined with respect to a reference layered architecture is the design of algorithms, protocols, or architectures that exploit or provide a set of inter-layer interactions that is a superset of the standard interfaces provided by the reference layered architecture [18]. In order to achieve high gains in the overall performance of WSN, cross-layer interaction is used in the design of proposed protocol. The concept of cross-layer design is about sharing of information among two or more layers for adaptation purposes and to increase the inter-layer interactions [19–21]. The proposed system uses interaction between physical layer, Mac layer and network layer in order to select the next optimal forwarding node. The process at the network layer optimizes the optimal forwarding decision based on the physical parameters translated as forwarding metrics. The physical parameters are remaining power and timestamp. The forwarding metrics is used to determine the next hop communication. At Mac layer a dynamic duty cycle is used which change according to required time for transmission to prevent network congestion and reduce energy consumption.

3.2 Routing Discovery and Data Transmission Using Improved ACO

In standard ACO algorithm, the ants carry nodes information from source node to destination and come back to the source node to confirm the selected path along the visited nodes. This technique cause heavy traffic and overhead which lead to exhaust energy very fast, furthermore is not suitable for WMSN which saving energy is very important factor, so we use the proposed algorithm in BIOSARP [16] but with different QoS forwarding metrics. In BIOSARP routing process is activated only when there is data to be transferred. To avoid traffic overhead in standard ACO only two ants are proposed which are, search ant $SA_{c \to n}^i$ and data ant $DA_{s \to d}^i$, c is current node, n is next node, s is the source node and d is the destination node. The routing process is explained more clearly as following:

In case there is data to be sent, the node call data ant $DA_{s \to d}^i$ to transfer data packets. The data ant checks the pheromone value in the neighbor table $p_{cv}^k(t)$ where c is the current node, v is neighboring nodes and k is the neighboring nodes ID. In case there is no pheromone value in the neighbor table the data ant calculates the best pheromone value. If there is no information in the neighboring tables, the search ant generates to search neighbors and fulfill the neighboring table with new records. The pheromone value depends only on the neighboring nodes vk QoS metrics (Queuing delay and remaining energy) which well be stored in the neighbor table R^{vk}. The probabilistic forwarding rule is expressed as following:

$$P_{cv}^k(t) = \frac{[Qd_{cv}(t)]^\beta \cdot [E_{cv}(t)]^\gamma}{\sum_{h \in v_c^k}[Qd_{cv}(t)]^\beta \cdot [E_{cv}(t)]^\gamma}.$$

Where P_{cv}^k is the main entry required to send data from node c to neighbor node v with help of k which is the neighboring node ID. β, γ are parameters weight that controls the priority according to the application needs.

3.2.1 Queuing Delay

Minimizing end-to-end delay is a typical goal of routing protocols. Out of the four components of the delay between two adjacent nodes, namely processing delay, queuing delay, transmission delay, and propagation delay, queuing delay is the most dynamic component and normally dominates all the other delay components. Queuing delay is determined by traffic load and available bandwidth, so queuing delay is a sign of bandwidth utilization and congestion level. Hence, queuing delay is a very important routing metric to be considered in order to enhance the adaptability of the routing protocol. The mean queuing delay is calculated by averaging the per-packet queuing delays over that time interval. The mean queuing delay over time interval i is calculated as following [22].

$$Qd_i = \frac{T_{b+} \sum_{i \in p} Qi/B}{Np_i} \times \frac{TQ_i}{L_i}$$

Where:

i = the index of the current time interval; Np_i = the total number of packets being queued in i.

Qi = the queuing size in terms of total bytes at the moment when packet i put in queue

B = the fixed bandwidth., P = the aggregation of all the packets queued in time interval i.

T_b = the total time the node backed off (suspended packet transmission) in i because of channel contention

TQ_K = the total time when the queue is not empty in T_k, L_k = the length of each time interval;

All the variables are locally available to the node conducting the calculation. The term TQ_K/L is a scalar to count in the utilization level of the routing queue. It scales down the mean queuing delay proportionally with the ratio of the total length of idle periods (i.e., no traffic) within a time interval to the length of the time interval.

3.2.2 Adaptive MAC with Dynamic Duty Cycle

To improve network performance, dynamic MAC protocol should be considered which responses to change in network conditions and adapts to the unstable nature of WMSN. Many sleep/wake schemes have been suggested. These schemes use: pre-defined duty cycle, differential duty cycle and adaptive duty cycle. Predefined duty cycle causes high energy wastage due to idle listening, low throughput and more delay where is not suitable for WMSN. Adaptive schemes utilize different metrics like traffic priority, traffic load, residual energy, and network topology and sensor density to adjust duty cycle on nodes. Majority of duty cycle aim to prolong network lifetime by conserve energy, but introduce multimedia transmission in WSN has led to other requirements like high throughput and low latency duty cycle scheme. In the following steps we are going to discuss a dynamic traffic-aware duty cycle which consider queue delay and traffic rate. The duty cycle changes according to the queue delay and traffic rate. We use the techniques used in [1] to calculate the transmission rate in each time interval then we use our scheme to control the change in duty cycle according to required transmission rate, if the transmission finishes with the time of duty cycle, then the queuing delay will be under control.

As in [1] sensor nodes in WSN has two duties source duty and router duty and there are two sources of traffic first traffic generated by node itself and second is relay traffic where the node receive packets from its neighbor to forwarded to sink due to multi-hop nature of WSN. To prevent congestion at a node, the generated and received packets should be transmitted during the time the node is active. Because of the duty cycle operation, on the average, a node is active σT_∞ seconds therefore:

$$\sigma T_\infty \geq \left[(1 + e_i)\lambda_{ii} + (2 + e_i)\lambda_{i,relay}\right] T_\infty T_{PKT}$$

Where T_∞ is a long enough interval, e_i is the packet error rate and $1 + e_i$ is used to approximate the retransmission rate, λ_{ii} is the generated traffic rate, $\lambda_{i,relay}$ is the relayed traffic rate, T_{PKT} is the average duration to transmit packet to other node including the medium access overhead.

Consequently, the input relay packet rate $\lambda_{i,relay}$ is bounded by:

$$\lambda_{i,relay} \leq \lambda_{i,relay}^{Th}$$

Therefore, we introduce a Duty cycle measurement D_m where is used to check the suitability of current duty Cycle to the required transmission time, so D_m will be calculated as

$$D_m = \frac{Dt_i}{\left[(1 + e_i)\lambda_{ii} + (2 + e_i)\lambda_{i,relay}\right] T_{PKT}}$$

Where Dt_i is the duty cycle time while the other samples as we defined in previous equation.

The duty cycle assignment algorithm strives to keep the value of D_m close to 1. The value of D_m is calculated at every i second. There are three possible values of D_m [23]:

1. ($D_m > 1$) which mean the duty cycle time bigger than required transmission time within current traffic rate, this can be result of:

   – The node away from the sink where the traffic load is low
   – The node don't generate data only relay data or the traffic relay rate is low

   In this situation the duty cycle can be safely decreased gradually to allow the value of the D_m to converge gracefully and save energy, this prompt us for liner decrease strategy to update the duty cycle as following:

$$D_{m(i+1)} = D_{m(i)} - Qd.\lambda_{ii} / v$$

   v is the transmission rate throttle factor, the duty cycle can be decreased only to specified minimum value which is permissible duty cycle value

2. ($D_m < 1$) which indicate that the duty cycle less than required time for transmission, and the transmission cannot be adjusted through current duty cycle. This happen due to following reasons:

   – The node close to the sink where the traffic load is high.
   – The node exists in high contention area which cause high channel access delay and packet drop.

In this case the duty cycle change according to two different scenarios:

a) If current $D_{m(i)} < D_{m(i-1)}$ which show that the time required for transmission continuously rising without change in duty cycle, which cause more queuing delay, this sharp rise in queuing delay require to increase the duty cycle as in following:

$$D_{m(i+1)} = D_{m(i)} \cdot \left(\frac{K}{\lambda_{ii}}\right)^{Qdi}$$

Where K is variable updated according to constant value.

b) If current $D_{m(i)} > D_{m(i-1)}$ which mean the time required for transmission less than in $(i-1)$ which indicate that the transmission rate gradually converging to $D_{m(i)}$ this prompt us to decrease the duty cycle according to following formula:

$$D_{m(i+1)} = D_{m(i)} \cdot (1 - Qdi)^{\lambda_{ii}}$$

In the last case the value of $D_{m(i)} = 1$ which indicates that the duty cycle value is suitable for current transmission rate which represent the ideal case and the duty cycle remain unchanged.

4 Implementation, Initial Results and Comparison

In this section, the proposed protocol is studied and analyzed through simulation implementation and results.

4.1 Network Model

The NS-2 based simulation has been conducted to simulate 49 nodes are distributed in 80 m x 80 m region grid topology based WMSN model to show the effect of CARPM (Table 1).

In the simulation study, NS-2 simulator is used to develop and evaluate the performance of CARPM. Figure 3 shows the countermeasures against abnormalities found in WSN. 49 wireless sensor nodes were deployed shown in Fig. 1, where node 0 is the source node. The application traffic used as in [24]. Packet delivery ratio and energy consumption are the metrics used to analyze the performance of CARPM and the baseline BIOSARP. Multimedia traffic is configured according to G.711 Codec rate given in [24], as constant bit rate (CBR) with an 80-byte using G.711 codec and data rate of 64kbps.

4.2 Results Comparison

The simulation results in Fig. 2 shows that CARPM increase the delivery ratio by 5 % as the packet rate is varied. Also the results shows in Fig. 3 that the CARPM consume less energy by 7 % than BIOSARP and prolongs network lifetime. The proposed

Table 1. Network parameters to simulate routing mechanism

Propagation Model	Shadowing
path loss exponent	2.45
Shadowing deviation	4.0 dB
Reference distance	1.0 m
Low Rate WPAN	IEEE 802.15.4
phyType	Phy/WirelessPhy/802_15_4
MacType	Mac/802_15_4
Antenna	OmniAntenna set X_ 0 OmniAntenna set Y_ 0 OmniAntenna set Z_ 1.5 OmniAntenna set Gt_ 1.0 OmniAntenna set Gr_ 1.0

Transport layer	UDP
Operation mode	Non Beacon (unslotted)
Acknowledgement	Yes
CSThresh_	1.10765e-11 Watts
RXThresh_	1.10765e-11 Watts
Initial Energy	3.6 Joule
Power transmission	1 mW
Physical (wirelessPhy)	set bandwidth_ 2e+6 set Pt_ 0.001 set freq_ 2.4e+9 set L_ 1.0
Traffic	CBR

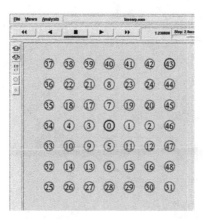

Fig. 1. Network model

parameters help real-time multi-media in achieving better throughput with lower energy consumption. The analysis shows that by considering queuing delay factor, we can enhance the multimedia traffic performance in WMSN.

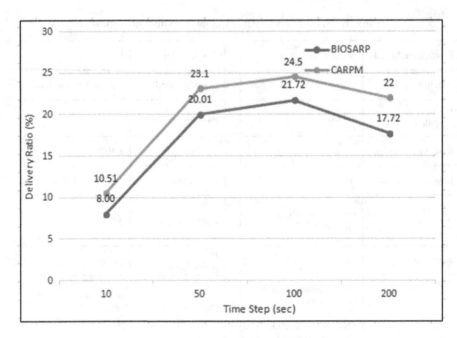

Fig. 2. Delivery ratio comparison

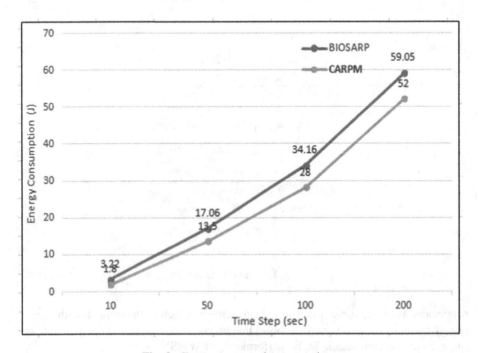

Fig. 3. Energy consumption comparison

5 Conclusion and Future Work

We have proposed a cross layer ant based routing protocol (CARPM) for Wireless Multimedia Sensor Network (WMSN). The routing decision is based on improved Ant Colony Optimization (IACO). The decisions depends on energy consumption and queuing delay. The adopted cross layer architecture helps WMSN in improving the overall data throughput, especially in the case of multimedia traffic. The cross layer design also assists WMSN to gain better delivery ratio while reducing energy consumption. Results shows that by giving weightage to queuing delay actually enhances the overall data routing efficiency. Hence, the outcomes clearly demonstrate that CARPM provides better delivery ratio and energy consumption as compared to BIOSARP for routing multimedia traffic in WSN.

In future, we will avail detailed results and comparisons by emphasizing more on dynamic duty cycle parameter as illustrated in this paper. Additionally, we will perform analysis by varying the duty cycle time according to transmission requirements to avail better performance. Furthermore, we will compare our proposed protocol with other state of the art routing protocols designed specifically for WMSN and finally, implement it on real WMSN testbed to see the actual behavior.

Acknowledgment. The authors wish to express sincere appreciation to Universiti Technology Malaysia (UTM), Malaysia for their support and special thanks to researchers at Center of Excellence in Information Assurance (CoEIA), King Saud University, Kingdom of Saudi Arabia. The authors would like to thank the anonymous reviewers for their helpful suggestions.

References

1. Vuran, M.C., Akyildiz, I.F.: XLP: A Cross-Layer Protocol for Efficient Communication in Wireless Sensor Networks. IEEE Trans. Mob. Comput. **9**, 1578–1591 (2010)
2. Fallahi, A., Hossain, E.: QoS provisioning in wireless video sensor networks: a dynamic power management framework. IEEE Wirel. Commun. **14**, 40–49 (2007)
3. Stutzle, T., Dorigo, M.: A short convergence proof for a class of ant colony optimization algorithms. IEEE Trans. Evol. Comput. **6**, 358–365 (2002)
4. Çelik, F., Zengin, A., Tuncel, S.: A survey on swarm intelligence based routing protocols in wireless sensor networks. Int. J. Phys. Sci. **5**, 2118–2126 (2010)
5. Saleem, M., Di Caro, G.A., Farooq, M.: Swarm intelligence based routing protocol for wireless sensor networks: survey and future directions. Inf. Sci. **181**, 4597–4624 (2011)
6. Ehsan, S., Hamdaoui, B.: A survey on energy-efficient routing techniques with QoS assurances for wireless multimedia sensor networks. IEEE Commun. Surv. Tutor. **14**, 265–278 (2012)
7. Almalkawi, I.T., Guerrero Zapata, M., Al-Karaki, J.N., Morillo-Pozo, J.: Wireless multimedia sensor networks:current trends and future directions. Sensors **10**, 6662–6717 (2010)
8. Cobo, L., Quintero, A., Pierre, S.: Ant-based routing for wireless multimedia sensor networks using multiple QoS metrics. Comput. Netw. **54**, 2991–3010 (2010)
9. Al-zurba, H.L.T., Hassan, M., Abdelaziz, F.: On the suitability of using ant colony optimization for routing multimedia content over wireless sensor networks. Int. J. Appl. Graph Theory Wirel. Ad Hoc Netw. Sens. Netw. **3** (2011)

10. Xiaohua, Y., Jiaxing, L., Jinwen, H.: An ant colony optimization-based QoS routing algorithm for wireless multimedia sensor networks. In: 2011 IEEE 3rd International Conference on Communication Software and Networks (ICCSN), pp. 37–41 (2011)
11. Zungeru, A., Ang, L.-M., Prabaharan, S.R.S., Seng, K.: Ant based routing protocol for visual sensors. In: Abd Manaf, A., Zeki, A., Zamani, M., Chuprat, S., El-Qawasmeh, E. (eds.) ICIEIS 2011, Part II. CCIS, vol. 252, pp. 250–264. Springer, Heidelberg (2011)
12. Camilo, T., Carreto, C., Silva, J., Boavida, F.: An energy-efficient ant-based routing algorithm for wireless sensor networks. In: Dorigo, M., Gambardella, L.M., Birattari, M., Martinoli, A., Poli, R., Stützle, T. (eds.) ANTS 2006. LNCS, vol. 4150, pp. 49–59. Springer, Heidelberg (2006)
13. Mohammed, B.M.a.F.: QoS Based on ant colony routing for wireless sensor networks. Int. J. Comput. Sci. Telecommun. 3 (2012)
14. Saleem, K., Fisal, N., Hafizah, S., Kamilah, S., Rashid, R.A.: Ant based self-organized routing protocol for wireless sensor networks. Int. J. Commun. Netw. Inf. Secur. (IJCNIS) 2, 42–46 (2009)
15. Saleem, K., Fisal, N., Hafizah, S., Kamilah, S., Rashid, R., Baguda, Y.: Cross layer based biological inspired self-organized routing protocol for wireless sensor network. In: TENCON 2009. IEEE, Singapore (2009)
16. Saleem, K., Fisal, N., Al-Muhtadi, J.: Empirical studies of bio-inspired self-organized secure autonomous routing protocol. IEEE Sens. J. 14, 1–17 (2014)
17. Ahmed, A.A., Latiff, L.A., Sarijari, M.A., Fisal, N.: Real-time routing in wireless sensor networks. In: The 28th International Conference on Distributed Computing Systems Workshops. IEEE (2008)
18. Jurdak, R.: Wireless Ad Hoc and Sensor Networks: A Cross-Layer Design Perspective. Signals and Communication Technology. Springer, New York (2007)
19. Costa, D.G., Guedes, L.A.: A survey on multimedia-based cross-layer optimization in visual sensor networks. Sensors 11, 5439–5468 (2011)
20. da Silva Campos, B., Rodrigues, J.J.P.C., Mendes, L.D.P., Nakamura, E.F., Figueiredo, C.M.S.: Design and construction of wireless sensor network gateway with IPv4/IPv6 support. In: 2011 IEEE International Conference on Communications (ICC), pp. 1–5 (2011)
21. Hamid, Z., Hussain, F.: QoS in wireless multimedia sensor networks: a layered and cross-layered approach. Wirel. Pers. Commun. 75, 729–757 (2014)
22. Zhihao, G., Malakooti, B.: Delay prediction for intelligent routing in wireless networks using neural networks. In: Proceedings of the 2006 IEEE International Conference on Networking, Sensing and Control, 2006. ICNSC '06, pp. 625–630 (2006)
23. Hamid, Z., Bashir, F.: XL-WMSN: cross-layer quality of service protocol for wireless multimedia sensor networks. EURASIP J. Wirel. Commun. Netw. 2013, 1–16 (2013)
24. Sun, Y., Sheriff, I., Belding-Royer, E.M., Almeroth, K.C.: An experimental study of multimedia traffic performance in mesh networks. Papers presented at the 2005 workshop on Wireless traffic measurements and modeling, pp. 25–30. USENIX Association, Seattle, Washington (2005)

Access and Resources Reservation
in 4G-VANETs for Multimedia Applications

Mouna Garai[1], Mariem Mahjoub[1], Slim Rekhis[1]([⊠]), Noureddine Boudriga[1],
and Mohamed Bettaz[2]

[1] Communication Networks and Security Research Laboratory,
University of Carthage, Tunis, Tunisia
{mouna.garai,mariem.mahjoub,slim.rekhis,noure.boudriga2}@gmail.com
[2] Methods of Systems Design Laboratory,
National School of Computer Science, Algiers, Algeria
m.bettaz@mesrs.dz

Abstract. The development of Vehicular Ad-hoc Networks (VANET)
has witnessed the release of various multimedia services and made it
important to develop architectures and routing protocols capable of
(a) handling the multimedia QoSs requirements and the real-time services' constraints; (b) maximizing the network coverage; and (c) managing resources on the Road Side Units (RSUs), in order to guarantee the
continuous delivery of real-time services.

In this work, we provide a novel 4G-based VANET heterogeneous
architecture, which integrates IEEE 802.11p and 3GPP LTE access networks, to provide access to multimedia services. A tree-based network
access scheme is developed, providing a rapid connection and handover
to highly mobility vehicles, and allowing to maximize at a large extent
the network coverage beyond the area uncovered by RSUs (in the form
of LTE eNodeBs). Techniques for the proactive resources provision on
the RSUs, and the management of QoS-aware vertical and horizontal
handovers, are developed.

Keywords: VANET · Multimedia · 4G · Tree-based access · QoS

1 Introduction

The development of Vehicular Ad-hoc Networks (VANET) has contributed to
the release of value added multimedia services that aims to promote the development of safe, secure and enhanced navigation. These services require: (a) the
design and development of appropriate models and techniques for the Quality of
Service (QoS) provision; (b) the resources reservation and management on the
Road Side Units (RSUs); (c) the efficient use of the limited wireless resources;
and (d) the achievement of good communication connectivity, and high network
bandwidth and coverage. Several issues make the achievement of aforementioned
requirements challenging, namely the high mobility of vehicles, the variation of
the wireless channel conditions, the limitation of the transmission radius between

© Springer-Verlag Berlin Heidelberg 2015
M. Garcia Pineda et al. (Eds.): ADHOC-NOW Workshops 2014, LNCS 8629, pp. 95–108, 2015.
DOI: 10.1007/978-3-662-46338-3_9

vehicles, and the limited availability of wireless resources. These problem are magnified by the continually increasing number of vehicles, and the diversity of multimedia services in VANets.

To cope with these challenges, recent researches proposed the design of wireless architecture integrating the fourth generation Long Term Evolution (LTE) networks with IEEE 802.11 VANETs. Authors in [5] proposed a VANET architecture integrating 3GPP LTE and 802.11p networks to provide a seamless connectivity. In [6], the authors optimized the construction of routes based on the lifetime, the distance, and quality of the links between vehicles. Nevertheless, these QoS parameters are related to low layers neglecting the users perception of QoS. In [7], a QoS-aware routing algorithm for VANETs, which constructs routes based on a grid approach, was proposed. All the aforementioned references do not consider QoS parameters that are appropriate for providing a high quality usage of multimedia services in VANets. In fact, requirements such as bandwidth, delay, and jitter are not considered during the generation of route. In [3], a tree-based access scheme for call admission in VANETs was proposed to. The proposed QoS model which does not consider the specific features of the access networks, and the lack of handover management make the proposal unsuitable for providing access to multimedia services by vehicles.

Authors in [8] proposed a user-oriented cluster-based multimedia delivery solution over VANETs that is able to offer personalized content to passengers according to their preferences but ignore the quality of links in the cluster heads election process. In [9], a grouping-based storage strategy was proposed based on a low layer VANET (vehicles access is done using WAVE interfaces) and an upper layer P2P Chord overlay on top of a cellular network (access is done using 4G interfaces). Authors in [4], proposed an approach that uses both IEEE 802.11p and LTE networks to periodically collect messages from vehicles and send them to a central server. In highways, where RSUs could be located far from each other and do not provide a full coverage of the roads, the approches proposed in [9] and [4] cannot guarantee the continuous delivery of multimedia services.

In this paper, we propose a heterogeneous VANET architecture integrating IEEE 802.11p and 3GPP LTE access networks. Connected vehicles are grouped in a tree topology to maximize the network coverage, make the global topology less dynamic, and provide a rapid and seamless data connectivity to vehicles, while taking into consideration the different QoS requirements of the multimedia services executed on them. Techniques for resources provision on the RSUs, the resources allocation to connected vehicles requesting services, and the management of QoS-aware vertical and horizontal handovers, are also proposed. The contribution of the paper is four-fold.

– First, the proposed tree-based connection scheme is rapid compared to the existing mechanisms. To access to the network, a vehicle has simply to listen to the messages broadcast by its neighbors, and select the route offering the best QoS. No broadcast nor flooding of the route requests are required.

- Second, the use of a tree-based topology allows to extend the network coverage beyond the area covered by the RSUs, allowing operators to reduce the cost associated to the deployment of a VANET.
- Third, we propose a technique for the QoS provision on 4G/VANET networks, which is based on a distributed and real-time computation of QoS metrics for multimedia services. QoS parameters are computed by the RSU announcing the route, and are then updated by each intermediate node part of the tree under construction.
- Fourth, we set up a technique for the proactive reservation of resources on LTE connections, and the execution of pre-handover procedure by vehicles that are susceptible to change their role (from a child vehicle to a root vehicle). We increase the likelihood of successful handovers.

This paper is structured as follows. The next section describes the Qos model and network architecture. Section 3 describes the all operations related to nodes attachment, resources allocation, and routes management. Section 4 describes the techniques proposed to manage QoS-aware horizontal and vertical handovers. Section 5 describes the simulation results. The last section concludes the paper.

2 System Model and Network Architecture

In this section we introduce our network architecture and we describe the proposed communication and QoS models.

2.1 Network Architecture

We consider an heterogeneous network architecture integrating IEEE 802.11p VANETs together with 3GPP LTE networks. A set of LTE Evolved Node B (eNB) are deployed as RSUs. They exchange data useful for QoS provision through the LTE backhaul network. In this work we mainly focus on the case of highway VANETs, where RSUs are deployed far from each other to minimize the cost associated to network deployment and maintenance. As shown by Fig. 1, a coverage hole exists between two successive RSUs. Vehicles connected to the network are organized into groups, creating a tree topology where, a root vehicle should be located under the coverage of a RSU, has an active connection to the LTE network, and acts as a gateway between the LTE and the 802.11 network by relaying messages sent to/from its child vehicles. A message generated by a mobile node is sent in multihop way to the root node, which in turn will forward the message to the RSU using a LTE connection. The techniques we provide are also usable in urban and freeway VANets.

- All mobile vehicles are equipped with two radio interfaces, one for 802.11p and another for LTE, and is equipped with a GPS receiver. The road map is supposed to be known by all vehicles. Only the gateway vehicle is expected to activate the two interfaces simultaneously. The other vehicles are connected

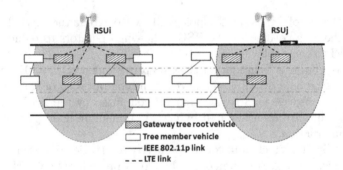

Fig. 1. Network architecture

using the 802.11p interface only.Vehicles can dynamically execute different multimedia services having different QoS requirements for a variable period of time. Furthermore, they change their roles dynamically depending on their positions, the quality of the provided service and radio communication, and the nature of service requests generated by child vehicles.

- Vehicles are able to simultaneously use different services that can be classified into two types: (a) safety services, which require the transmission of emergency notification and signaling message Each vehicle connected to the network is supposed to be constantly in use of such a type of service, requiring the reservation of resources on the tree topology connecting it; and (b) Multimedia services, which generate various QoS requirements. Each vehicle could dynamically start and stop several instances of these services at any time during navigation. The bandwidth consumed by the first type of services is very low in comparison with the second type.

2.2 Communication Model

We propose to use the Self-Organized Time Division Multiple Access (STDMA) technique [2] to manage the access to the service channel frequency on the tree, enabling the connected vehicles to use the same frequency with an alternate access according to the Time Slots they use. The STDMA is a decentralized and self-organizing Channel Access Method, which does not require a controlling via a central station. In this proposal, we use the Dedicated Short Range Communication (DSRC) architecture. We propose to divide the seven DSRC channels into: (1) A control channel used by all the nodes in the network to broadcast their QoS offers, announcing the presence of a tree route toward an RSU. A vehicle should always listens to alternative offers through this channel. (2) A set of service channels used for routing datagrams related to the use of multimedia services. Each tree in the network uses a different channel, and vehicles part of a same tree use STDMA for managing access to the same channel.

Before connection, a new vehicle should listen on the control channel for route announcements providing different QoS offers, select a tree together with a point of attachment, and transmit on the service channel related to the selected tree. The LTE radio access is assumed to use the OFDM technique and support different carrier frequency bandwidths (1.4–20 MHz).

2.3 QoS Model

We provide in this subsection the QoS model, taking into consideration the design difference between LTE networks and IEEE 802.11 VANETs.

Let v_i be a vehicle in the network part of a tree topology τ. We denote by α the route in the tree τ connecting v_i to the gateway vehicle. It is described as $\alpha = \langle v_i, .., v_n \rangle$, where: (a) v_n is the vehicle gateway of the tree which is connected to the RSU through the LTE network; and (b) each vehicle v_j ($j \in [i..n]$) is an intermediate node in the route.

We denote by Q_{v_i} the QoS vector computed by the vehicle v_i. It is expressed as $Q_{v_i} = \langle D_{v_i}, L_{v_i}, T_{v_i}, R_{v_i}, S_{v_i} \rangle$, where:

- D_{v_i} is the route delay computed by vehicle v_i. It is defined as: $D_{v_i} = \sum_{j \in [i,n-1]} d_{(v_j, v_{j+1})} + d_{(v_n, RSU)}$
  where (a) $d_{(v_n, RSU)}$ is the traffic residence delay in the buffer of the gateway vehicle v_n, in addition to the transmission delay between v_n and the RSU; and (b) $d_{(v_j, v_{j+1})}$ is the delay between two successive nodes $(j, j+1)$, which is equal to the sum of the transmission delay, the channel access delay according to the STDMA method, and the decoding delay.
- L_{v_i} is the average packet loss computed by vehicle v_i. It is equal to: $L_{v_i} = max\{l_{(v_i, v_{i+1})}, ..., l_{(v_{n-1}, v_n)}, l_{(v_n, RSU)}\}$
  where: $l_{(v_j, v_{j+1})}$ is the packet loss of the link connecting vehicles v_j and v_{j+1}. A loss $l_{(v_j, v_{j+1})}$, which is computed by vehicle v_j, represents the percentage of frames that are dropped by the decoder in v_{j+1} if the packet arrival time exceeds the playback deadline. $l_{(v_n, RSU)}$ is the packet loss of the link connecting the RSU to the gateway vehicle v_n.
- T_{v_i} is the available throughput computed by vehicle v_i. It is equal to the difference between the bandwidth of the LTE link connecting the gateway vehicle to the RSU, and the total bandwidth consumed by all vehicles in the tree. T_{v_i} is expressed as: $T_{v_i} = T_{(v_n, RSU)} - \sum_{v \in \tau}(c_v + \rho_v)$
  where: (a) $T_{(RSU, v_n)}$ is the maximum throughput supported by the LTE link connecting the gateway vehicle v_n to the RSU; (b) c_v is the flow peak rate related to the traffic sent by vehicle v; and (c) ρ_v is the average bandwidth of the traffic flow sent by vehicle v. The value of $T_{(RSU, v_n)}$ is estimated by the RSU based on the channel quality indicator (CQI) forwarded by the gateway vehicle v_n and the resource blocks already allocated by the RSU.
- R_{v_i} is the route lifetime observed at vehicle v_i. It is expressed as: $R_{v_i} = min\{r_{(v_i, v_{i+1})}, ..., r_{(v_{n-1}, v_n)}, r_{(v_n, RSU)}\}$ where $r_{(v_j, v_{j+1})}$ is the lifetime of the link connecting vehicle v_j to vehicle v_{j+1}; and $r_{(v_n, RSU)}$ is the lifetime of the link connecting the gateway vehicle v_n to the RSU. We denote by the lifetime of a link (v_j, v_{j+1}) the remaining time before v_j becomes out of coverage of v_{j+1}. It is equal to: $r_{(v_j, v_{j+1})} = (TR_{v_j} - d_{j,j+1})/ \mid s_{v_j} - s_{v_{j+1}} \mid$

where TR_{v_j} is the transmission range of the vehicle v_j, $d_{j,j+1}$ is the distance between the vehicle v_j and v_{j+1}, and s_{v_j} is the speed of the vehicle v_j.

- S_{v_i} is the signal-to-interference-plus-noise ratio (SNIR) related to the route connecting v_i to the RSU. It is equal to: $S_{v_i} = min\{S_{(v_i,v_{i+1})}, ..., S_{(v_{n-1},v_n)}, S_{(v_n,RSU)}\}$ where $S_{(v_j,v_{j+1})}$ is the SNIR of the link connecting vehicles v_j and v_{j+1} in the 802.11 VANET, and $S_{(v_n,RSU)}$ is the SNIR of the LTE link connecting the vehicle gateway v_n to the RSU.

3   Nodes Attachment and Resources Allocation

In this section, we develop the techniques and mechanisms related to routes announcement, nodes attachment, and resources allocation and management.

3.1   Routes Announcement by an RSU

To announce its availability, each RSU is required to generate and broadcast, over a control channel, a Connection Advertisement message (CAD) containing a description of the parameters related to the QoS that it can offer. A CAD can be: (a) sent to an already connected gateway vehicle in order to indicate the remaining throughput that can be offered through the tree it is heading, and also to acknowledge previous connection requests received from child vehicles connected to that tree; or (b) broadcast to all vehicles in the transmission coverage of the RSU cell to inform them about the availability of an LTE link toward the RSU and the QoS that it can support.

Each connected vehicle that receives the CAD message (described in Table 1), should update it (i.e., re-compute the QoS vector, modify the sender identity, adjust the handover flag if the vehicle is performing a handover, set the mobility parameters, add the vehicle identity to the field Route, specify the buffer status and the CQI parameters if the vehicle is a gateway) and perform a one-hop broadcast of that message to its child vehicles in the tree. If the receiving node is a gateway vehicle, it should translate the LTE QoS vector received from the RSU to a 802.11 QoS vector before broadcasting the message. In order to keep information about the connection state, each connected vehicle will periodically send a Keep Alive (KA) message to its parent vehicle. Such an information should reach the RSU. To avoid relaying each KA message separately, each time that a vehicle generates a KA message toward its parent, it includes the set of identities of its child vehicles from which it already received a KA message. Both vehicles and RSUs are required to maintain an updated list of their child vehicles. When a vehicle does not forward a KA message within a predefined period of time, its parent vehicle, together with the RSU will detect that it is no longer reachable, and will release the resources it was using.

3.2   Resources Reservation and Nodes Attachment

When a vehicle needs to establish a new connection or request a new service, it should firstly formulate its own needs in term of QoS by generating the three-tuple information $QV^* = \langle D^*, L^*, T^* \rangle$ describing the maximum allowed delay,

Table 1. Content of a CAD message

CAD parameter	Description
ID-Sender	The identity of the vehicle broadcasting the message
ID-RSU	The identity of the RSU serving connection to the vehicles member of the tree
ID-Gw	The identity of the gateway vehicle connecting the tree to the RSU
Serial Number	A number used to differentiate between an old and a new copy
Mobility parameters	The Position, the average speed, and the direction of the vehicle forwarding the message
QoS Vector	The QoS vector computed by the RSU (if the sender is an RSU), or the vehicle forwarding the message (if the sender is a vehicle), as described is Subsect. 2.3
Route	The route from the vehicle forwarding the message to the Gateway Vehicle (defined by ID-Gw)
New Connection flag	If set to 1, this flag indicates that the current CAD message is acknowledging a previous connection request sent from a vehicle connecting to the current tree
Acknowledged connection	The identity of the vehicle whose connection requests are being acknowledged
Handover status flag	If set to 1, this flag indicates that the vehicle sending the current message has started a handover procedure
Buffer status	The occupation rate of the buffer of the Gateway vehicle
CQI	The Channel Quality Indicator of the LTE link connecting the Gateway vehicle to the RSU

the maximum acceptable packet loss, and the requested throughput. After that, it listens during a period of time Δt to the different CAD messages broadcast by its neighbor vehicles, and also by the RSU if it is located within an LTE cell. The neighbor vehicles could be connected through different tree topologies. From each CAD message, the vehicle extracts the four values $[R, T, D, L]$ from the QoS vector, and proceeds as follows (after eliminating the offers provided by nodes in the opposite direction) to select the vehicle to which it will connect. First, it selects offers having a received SNIR that exceeds a predefined threshold. Second, it eliminates any offer that: (a) can not provide the requested throughput ($T > T^*$); (b) is unable to guarantee the maximum allowed delay ($D < D^*$); or (c) provides a packet loss rate higher than the accepted value ($L < L^*$). Third, it retains the offer providing the highest value of R.

From the selected offer, the vehicle extracts the identity of the related neighbor vehicle and the route connecting it to the RSU, and generates an Attachment Request (AR) message containing: (a) the requested QoS parameters (QV^*); (b) its position and direction on the map; (c) the identity of the vehicle whose offer was selected; the identity of the RSU toward which the message will be forwarded;

and (d) its identity. To differentiate between handover requests, new connection requests and a service update, a two-bit state flag is used. The generated AR message is source routed to the RSU (the reverse of the route extracted from the CAD is specified) through the existing tree. Once received, the RSU extracts the identity of the new connected vehicle together with the requested QoS, adds the new connected vehicle to the database, generates an updated QoS vector after computing the new available throughput, and down-forwards an updated CAD (containing the new QoS vector) to the gateway vehicle. The vehicle starts using the requested service after receiving the new CAD message acknowledging its request. If no offer can satisfy the vehicle's QoS requirements (QV^*) the vehicle should either wait for new offers of reformulate its QoS requirements.

3.3   Resources Allocation by the RSU

Resources allocation is made each time a new connection is served, a vehicle is disconnected (i.e., its identity is no longer received in the KA message on the current tree), or a vehicle updates its QoS requirements. Thus, the eNB identifies the CQI based on the QoS factors sent by the vehicle (i.e., throughput, delay, and loss), and allocates the physical resources (i.e., resource blocks) to it through a scheduling process that guarantees fairness.

For each tree, the RSU computes the maximum available throughput, that can be allocated in the future to vehicles connecting to that tree, on the basis of the CQI report controlled by the eNB and periodically transmitted by the gateway vehicle on the Physical Uplink Shared Channel. The CQI reference resources is defined by the group of downlink physical resource blocks corresponding to the band to which the derived CQI value relates [1]. The computed throughput is integrated in the QoS vector sent in the new CAD message. This allocation is dynamic, in fact, it is updated every predefined period of time, upon reception of a vehicle request throughout the gateway vehicle, or upon reception of a pre-handover registration sent by vehicles child of the gateway (Subsect. 4.3). In this work, the RSUs are supposed to use a de-jitter buffer in order to compensate the delay variation caused by the different potential Handovers, and make the video stream displayed at the end user continuous.

3.4   Updates of CAD Messages Announcing Available Routes

Each vehicle is assumed to maintain an updated value regarding the average packet loss, and the SNIR related to the link connecting it to its parent vehicle on the tree. Every vehicle, which receives a new CAD message over that link, proceeds as follow (formulas provided in Subsect. 2.3 are applied): (a) it extracts the position and velocity of the vehicle sending the CAD message, identifies its own position and velocity, computes the lifetime of the link over which the CAD message is being received, and computes the new value of the route lifetime; (b) it extracts the timestamp of the received CAD message, identifies the current time value, and computes the new value of the route delay; and (c) it computes the new value associated to the average packet loss, and SNIR. The updated CAD message is broadcast locally to child vehicles in the tree.

4   QoS Aware Handover Management

The heterogeneity of the access networks used to connect vehicles, and their high mobility, makes it necessary to manage two types of handovers: (1) Horizontal Handover: the switching of a child vehicle from its point of attachement to another node in the same tree or in another tree. The vehicle keeps using a 80.211 connection, (2) Vertical Handover: Such an event occurs in two situations: (a) the switching of a child vehicle from a 802.11 access network to a LTE access network; and (b) the switching of a gateway from a LTE access network to a 802.11 access network.

In our proposal, we aim to: (a) minimize the handover delay, (b) reduce to the maximum the number of handovers to prevent wasting resources and degrading the QoS; and (c) reduce the handover failure probability due to insufficiency of resources. A vehicle initiates a handover in two situations. First, when it notices a degradation in the QoS parameters it initially requested, especially if: (a) the RSS of the link connecting it to its parent, or the route lifetime, drops below a predefined handover threshold; or (b) the route delay or the packet loss rate drops below the requested value. Second, when it requests a new service with a set of QoS parameters that cannot be satisfied by the already available route. Third, when it does not receive any message from its parent for a threshold period of time TO. To reduce the number of handovers, we propose to prevent a vehicle to change its point of attachement in the tree if its requested QoS is still satisfied (e.g., a tree member vehicle, which enters in the coverage of a new RSU, will not automatically become a gateway vehicle or change its point of attachement).

4.1   Vertical Handover

When a gateway vehicle detects a degradation of the CQI indicator, or when the lifetime of the LTE link falls below a handover threshold, it performs a vertical handover following two steps: (1) it stops sending the CAD messages to its child vehicles to inform them that it will leave the tree; and (2) it looks for other alternative paths(satisfying its QoS requirements) announced by its neighbor vehicles to become a member of an existing tree. If not, it waits until losing its connection.

To avoid that all the vehicles of the tree, which do not receive the CAD message, trigger a handoff procedure at the same time, we increase the timeout value TO_v as long as the depth of the vehicle v in the tree increases. Therefore, we have $TO_v = TO^* + (\epsilon \times dpt_v)$, where TO^* is a nominal timeout value, ϵ is a very low constant value, and dpt_v is the depth of the vehicle v in the tree.

Each vehicle immediate child of a gateway, which does not receive the CAD message for a period of time TO, calculates the Remaining Resident Time (RRT)

under the RSU coverage based on its position, velocity, and the RSU position and transmission radius. If the *RRT* is less than a predefined value, called Handover Vehicle Threshold, the node must start a vertical handover. In order to prevent its child vehicles to execute at the same time a vertical handover, it sends a copy of the previous received CAD, while setting the Handover flag to 1. If the *RRT* is higher than the threshold, the vehicle does not start a vertical handover, nor it sends a copy of the CAD message. Therefore, its children will detect the timeout, and proceed with same manner.

4.2   Horizontal Handover

A vehicle performs a horizontal handover if its detects, upon the reception of the new CAD message, that the new computed route lifetime is lower than the threshold T_1, or if the initially required QoS is no longer satisfied. In that case, the vehicle has to find another offer from its neighbor vehicles that can satisfy its requirements. If it is the case, it connects to it through the 802.11p link. The vehicle notifies its child vehicles using the same way mentioned in the previous subsection (it sends a CAD message with a handover flag set to 1). The vehicle executing an horizontal handover sends to its new parent vehicle an Attachment Request message containing a handover flag set to 1 and the identity of its previous RSU. The receiving parent will forward the request to the RSU which will update its database. If the vehicle, which executed the handover, is connecting to a new tree, the receiving RSU informs the previous RSU that the vehicle has changed its point of attachment to stop forwarding messages to it, and updates the available resources in the CAD message. The vehicle disconnection will be detected due to the absence of KA messages.

4.3   Node Pre-handover Process

Once the gateway vehicle forwards the CAD message, every direct child vehicle, checks if there is a degradation in the CQI indicator, or if the route lifetime is becoming lower than a pre-handover threshold (selected to be higher than the handover threshold). If its is the case, it registers to the LTE network and immediately switches to the idle mode. Later, a vertical handover could be executed with a reduced delay, while minimizing the QoS degradation.

4.4   Tree Splitting

Each RSU sends to the upcoming RSU on the road map a local snapshot showing the positions and speeds of vehicles, together with their positions in the tree. The aim is to let RSUs release resources on the trees so that new upcoming vehicles can find resources to hand over successfully. In this context, it would be necessary to split a tree so that the route delay can be reduced, and the available throughput can be increased.

Let $\alpha = \langle v_0, ..., v_n \rangle$ be a route in a tree τ connected to a RSU ρ, and let $v \in \alpha$ be a neighbor of a vehicle, which is connected to another RSU and is susceptible to generate a handover. The RSU ρ will split the route α into $\alpha_1 = \langle v_0, ..., v_i \rangle$ and $\alpha_2 = \langle v_{i+1}, ..., v_n \rangle$ where $v \in \alpha_1$, and v_i is the closest vehicle to v satisfying these conditions: (a) v_i is in the coverage of the RSU ρ; (b) the SNIR of the LTE link, that will connect v_i to the RSU when it becomes a gateway vehicle, is higher than a predefined threshold; (c) and the new CAD message to be generated by v once the tree is split will provide a QoS offer satisfying the requirements of the vehicle in handoff.

5 Simulation

We simulate the proposed solution in a 3 lanes based Highway where each lane is characterized by a different speed (19.44, 22.22 and 25 m/s). The number of simulated vehicles varies between 100 and 400. They are randomly introduced to the highway during a period of time varying between 6,66 and 8.57 min. We set the vehicle and the RSU coverage radius equal to 200 and 5000 m, respectively. At each time-slot (set to 4 s) a vehicle can request an additional multimedia service with a probability equal to 10 %. Each one of these services, requiring a bandwidth of 5 Mbps, has a duration equal to 200 s. In our simulation, we varied the uncovered distance between two successive RSUs from 3000 to 11000 m and the number of vehicles from 100 to 400. We do not compare our solution with the works referenced in the introduction, especially as their QoS model is not suitable for multimedia applications.

The connection/handover failure ratio with respect to the uncovered distance between two successive RSUs was firstly evaluated. Figure 2a shows a failure ratio inversely proportional to the number of vehicles in the network. In fact, as long as the network becomes dense, vehicles are likely to find new neighbor nodes that satisfy the requested QoS parameters, increasing the likelihood of a successful

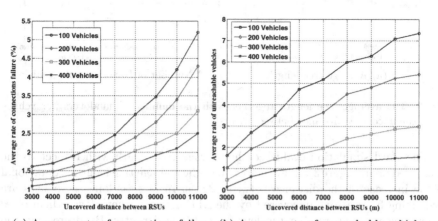

(a) Average rate of connections failure   (b) Average rate of unreachable vehicles

Fig. 2. Average rate of connections failure together with the Average rate of unreachable vehicles w.r.t uncovered distance between RSUs

connection. Besides, the more the distance between RSU is, the higher will be the connection/handover failure ratio and, the difference between the failure rate, for the different simulated scenarios, decreases as long as the uncovered area decreases. In fact, as long as vehicles become far from the gateway, the number of vehicles connected to the same tree increases, reducing the availability of free bandwidth and the likelihood of connection or handover failure.

Secondly, we estimated the average rate of unreachable vehicles with respect to the uncovered distance between RSUs. Figure 2b shows a growth of the rate of unreachable vehicles when the number of vehicles decreases. Besides, the higher is the distance between RSUs, the higher will be the rate of unreachable vehicles. In addition, the difference between the rates of unreachable vehicles in the different scenarios, decreases as long as the distance between RSUs decreases. In fact, as long as vehicles become far from the gateway, the distance between vehicles increases with the decrease of the number of simulated vehicles, making it more difficult for a vehicle to find a neighbor satisfying its QoS requirement. In our proposal, since a route is generated considering the QoS constraints required by the multimedia services, including bandwidth, disconnections could happen even if a vehicle is close to connected neighbors. In fact, as new services are executed on vehicles, the bandwidth remaining on the available trees could be insufficient.

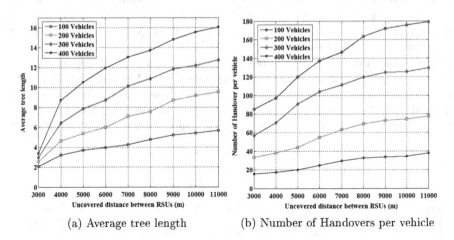

(a) Average tree length   (b) Number of Handovers per vehicle

Fig. 3. The average tree length together with the number of handovers per vehicle w.r.t. the uncovered distance between RSUs

The third simulation evaluated the average tree length with respect to the uncovered area. Figure 3a shows a growth of the average tree length proportional to the increase of the distance between RSUs. In fact, as long as vehicles are connecting from a zone uncovered by the LTE network, they attach themselves to existing trees, increasing the length of the used routes. By decreasing the number of vehicles in the network, the average tree length decreases since most of vehicles located beyond the RSUs coverage become unreachable (i.e., unable to attach themselves to an existing tree).

The fourth simulation, presented in Fig. 3b, shows the evolution of the number of handovers per vehicle, with respect to the uncovered area. That number increases as long as the number of vehicles in the network decreases. In fact, when the distance between the RSUs increases, the route length increases, decreasing the available resources on them. As vehicles dynamically generate new service requests, their QoS requirements are likely to be unsatisfied due to the unavailability of resources on the used long routes. The decrease of the number of vehicles in the network reduces more and more the number of available routes and resources, increasing the number of handovers. In our proposal, a handover can not only be executed due to unreachability of neighbors, but also due to the impossibility to satisfy the QoS constraints of a new multimedia service.

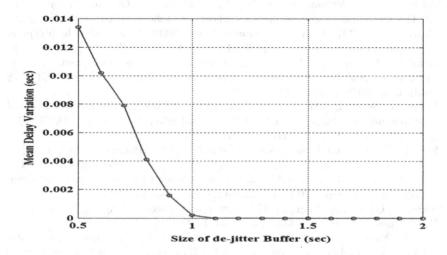

Fig. 4. Mean delay variation of video stream received by vehicles

The last simulation evaluated the average time variation of the video stream received by the vehicle. In this simulation, we consider 400 vehicles introduced randomly to the highway during a period equal to 8.57 min, and we set the distance between RSUs equal to 5000 m. We assume that a vertical handover requires a delay of 0.7 s. We also assume that a horizontal handover, which is consists in changing the point of attachement of a vehicle in the tree, costs 0.1 s. Figure 4 shows a decrease of the mean change delay proportionally to the increase of the size of the de-jitter buffer. We notice that the delay becomes very low (less than 2×10^{-3}) if the de-jitter buffer becomes higher than 0.7 (this value corresponds the vertical handover delay).

6   Conclusion

In this paper we presented a heterogeneous VANET architecture integrating LTE and 802.11p networks. We propose an access scheme based on a tree topology to maximize the network coverage, make the global topology less dynamic, and provide a rapid and seamless data connectivity to vehicles. A set of mechanisms for QoS provision (suitable for multimedia services) and resources reservation, are developed. The architecture enables a rapid access of vehicles to the network and a low rate of connection failure thanks to the availability of pre-built routes, the continuous updates of QoS offers on these routes, and the use of a pre-handover mechanism.

References

1. Physical layer procedures: 3GPP TS 36.213, ETSI (2009)
2. Bilstrup, K., Uhlemann, E., Ström, E., Bilstrup, U.: On the ability of the 802.11p mac method and stdma to support real-time vehicle-to-vehicle communication. EURASIP J. Wirel. Commun. Netw. **2009**, 1–13 (2009). http://jwcn.eurasipjournals.com/content/pdf/1687-1499-2009-902414.pdf
3. Garai, M., Boudriga, N.: A novel architecture for qos provision on vanet. In: High-Capacity Optical Network and Emerging/Enabling Technologies (HONET-CNS 2013), Cyprus, December 2013
4. Rémy, G., Senouci, S.M., Jan, F., Gourhant, Y.: Lte4v2x: Lte for a centralized vanet organization. In: IEEE Global Telecommunications Conference (GLOBECOM 2011), Houston, TX, USA, December 2011
5. Sivaraj, R., Gopalakrishnay, A.K., Chandraz, M.G., Balamuralidhar, P.: Qos-enabled group communication in integrated vanet-lte heterogeneous wireless networks. In: 7th IEEE Wireless and Mobile Computing, Networking and Communications Conference (WiMob), Wuhan, China, October 2011
6. Sofra, N., Gkelias, A., Leung, K.K.: Route construction for long lifetime in vanets. IEEE Trans. Veh. Technol. **60**(7), 3450–3461 (2011)
7. Sun, W., Yamaguchi, H., Yukimasa, K., Kusumoto, S.: Gvgrid: a qos routing protocol for vehicular ad hoc networks. In: 14th IEEE International Workshop on Quality of Service, New Haven, CT, USA, June 2006
8. Tal, I., Muntean, G.M.: User-oriented cluster-based solution for multimedia content delivery over vanets. In: 2012 IEEE International Symposium on Broadband Multimedia Systems and Broadcasting (BMSB), Seoul, South Korea, June 2012
9. Xu, C., Zhao, F., Guan, J., Zhang, H., Muntean, G.M.: Qoe-driven user-centric vod services in urban multi-homed p2p-based vehicular networks. IEEE Trans. Veh. Technol. **62**(5), 2273–2289 (2013)

Security in Ad Hoc Networks, SecAN 2014

Detection and Prevention of Black Hole Attacks in Mobile Ad hoc Networks

Muhammad Imran[1], Farrukh Aslam Khan[1,2(✉)], Haider Abbas[2,3], and Mohsin Iftikhar[4]

[1] Department of Computer Science, National University of Computer and Emerging Sciences, Islamabad, Pakistan
mimran181@gmail.com
[2] Center of Excellence in Information Assurance (CoEIA), King Saud University, Riyadh, Saudi Arabia
{fakhan,hsiddiqui}@ksu.edu.sa
[3] National University of Sciences and Technology, Islamabad, Pakistan
[4] Department of Computer Science, College of Computer and Information Sciences, King Saud University, Riyadh, Saudi Arabia
miftikhar@ksu.edu.sa

Abstract. Mobile Ad hoc Networks (MANETs) are vulnerable to external threats due to their open access and lack of central point of administration. Black hole attack is a well-known routing attack, in which an attacker node replies to the Route Requests (RREQs) by pretending itself as a neighbor of the destination node in order to get the data. These days, it has become very challenging to secure a network from such attacks. In this paper, we propose a Detection and Prevention System (DPS) to detect black hole attack in MANETs. For this purpose, we deploy some special nodes in the network called DPS nodes, which continuously monitor RREQs broadcasted by other nodes. DPS nodes detect the malicious nodes by observing the behavior of their neighbors. When a node with suspicious behavior is found, DPS node declares that suspicious node as black hole node by broadcasting a threat message. Hence, the black hole node is isolated from the network by rejecting all types of data from it. The simulations in NS-2 show that our proposed DPS mechanism considerably reduces the packet drop ratio with a very low false positive rate.

1 Introduction

Mobile Ad hoc Networks (MANETs) are infrastructure-less networks in which nodes are free to move arbitrarily in any direction. These mobile nodes have limited transmission range; consequently they need the assistance of their neighboring nodes in order to transmit a message to a node away from their transmission range. For this purpose, specific routing protocols are used that have the ability to establish a route between nodes that are not in the transmission range of each other. Ad hoc On-demand Distance Vector (AODV) [1] routing protocol is one of such protocols, which is widely used in MANETs. As being easily configurable, MANETs are mostly used in the areas where infrastructure is not available, such as military and rescue operations etc. Due to having open access, MANETs are always vulnerable to external and internal attacks

© Springer-Verlag Berlin Heidelberg 2015
M. Garcia Pineda et al. (Eds.): ADHOC-NOW Workshops 2014, LNCS 8629, pp. 111–122, 2015.
DOI: 10.1007/978-3-662-46338-3_10

such as Denial of Service (DoS), Flooding, Wormhole, Black hole, Gray hole, and Sinkhole etc.

Black hole attack is a well-known attack in wireless ad hoc networks that can occur especially in case of on-demand routing protocols such as AODV [19]. It is an attack in which a malicious node acquires route from a source node to a destination node by falsification of sequence number or hop count or both [2, 3, 17, 18]. A black hole node builds a route reply with fake larger sequence number and shorter hop count (usually 1) of a routing message in order to forcefully acquire the route and then listen or drop all data packets that pass through that route. The original AODV had a feature that any intermediate node in an ad hoc network could respond back to a route request if it had a fresh route to the destination in order to decrease the routing delay in the network. However, the original AODV protocol assumed that all nodes in the network were trusted nodes. If this is not the case then it would be very easy for any malicious node to destroy the network partly or as a whole by responding to the route request. Moreover, the malicious node does not need to look into its routing table to reply to the route request, therefore, the route reply from the malicious node would be much faster than the reply from a normal node. After receiving reply from the malicious node, the source node would start sending the data assuming that the route discovery process is complete. Consequently, all the packets transferred through that malicious node are lost [20].

Figure 1 shows the behavior of a black hole attack. In this figure, the source node S wants to establish a route to the destination node D. In case of AODV, node S broadcasts a Route Request (RREQ) packet to search for the destination node D. The intermediate nodes I, J, K, L, M and N will receive and rebroadcast the RREQ, whereas the black hole node B will send a RREP with a large sequence number or hop count of 1 to the source node S pretending that it is a neighbor of the destination node D. Actual RREP from the destination node D containing route S-J-M-D will be discarded by the source node S due to having more hop count value i.e. 3 as compared to RREP sent by the black hole node B whose hop count value is 1.

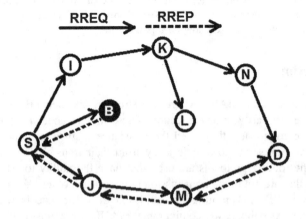

Fig. 1. Black hole attack

Therefore, according to AODV, a source node would select the largest sequence number and shortest route with minimum hop count to send data packets upon receiving multiple RREPs. Hence, a route would be selected by node S via a black hole node. The black hole node will drop all the data packets that it receives. Two or more black hole nodes can collaborate to launch an advanced form of attack, known as cooperative black hole attack [4].

In this paper, we propose a Detection and Prevention System (DPS) to detect black hole attacks in MANETs. Some special nodes called DPS nodes are deployed in the network, which continuously monitor RREQs broadcasted by other nodes. The DPS nodes detect malicious nodes by observing the behavior of their neighbors. When a node with suspicious behavior is detected, DPS nodes declare that suspicious node as black hole node and broadcast a threat message. The black hole nodes are isolated from the network by rejecting all types of data from them. Simulations are performed in Network Simulator 2 (NS-2) to check the performance of the proposed technique. The results show that the proposed DPS mechanism considerably reduces the packet drop ratio with a very low false positive rate. Our proposed scheme detects and isolates all forms of black hole attacking nodes.

The remainder of the paper is organized as follows: In Sect. 2, the related work is presented. Section 3 describes the details of our proposed scheme, while in Sect. 4, experiments and results are presented. Finally, Sect. 5 concludes the paper.

2   Related Work

Several researchers have proposed different solutions to countermeasure black hole attacks in MANETs. Ramaswamy et al. [5] proposed a technique to detect multiple and coordinated black hole attacks working in a group by adding a Data Routing Information (DRI) table in each node. This table contains information of data sent and received by a node to and from its neighboring nodes respectively. Malicious nodes are detected on the basis of information contained in DRI table. This technique adds some delay in route discovery process due to cross checking of intermediate nodes. Kurosawa et al. [6] presented an anomaly detection technique using dynamic training method in which the training data is updated at regular time intervals. This scheme required to check whether the characteristic change of a node exceeds the threshold within a specified period of time. If yes, this node is considered as a black hole node, otherwise, latest observation data is added to the dataset for the purpose of dynamic updating. The characteristics under observation are the number of RREQs sent, the number of RREPs received, and the mean destination sequence number of observed RREQs and RREPs. This scheme requires additional processing as values are updated after specific time interval. So shorter updating time interval requires more processing overhead otherwise detection accuracy will decrease.

Tamilselvan and Sankaranarayanan [7] proposed a solution called Prevention of a Co-operative Black Hole Attack (PCBHA) to prevent the cooperative black hole attack. The authors used a table called Fidelity Table, where each participating node is assigned with a fidelity level, which acts as a reliability measure of that node. In the beginning, a default fidelity level is assigned to each node. After broadcasting a route request, a source node waits to receive route replies from the neighboring nodes and then selects a node with a higher fidelity level to transmit data to the destination node. The destination node will return an acknowledgement (ACK) after receiving the data packets and the source node adds 1 to the fidelity level of the neighboring node upon receiving an ACK response. If no ACK is received by the source node, the value of 1 is subtracted from the fidelity level, which shows the possibility of a black hole node on this route. When the fidelity level becomes equal to 0, it is declared as black hole node. This solution adds more traffic to the network while exchanging fidelity table within the node and sending ACK message for each data packet. Another solution was presented by Weerasinghe and Fu [8] to countermeasure cooperative black hole attacks. This solution was basically an enhanced form of a previous solution [5], which uses the Data Routing Information (DRI) table to detect wormhole attacks. The problem of delay in route discovery process still exists in the solution.

Su et al. [9] proposed an intrusion detection system which runs Anti-Black hole Mechanism (ABM) to detect malicious nodes. ABM increases the suspicious value of a node on the basis of abnormal difference between routing packets transmitted from that node. When the suspicious value exceeds a specified threshold, the node is declared as black hole by broadcasting a block message. Due to the restriction that an intermediate node cannot reply to the RREQ, there is a delay in finding the destination node and rediscovering same destination by neighboring source nodes. Gupta et al. [10] presented a Black hole Attack Avoidance Protocol (BAAP), which avoids malicious nodes with the help of a legitimacy table, which is maintained by each node in the network. The black hole nodes are isolated from the network on the basis of the values in the legitimacy table. This technique requires additional fields in the routing table, RREQ and RREP, and also causes additional processing at each node. In [11], Su used a technique to prevent selective black hole attack in MANETs.

Jhaveri et al. [12] presented a novel approach in which intermediate node calculates the peak value from RREP sequence number, routing table sequence number, and number of RREP received. On the basis of peak value, the malicious nodes are detected. Chatterjee and Mandal [13] proposed a black hole detection method using triangular encryption. On receiving the RREQ, an intermediate node encrypts the plain text in the packet using partition and key on which the sender and receiver are agreed. The requirement of same partition and encryption key makes this approach complex and difficult to implement in large networks.

Tan and Kim [14] proposed a mechanism that provides Secure Route Discovery for AODV protocol (SRD-AODV) to prevent black hole attacks. It requires source and destination nodes to verify the sequence numbers in the Route Request (RREQ) and Route Reply (RREP) messages based on defined thresholds before establishing a connection with a destination node for sending the data. Thachil and Shet [15] presented a trust-based approach to mitigate blackhole attack in MANETs. In this approach, each node listens its neighbors promiscuously and calculates their trust value as a ratio of the number of packets dropped to the number of packets forwarded. If the trust value goes below a threshold, it is assumed as malicious node and avoided. This approach is difficult to implement in large networks where nodes change their positions rapidly. Zhang et al. [16] presented a technique based on sequence number to overcome black hole attack. In this technique, each intermediate node that forwards the route reply back to the source node also sends a message containing sequence number to the destination node. The destination node sends the updated sequence number to the source node. So the source node checks the sequence number to detect fake route replies having larger sequence number. This technique adds more traffic to the network and causes delay in route discovery process due to additional messages.

3 Proposed Scheme

As the black hole node does not forward the RREQ broadcasted by other nodes, therefore, the number of RREQs broadcasted by a black hole node is always less as compared to its neighbors. The proposed DPS works on the basic principle that "the black hole node broadcasts either no route requests (single black hole attack) or very few route requests (cooperative black hole attack) as compared to the normal nodes".

3.1 Assumptions

We have made two assumptions for our solution:

- All the DPS nodes are set in promiscuous mode.
- The network is secured from impersonation attack.

The promiscuous mode is necessary for DPS nodes to detect black hole attack because black hole node in single form of attack does not broadcast RREQ, so their presence can be detected through RREP as they respond to each route request. It is also necessary to keep the network secure from impersonation attack to prevent adversary to send fake threat messages.

3.2 Types of Nodes

Our detection and prevention system has three different types of nodes, which perform different tasks according to their role.

Normal Nodes: These are the common nodes in the network, each of which maintains a list (block list) of malicious nodes that only updates on receiving a block message from a DPS node. These nodes simply drop all packets received from the malicious nodes.

Table 1. DPS analysis table

Status	Node ID	RREQ count	Suspicious value	Blackhole confirmed
Active	43	0	3	No
Active	31	4	0	No
Inactive	41	3	0	No
Active	35	4	1	No

Malicious (Black hole) Nodes: These nodes respond to each RREQ with a RREP with greater sequence number and hop count value equal to 1, so that the source node considers them as neighbor of the destination node and starts sending the data packets.

DPS Nodes: These are the detective nodes that only sniff RREQs and RREPs to maintain an analysis table as shown above (i.e. Table 1). These nodes only broadcast block messages and do not involve in normal data transfer. The analysis table has the information about node's status, RREQ broadcasted, suspicious value, and black hole status. Whenever a DPS node receives a RREQ or RREP from a node, it adds that node into its analysis table and changes that node's status to active.

3.3 System Parameters

In DPS we use two system parameters for different purposes whose values are predefined.

Max_Req_Count: When the *RREQ Count* of a single node reaches the Max_Req_Count, the DPS node initiates the process of calculating suspicious value for each node in its analysis table.

Threat_Value: When the suspicious value of a node becomes equal to Threat_Value, the DPS node broadcasts a block message to alert the normal nodes and other DPS nods.

3.4 DPS Processes

The DPS nodes in the system perform three functions, which are described below.

RREQ Counting: Whenever a DPS node receives a RREQ from a neighboring node, it increments the corresponding node's *RREQ Count* by one. If the new value is equal to the Max_Req_Count then the suspicious value calculation process starts.

Suspicious Value Calculation: This process increases the *Suspicious Value* of all those nodes in the analysis table whose *RREQ Count* is zero and *Status* is active. To avoid false positive detection, this process also decreases the *Suspicious Value* of nodes (with non-zero suspicious value) whose *RREQ Count* is greater than zero and *Status* is active. If the new *Suspicious Value* of a node becomes equal to Threat_Value and its *Black hole Confirmed* field is No then the DPS node will broadcast a block message containing the ID of the malicious node. After sending the block message, the *Black hole Confirmed* field is set to Yes. At the end of the suspicious value calculation process, the status of all the nodes in the analysis table are set to inactive and the *RREQ count* is set to zero.

Block Message Broadcasting: When the *Suspicious Value* of a node reaches Threat_Value then the DPS node broadcasts a block message if not done before. On receiving a block message, the other DPS nodes will rebroadcast it, while the normal nodes add malicious node ID into their block lists.

A flowchart of the proposed DPS is shown in Fig. 2.

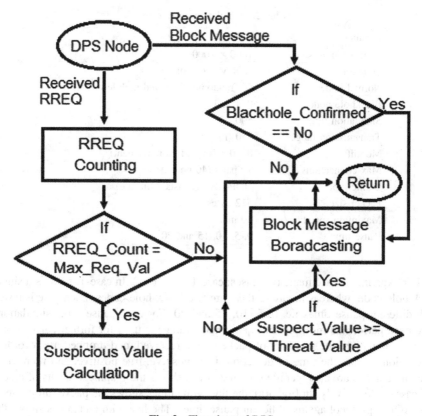

Fig. 2. Flowchart of DPS

4 Experiments and Results

To evaluate the performance of our proposed DPS mechanism, we conducted experiments in NS-2 version 2.34. The parameter values used in the experiments are given below in Table 2.

4.1 Experimental Detail

To measure the performance of the proposed DPS, we have taken into account 50 normal nodes in an area of 1000 × 1000 m and 9 DPS nodes at fixed locations so that they can cover the whole area. The black hole attack is implemented using the minimum hop count in RREP, in which a malicious node responds to all RREQs with a reply having hop count value 1. Since this RREP has less hop count as compared to others, therefore, the other nodes consider the malicious node as the neighbor of the destination. In this way, the black hole node gets involved in routes and starts dropping data packets that it receives.

Table 2. Simulation parameters

Parameter	Value
Simulation area	1000 × 1000
Protocol	AODV Protocol
Normal nodes	50 (randomly deployed mobile nodes)
Blackhole nodes	0, 1 and 2
Simulation time	500 (s)
Transmission range	250 (m)
Mobility	0–20 m/s (random movement)
Max connections	20 pairs (40 nodes)
Traffic type	UDP-CBR (constant bit rate)
Packet size	512 bytes
Maximum speed	20 m/s
Pause time	0, 5, 10, 15 and 20 s

For experiments, we made two cases; case-1 and case-2. In case-1, there is a single black hole node, whereas in case-2 there are two black hole nodes. Each case is tested with different pause times i.e. 0, 5, 10, 15 and 20. For each pause time, simulations have been performed multiple times and their average is used for further calculations. In each simulation, the numbers of packets sent, received, and dropped are recorded. In addition to that, the time of detection, false positive, true positive, and number of black hole nodes are also recorded. Figure 3(a-b) shows the graphical results of case-1 and case-2 respectively. In Fig. 3(a), the line with circles shows the packet drop rate of pure AODV protocol against different pause times. The line with rectangles shows the packet drop rate of AODV with one black hole node. Whereas, the dotted line with circles represent the packet drop rate of AODV with DPS and one black hole node. In Fig. 3(b), the line with circles shows the packet drop rate of AODV against different pause times. The line with rectangles shows the packet drop rate of AODV with two black hole nodes. Whereas, the dotted line with circles represent the packet drop rate of AODV with DPS and two black hole nodes.

4.2 Performance Metrics

The results of experiments are analyzed on the basis of packet drop rate, transmission delay, detection time, and false positive rate. Table 3 shows the detailed experimental results including packet drop rate with and without DPS, detection time of all black hole nodes in the system, and false positive rate.

Packet Drop Rate: Packet drop rate is the difference rate between packets sent by the source node and received by the destination node. By using DPS with AODV protocol, there is about 13 %–47 % decrease in packet drop rate as compared to AODV without DPS against different pause times in case of one black hole node. In case of two black hole nodes, packet drop rate reduces to 28 %–45 % against different pause times by using AODV with DPS.

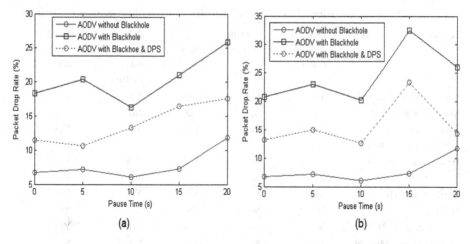

Fig. 3. Packet drop rate (a) One black hole node (b) Two black hole nodes

False Positive Rate: False positive rate is the rate of declaration of normal nodes as malicious nodes. As shown in Table 3, the proposed DPS has very low false positive rate. During simulations, sometimes a normal node was also declared as black hole node when it was almost isolated from the other network nodes near the boundary.

Transmission Delay: It is the delay that is caused due to finding a valid route to the destination before sending data packets. By implementing the proposed DPS, there is no further delay added to route discovery process.

Routing Overhead: Routing overhead is the extra amount of data, which is required to transmit other than actual data. There is no overhead data in DPS except the threat message that is broadcasted only when a black hole node is found.

Table 3. Experimental results

Pause time (s)	No. of blackhole node s	Packet drop rate without DPS (%)	Packet drop rate with DPS (%)	Detection time (s)	False positive rate (%)
0	1	18.35	11.47	57	2
5	1	20.36	10.59	69	0
10	1	16.28	14.30	72	2
15	1	20.99	16.48	89	0
20	1	25.93	17.59	54	2
0	2	20.76	13.20	75	0
5	2	23.00	15.02	54	2
10	2	20.25	12.60	75	1
15	2	32.53	23.36	88	2
20	2	25.96	14.26	66	2

Table 4. Comparison of the proposed technique with existing techniques

Techniques proposed by	Delay in route discovery	Routing overhead	Handle node mobility	Additional packet for route discovery	Block blackhole node
Ramaswamy et al. [5]	Yes	Yes	Yes	Yes	Yes
Kurosawa et al. [6]	Yes	No	Yes	No	Yes
Tamilselvan and Sankaranarayanan [7]	No	Yes	Yes	Yes	Yes
Weerasinghe and Fu [8]	Yes	Yes	Yes	Yes	Yes
Su et al. [9]	Yes	No	No	No	Yes
Gupta et al. [10]	Yes	Yes	No	Yes	No
Su [11]	No	No	No	No	Yes
Jhaveri et al. [12]	No	No	Yes	No	Yes
Chatterjee and Mandal [13]	Yes	Yes	Yes	No	No
Tan and Kim [14]	No	No	Yes	No	No
Thachil and Shet [15]	No	No	No	No	No
Zhang et al. [16]	Yes	Yes	Yes	Yes	Yes
Proposed Technique	**No**	**No**	**Yes**	**No**	**Yes**

Table 4 shows a general comparison of our proposed technique with some previous techniques discussed in the related work section on the basis of five factors. These factors include delay in route discovery process, routing overhead, either the technique handles node mobility or not, additional packets for route discovery process, and whether the technique blocks the black hole node or not.

5   Conclusion

In this paper, we proposed a detection and prevention system (DPS) for black hole attacks in Mobile Ad hoc Networks (MANETs). The proposed system is based on the fact that the black hole nodes do not forward the route requests. For detection purposes, some extra nodes are used, which are not involved in normal routing. These nodes (DPS Nodes) monitor the route request and route replies from their neighboring nodes and manage their suspect value. The suspect value of nodes having very low RREQ broadcasting rate is increased gradually. When the suspicious value of a node reaches a predefined threshold, it is declared as black hole node. After that all nodes in the network add it (black hole node) into their block list and ignore all the traffic coming from these nodes. There is no routing overhead and no delay in transmission. Furthermore, the proposed DPS is equally effective for cooperative black hole attacks. The NS-2 simulation results show that the proposed system detects all actual black hole nodes that are present in the range of DPS nodes. This increases the throughput of the network by reducing the packet drop rate, with very low false positive rate.

In future, we plan to implement our proposed system for the detection and prevention of other attacks (e.g., wormhole attack) with necessary modifications.

Acknowledgement. This work was supported by the Research Center of College of Computer and Information Sciences, King Saud University, through Grant Number RC131028. The authors are grateful for this support.

References

1. Perkins, C.E., Royer, E.M.: Ad-hoc on-demand distance vector routing. In: Second IEEE Workshop on Mobile Computing Systems and Applications (WMCSA 1999), New Orleans, LA, USA, pp. 90–100 (1999)
2. Mohebi, A., Scott, S.: A survey on detecting black-hole methods in mobile ad hoc networks. Int. J. Innovative Ideas. **13**(2), 55–63 (2013)
3. Mandala, S., Abdullah, A.H., Ismail, A.S., Haron, H., Ngadi, M.A., Coulibaly, Y.: A review of blackhole attack in mobile ad hoc network. In: 3rd International Conference on Instrumentation, Communications, Information Technology, and Biomedical Engineering (ICICI-BME), Bandung, pp. 339–344 (2013)
4. Tseng, F.-H., Chou, L.-D., Chao, H.-C.: A survey of black hole attacks in wireless mobile ad hoc networks. Hum.-Centric Comput. Inf. Sci. **1**(4), 1–16 (2011)
5. Ramaswamy, S., Fu, H., Sreekantaradhya, M., Dixon, J., Nygard, K.: Prevention of cooperative black hole attack in wireless ad hoc networks. In: International Conference on Wireless Networks (ICWN 2003), Las Vegas, Nevada, USA (2003)
6. Kurosawa, S., Nakayama, H., Kato, N., Jamalipour, A., Nemoto, Y.: Detecting blackhole attack on AODV-based mobile ad hoc networks by dynamic learning method. Int. J. Netw. Secur. **5**(3), 338–346 (2007)
7. Tamilselvan, L., Sankaranarayanan, V.: Prevention of co-operative black hole attack in MANET. J. Netw. **3**(5), 13–20 (2008)
8. Weerasinghe, H., Fu, H.: Preventing cooperative black hole attacks in mobile ad hoc networks: simulation implementation and evaluation. Int. J. Softw. Eng. Appl. **2**(3), 39–54 (2008)
9. Su, M.-Y., Chiang, K.-L., Liao, W.-C.: Mitigation of black-hole nodes in mobile ad hoc networks. In: International Symposium on Parallel and Distributed Processing with Applications (ISPA), Taipei, Taiwan, pp. 162–167 (2010)
10. Gupta, S., Kar, S., Dharmaraja, S.: BAAP: blackhole attack avoidance protocol for wireless network. In: International Conference on Computer and Communication Technology (ICCCT), Allahabad, India, pp. 468–473 (2011)
11. Su, M.-Y.: Prevention of selective black hole attacks on mobile ad hoc networks through intrusion detection systems. Comput. Commun. **34**(1), 107–117 (2011)
12. Jhaveri, R.H., Patel, S.J., Jinwala, D.C.: A novel approach for GrayHole and BlackHole attacks in mobile ad-hoc networks. In: Second International Conference on Advanced Computing and Communication Technologies (ACCT), Haryana, India, pp. 556–560 (2012)
13. Chatterjee, N., Mandal, J.K.: Detection of blackhole behaviour using triangular encryption in NS2. In: 1st International Conference on Computational Intelligence: Modeling Techniques and Applications (CIMTA), Procedia Technology, vol. 10, pp. 524–529 (2013)
14. Tan, S., Kim, K.: Secure route discovery for preventing black hole attacks on AODV-based MANETs. In: International Conference on ICT Convergence (ICTC), Jeju, Korea, pp. 1027–1032 (2013)

15. Thachil, F., Shet, K.C.: A trust-based approach for AODV protocol to mitigate blackhole attack in MANET. In: International Conference on Computing Sciences (ICCS), Phagwara, pp. 281–285 (2012)
16. Zhang, X.Y., Sekiya, Y., Wakahara, Y.: Proposal of a method to detect black hole attack in MANET. In: International Symposium on Autonomous Decentralized Systems, Athens, Greece, pp. 1–6 (2009)
17. Hu, Y.-C., Perrig, A.: A survey of secure wireless ad hoc routing. IEEE Secur. Priv. 2(3), 28–39 (2004)
18. Kant, R., Gupta, S., Khatter, H.: A literature survey on black hole attacks on AODV protocol in MANET. Int. J. Comput. Appl. 80(16), 22–26 (2013)
19. Ehsan, H., Khan, F.A.: Malicious AODV: implementation and analysis of routing attacks in MANETs. In: 11th IEEE International Conference on Trust, Security and Privacy in Computing and Communications (TrustCom), Liverpool, UK, pp. 1181–1187 (2012)
20. Patcha, A., Mishra, A.: Collaborative security architecture for black hole attack prevention in mobile ad hoc networks. In: IEEE Radio and Wireless Conference, Boston, MA, USA, pp. 75–78 (2003)

A Novel Collaborative Approach for Sinkhole Detection in MANETs

Leovigildo Sánchez-Casado[1]([✉]), Gabriel Maciá-Fernández[1],
Pedro García-Teodoro[1], and Nils Aschenbruck[2]

[1] Department of Signal Theory, Telematics and Communications,
School of Computer Science and Telecommunications, CITIC-UGR,
University of Granada, C/Periodista Daniel Saucedo Aranda S/n,
18071 Granada, Spain
{sancale,gmacia,pgteodor}@ugr.es

[2] Distributed Systems Group, Institute of Computer Science,
University of Osnabrück, Albrechtstr. 28, 49076 Osnabrück, Germany
aschenbruck@uos.de

Abstract. This paper presents a novel approach intended to detect *sinkholes* in MANETs running AODV. The study focuses on the detection of the well-known sinkhole attack, devoted to attract most of the surrounding network traffic by providing fake routes, and thus, invalidating alternative legitimate routes and disrupting the normal network operation. Our detection approach relies on the existence of "contamination borders", formed by legitimate nodes under the influence of the sinkhole attack and, at the same time, neighbors of non-contaminated legitimate nodes. Thus, by collecting the routing information of the neighbors, these nodes are likely to be able to properly detect sinkholes. We evaluate our approach in a simulation framework and the experimental results show the promising nature of this approach in terms of detection capabilities.

Keywords: AODV · Intrusion detection systems · MANETs · Poisoning attacks · Sinkhole

1 Introduction

MANETs are a particular type of infrastructure-less networks composed of mobile devices communicating via a multi-hop strategy, *i.e.*, a given node can directly communicate with those within its communication range, but it makes use of other nodes to relay its messages to out-of-range destinations. These inherent characteristics make this kind of networks a particularly useful candidate in certain areas, such as military applications, disaster management, etc. [1]. As MANETs proliferate, specific security issues become more relevant and need to be appropriately addressed for these environments. Different factors, usually referred to the constrained nature of nodes (reduced bandwidth, battery lifetime, etc.), must be taken into account in the mentioned security related aspects.

Among others on the networking layer, *route poisoning* attacks [2] are among the most potentially disruptive threats in MANETs. The present work focuses on

© Springer-Verlag Berlin Heidelberg 2015
M. Garcia Pineda et al. (Eds.): ADHOC-NOW Workshops 2014, LNCS 8629, pp. 123–136, 2015.
DOI: 10.1007/978-3-662-46338-3_11

the study of the *sinkhole* attack, possibly the most representative route poisoning attack. Nodes exhibiting this malicious behavior attempt to forge the source-destination routes in order to attract through them the surrounding network traffic. For this purpose, sinkhole nodes modify the control packets of the routing protocol and publish fake routing information that makes them appear as the best path to some destinations. In this manner, they achieve to be selected by other legitimate nodes as next hop on the forged route.

Focused on detecting sinkhole attacks, this work proposes an intrusion detection system (IDS) that relies on the existence of "contamination borders". These borders are formed by legitimate nodes under the influence of the sinkhole attack but with other neighbors which are not. We hyphotesize that by collecting and analyzing part of their own routing information and those belonging to their neighbors, these frontier nodes can precisely determine the existence of sinkhole behaviors. Based on this hypothesis, we suggest an IDS for the detection of sinkhole attacks which performs a collaborative process that collects from the neighbors the features for estimating the malicious behavior of a given node. The detection capabilities of our approach are enhanced regarding previous approaches due to the employment of information gathered from the contamination borders. These capabilities are proven in AODV (*Ad hoc On-Demand Distance Vector*) [3], one of the most representative and studied routing protocols in MANETs, obtaining promising results.

The rest of the paper is organized as follows. Section 2 describes the implementation of a sinkhole attack in AODV. Section 3 provides some related work regarding fighting against sinkhole attacks in MANETs. The existence of "contamination borders" and their utility as the basis for our detection approach is proven in Sect. 4. Our IDS is explained in Sect. 5, while Sect. 6 describes the experimental environment to evaluate the approach and the results obtained. Finally, main conclusions and future work are presented in Sect. 7.

2   Sinkhole Attacks in AODV

Among various routing protocols for MANETS, AODV is perhaps the most well-known and one of the most widely used ones. This is mainly due to its many useful characteristics. AODV is a reactive routing protocol for MANETs, *i.e.*, routes to a given destination are established on demand. If a source node N_s needs a connection with a destination node N_d and it does not have a valid route towards it, N_s initiates a route discovery process by broadcasting a *route request* message (RREQ). Upon receiving this RREQ, intermediate nodes forward it to their own neighbors, repeating the process until the RREQ reaches the intended destination. Once N_d receives the first RREQ, it sends a *route reply* message (RREP) backwards via the inverse route. Besides, AODV permits that intermediate nodes having a valid route to the destination generate RREP messages as a response to the received RREQ messages. Therefore, source and intermediate nodes are responsible for managing the routing information related to the next hop for every communication flow.

To avoid routing loops, AODV employs *destination sequence numbers*. These monotonically increasing numbers allow the nodes to determine the freshness of their information. Sequence numbers are updated whenever a node receives new (*i.e.*, not stale) information from control messages. This way, a node updates its routing information if the sequence number received in the RREP message is greater than the last stored sequence number. Given the choice between two routes, a node selects the one with the greatest sequence number. This fact can be exploited by malicious nodes to introduce themselves in the path.

Routing tables of the nodes in AODV are composed of the following fields: destination, next hop, distance to the destination measured in number of hops (*HopCount*), status (VAL -valid- or INV -invalid-) and sequence number (*SeqNum*), as well as other fields, like the lifetime of the route, several flags, the output interface, etc.

Once the very basics of AODV are known, it is easy to understand how a malicious node can carry out a sinkhole attack. It could modify or create a RREP message which announces an optimal metric, *i.e.*, a sequence number greater than the one received in the RREQ. If the sequence number is large enough, all other alternative routes will be invalidated. As a consequence, the malicious node guarantees that the requesting node will learn the route through the former, which will be selected as the next hop on the path. If the sinkhole node replies with fake RREP messages to every received RREQ packet, it will eventually become a sink of all data packets. Having achieved that, the malicious node will be able to apply different actions over the collected traffic, such as extracting sensitive information, modifying or discarding packets or carrying out more sophisticated attacks.

Figure 1 shows an example of a sinkhole attack. Here, the source node N_s broadcasts a RREQ message (1) asking for a route towards the destination N_d, this message being forwarded by the intermediate nodes. When the RREQ packet reaches the malicious node N_m, it replies with a fake RREP message (2a) claiming to have a shorter (*HopCount* = 1) and fresher (*SeqNum* = 37) route. At the same time, N_d is replying with a RREP message (2b) that includes the legitimate values for *HopCount* and *SeqNum* (3 and 8 respectively). Therefore, despite receiving other legitimate replies, N_s will choose the route through N_c, considered the most recent. Thus, the traffic from N_a towards N_d will eventually go through the malicious node N_m.

3   Related Work

Intrusion detection techniques have been recurrently used to determine the potential existence of non-legitimate events in a communication environment [4]. Consequently, in the literature a wide variety of IDS schemes was already proposed to detect sinkhole attacks in MANETs. Typically, they are classified as network-based IDS (NIDS) or host-based IDS (HIDS) dependending on the source of the features that support the detection process [4]. In what follows, we show that most of the IDS solutions adopted at present to detect sinkhole attacks are NIDS-like, that is, network parameters are monitored to determine the potential occurrence of malicious events.

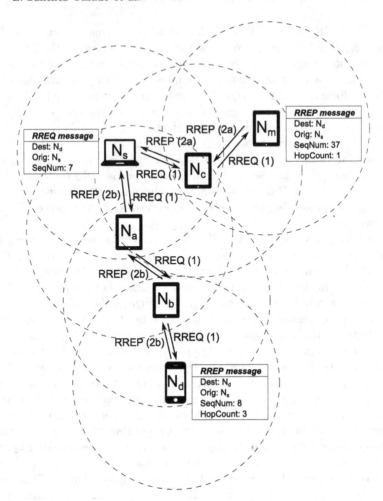

Fig. 1. Example of sinkhole node, N_m, replying with a fake RREP for a destination N_d.

Machine learning approaches are used in many approaches. Zhang *et al.*, in [5], introduce a local and cooperative scheme in which each mobile node runs a SVM-based IDS agent that monitors local traces, being responsible for locally detecting signs of intrusions. However, if an anomaly is detected among the local data, or if an evidence is inconclusive, neighboring IDS agents will collaboratively investigate, participating in a global detection procedure. A cross-feature method is described in [6], where a total of 141 traffic and topology related features are defined. This method also analyses correlations between features, in order to reduce the number of them. Then, a classifier like C4.5, RIPPER or Naïve-Bayes is used to carry out the anomaly detection procedure.

Other approaches perform some sort of matching techniques. For instance, IDAD [7] is an IDS solution to detect both single and multiple sinkholes.

This scheme compares every network activity of a host with a pre-collected set of anomaly and attack activities. The parameters used are obtained from each anomaly RREP packet: destination sequence number, hop count, route lifetime, destination IP address and timestamp. This way, IDAD is able to differentiate normal from abnormal RREP packets just by checking resemblances among them.

Finally, most of the techniques simply monitor the target environment, comparing the value of the collected features with a given threshold, which could be adaptive or not. Kurosawa *et al.* [8] introduce an anomaly detection scheme which uses a dynamic training method. They consider the number of RREQ packets sent and RREP packets received, as well as the average of the differences between the destination sequence numbers sent in RREQ packets and the ones received in RREP packets. Thus, this training set of features is employed to calculate the detection threshold based on the normal state of the network, which is dynamically adapted at regular time intervals to improve the detection accuracy. For the detection process, every sample in the data set is compared with the threshold to detect deviations from the normal network state. Similarly, in [9], the authors propose DPRAODV, in which the node receiving a RREP message checks whether the sequence number value exceeds a given threshold. To reduce inaccuracies which can lead to false alarms, this threshold value is dynamically updated at every time interval. If the sequence number is higher than the threshold, the intermediate node is suspected to be malicious.

Furthermore, a number of slight variations that also follow the approach of comparing the sequence number received in the RREP packet with the sequence number sent in the RREQ can be found in [10–13].

These schemes only consider the behavior of the sequence numbers in a local way, *i.e.*, without taking into account information of the network vicinity. By considering this behavior in a more global way, we will demonstrate that it is possible to improve the detection capabilities.

4   "Contamination Borders" in the Sinkhole Attack

Let us consider the existence of a MANET composed of L legitimate nodes $\{N_1, ..., N_L\}$. For every node N_i in the network, we extract some features following a time-based procedure, by using non-overlapping windows of δ seconds. As we assume mobility of the nodes, every node N_i has different sets of neighbors $NB_i(\omega)$ at the time of study $\omega \in \mathbb{N}$. Nodes can generate different traffic flows and they communicate by using AODV. We use the notation $R_{i,j}$ to refer to the route learned by node N_i towards a given destination N_j. Routes are composed, among other fields, by the following information: $R_{i,j}(\omega) = \{SN_{i,j}(\omega), NH_{i,j}(\omega)\}$, where $SN_{i,j}(\omega)$ is the sequence number learned for the route $N_i \rightarrow N_j$ and $NH_{i,j}(\omega)$ represents the next hop towards the destination at time $\omega \cdot \delta$.

In this general scenario, we additionally consider the existence of M malicious nodes behaving as sinkhole nodes, *i.e.*, nodes that reply to every RREQ packet with a forged RREP message, trying to include themselves as the next hop in the path to the destination.

4.1 Existence of "Contamination Borders"

In the above scenario, our approach relies on the existence of contamination zones, formed by legitimate nodes which are under the influence of the attack. Some of these nodes conform the "contamination border". The peculiarity of these last nodes is that they are simultaneously neighbors of contaminated nodes and nodes which are not under the influence of the sinkhole (*i.e.*, those that have the knowledge about the legitimate routes).

The nodes at the "contamination zone" forward traffic through the sinkhole node. At the same time, when a non-contaminated node requests to one of the contamination border nodes a route that has been compromised, it will reply with fake information, *i.e.*, the border node will unintentionally publish fake learned routes when asked for them. In such a situation, the border nodes behave in a similar way to how a malicious node would, being indistinguishable from actual sinkholes.

Let us illustrate this idea with the example shown in Fig. 2. Let us assume first that, at a given time t_0, node N_c has a legitimate route to a destination N_d with sequence number 35. At t_1, N_b needs a route towards N_d and generates a RREQ which is forwarded by N_a. As a consequence, at t_2, N_m replies with a fake RREP including an increased sequence number (for instance, 100) and N_c replies with its legitimate RREP. Since the sequence number in N_c is smaller, N_a learns the route through N_m and forwards that RREP to N_b. The routes are updated at t_3.

In such a situation, N_a will become a contamination border node, since it sends a fake RREP to N_b without a malicious intention. Thus, the contaminated area will be formed by N_a and N_b, N_m being the malicious sinkhole. Nodes N_c and N_d will remain without being contaminated.

Under these circumstances, the only difference between a sinkhole node and a contaminated node is that sinkhole nodes deliberately try to attract most of the surrounding traffic, whereas contaminated ones only act like the sinkhole for those requests related to fake routes learned from it, and not for every request they receive.

4.2 Use of "Contamination Borders" to Detect Sinkhole Behaviors

As shown in Sect. 3, most of the IDS schemes consider information directly extracted from the node carrying out the detection process, *i.e.*, they basically employ some metric related to the difference between sent and received sequence numbers. However, this approach suffers from some flaws which can lead to errors in the detection process.

The first weakness is related to the fact that these approaches provide good results as long as the increased sequence numbers published by the sinkholes are high. That is, the difference between the sent and received sequence numbers is noticeable. However, if the sinkhole node is somehow smart, it will publish fake sequence numbers moderately high, thus assuring that it is selected as the next hop whereas hindering the detection process. On the other hand, legitimate nodes learning fake routes are able to publish them and, as seen, they are

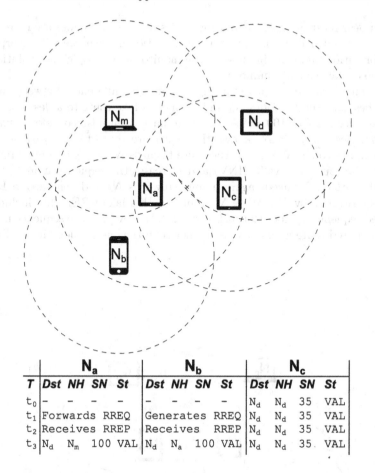

T	N_a				N_b				N_c			
	Dst	NH	SN	St	Dst	NH	SN	St	Dst	NH	SN	St
t_0	-	-	-	-	-	-	-	-	N_d	N_d	35	VAL
t_1	Forwards	RREQ			Generates	RREQ			N_d	N_d	35	VAL
t_2	Receives	RREP			Receives	RREP			N_d	N_d	35	VAL
t_3	N_d	N_m	100	VAL	N_d	N_a	100	VAL	N_d	N_d	35	VAL

Fig. 2. Existence of contamination zones and border nodes.

prone to be erroneously detected as sinkhole as well. Therefore, both facts can degrade the detection capabilities of these schemes.

Our approach is based on the fact that, due to the existence of "contamination borders", if a border node compares the received sequence number for a given route not only with the sequence number sent, but also with the sequence numbers stored by their neighbors, the dynamic range of the detection will be increased, thus leading to a better performance of the IDS scheme. Therefore, by collaborating with their neighbors and sharing the features of interest, these border nodes are able to perform a better detection, properly distinguishing between sinkhole nodes and legitimate nodes.

Besides, in a network with sinkhole nodes, it is expected that contaminated nodes being neighbors of a sinkhole node have in their routing tables many entries whose next hop is the given sinkhole, and not that many whose next hop is other contaminated node. For this reason, we hyphotesize that those nodes which appear most often in the routing tables of the other nodes are more likely to be considered malicious. Thus, in our detection system, we will incorporate this information and combine it with the monitoring of suspicious evolutions in the value of the sequence number.

The simple example depicted in Fig. 3 shows the differences between the two approaches. At time t_0, nodes N_a and N_b have a fake route to a destination N_d with sequence number 100, since this route have been falsified before (example in Fig. 2). Besides, node N_c knows the legitimate route to N_d, with sequence number equals to 20. At time t_1, the route towards N_d in N_b becomes stale and it marks the route as invalid (INV) and increases the sequence number in one unit (101). At t_2, N_b needs again a route towards N_d and generates a RREQ which is forwarded by N_a. At t_3, N_m replies with a fake RREP that includes an increased sequence number (for instance, 121). However, as the sequence number for the required route in node N_c is smaller than the one included in the RREQ,

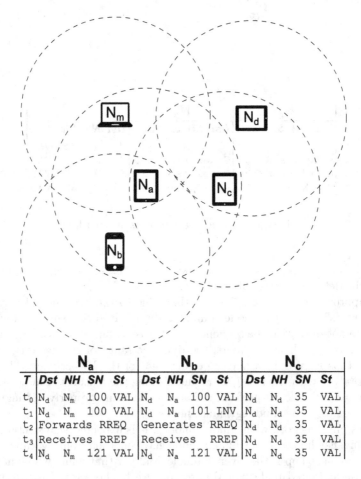

	N_a				N_b				N_c			
T	*Dst*	*NH*	*SN*	*St*	*Dst*	*NH*	*SN*	*St*	*Dst*	*NH*	*SN*	*St*
t_0	N_d	N_m	100	VAL	N_d	N_a	100	VAL	N_d	N_d	35	VAL
t_1	N_d	N_m	100	VAL	N_d	N_a	101	INV	N_d	N_d	35	VAL
t_2	Forwards	RREQ			Generates	RREQ			N_d	N_d	35	VAL
t_3	Receives	RREP			Receives	RREP			N_d	N_d	35	VAL
t_4	N_d	N_m	121	VAL	N_d	N_a	121	VAL	N_d	N_d	35	VAL

Fig. 3. Utility of a "contamination border" node, N_a, in the detection process.

N_c does not reply. The detection schemes compute the values at t_4, when routes have been updated.

In previous approaches, like [11] or [13], N_a would obtain the difference between the sent and received sequence numbers, resulting in $121 - 101 = 20$ units, which can be enough to attract the route but not to be detected by N_a. By using our approach, N_a computes the difference by comparing the sequence number received in the RREP with the minimum sequence number for the required route in its neighbors, giving as a result a difference of $121 - 35 = 86$ units.

As it has been explained by the static and straightforward scenario depicted in the example, by gathering very little information from the neighbors (basically the sequence number for some required routes), border nodes are able to increase the dynamic range of the metric usually employed to detect sinkhole attacks. This allows our approach to raise the detection threshold and therefore, to improve the detection rate whereas the misclassification rate remains low.

5   Deploying the Sinkhole Detection Scheme

This section presents the specific implementation of the proposed network-based intrusion detection system, which employs a simple heuristic to obtain an indicator value for the detection of sinkhole attacks. The IDS computes the heuristic by collecting information related to the routing tables of the node running the IDS and its neighbors. Even though the detection process is locally performed by each node running the IDS, the features involved in such a process are collaboratively gathered from the node itself and its neighbors.

This heuristic relies on the hypothesis that there are border nodes being under the influence of the sinkholes that have neighbors which are not under the influence and know the legitimate routes. The sequence number information provided by the neighbors allows to improve the detection capabilities in these border nodes. Besides, those nodes appearing most often in the routing tables as next hop are more likely to be considered malicious, since sinkhole nodes attract most of the surrounding traffic, and this fact must be taken into account in the heuristic.

It must also be noted that only those nodes that are neighbors of the actual sinkhole will be able to detect it, since non-neighbor nodes will detect as malicious those frontier nodes unintentionally sharing fake routes.

5.1   Overview of the Detection Approach

Our approach follows a window-based procedure to detect malicious nodes discretely over time. Every node N_i will run the IDS, and will check during each window if any of its neighbors is malicious or not. Thus, for every next hop node (NH), the following features are collected:

- $D_{i,NH}(\omega)$: the set of all destinations for the routes in the routing table of N_i which use NH as next hop, at time $\omega \cdot \delta$. Only valid routes with $HopCount$ greater than 1 are taken into account, since routes with $HopCount = 1$ indicate neighbors and do not have to be published, so they will not indicate whether or not a node is publishing false routes.

- $\boldsymbol{SN_{i,j}(\omega)}$: sequence number at node N_i for every destination N_j, at time $\omega \cdot \delta$.
- $\boldsymbol{NB_i(\omega)}$: set of neighbors of node N_i, at time $\omega \cdot \delta$.

Taking the above into account, we apply a heuristic to obtain an indicator value about the node's behavior as sinkhole. For that, the following procedure is executed:

(1) The IDS at node N_i obtains, for each NH in its routing table, a set of destinations N_j in $D_{i,NH}(\omega)$.
(2) Then, it requests to its neighbors theirs sequence numbers for those destinations N_j.
(3) After gathering the information from all the neighbors, N_i obtains the minimum sequence number of their neighbors for each destination N_j, and computes the difference between their own sequence numbers and these minimums.
(4) Finally, the malicious value for the NH is obtained as the summatory of these differences, thus considering that nodes NH with more poissoned routes are more likely to be a sinkhole node than a contaminated node:

$$MV_{i,NH}(\omega) = \sum_{j \in D_{i,NH}(\omega)} \left(SN_{i,j}(\omega) - \min_{v \in NB_i(\omega)} SN_{v,j}(\omega) + 1 \right) \quad (1)$$

Since, for a given destination, the computed difference between sequence numbers can be zero, we add 1 unit, thus taking every possible compromised destination into account in the summatory.

(5) After the calculation of the $MV_{i,NH}$, if it exceeds a given threshold, θ, the node NH is classified as a malicious sinkhole node:

$$class(NH) = \begin{cases} malicious, & \text{if } MV_{i,NH}(\omega) \geq \theta \\ legitimate, & \text{otherwise} \end{cases} \quad (2)$$

It can be observed that the calculation of the malicious value is a simple process with low computational cost once all the neighbors' information have been gathered.

(6) After the classification of NH as sinkhole, N_i could apply some response mechanism, like that of including NH in a blacklist or notifying all the nodes in the network about the malicious behavior of NH. These and other possible reaction schemes are out of the scope of this detection-oriented contribution.

6   Experimental Results

This section presents the description of the experimental environment used to evaluate the proposed scheme. Besides, some tests have been performed to prove the proper performance of the IDS, the experimental results being discussed.

6.1 Experimental Environment

In this work we have simulated some MANET deployments by using the network simulator OMNeT++ [14]. To simulate the sinkhole nodes, we have used NETA [15], a framework built on top of OMNeT++ that allows to simulate different network attacks in a simple manner and permit to apply several configuration parameters over them.

The simulation area is restricted to a $1000\,m \times 1000\,m$ square, with each node having a communication range of $250\,m$. As MAC and network layer protocols we have chosen 802.11-g and AODV. The simulation time is set to 300 s and the duration of the temporal window ω used for collecting the features is 1 s.

The total number of nodes is 25, with 24 legitimate nodes and only 1 sinkhole node. The attack is performed during the whole simulation by replying with false RREP every received RREQ, even if the sinkhole does not know a valid route. A value following a uniform distribution between 20 and 30 units is added to the one observed from the RREQ, giving the false increased sequence number. It must be noted that, in the literature, most of the works simply set the false sequence number to the maximum possible ($2^{31} - 1 = 4294967295$), meanwhile other works adds relatively high values, for instance, uniform values between 15 and 200 units. We consider a more realistic sinkhole which tries to hinder the detection process but assures being selected as the next hop.

To model the movement of the nodes the popular Random Waypoint Model (RWP) [16] has been chosen. In this model the node selects random destination and speed. When the node reaches the destination, it waits for a pause time before choosing a new random destination and speed and repeats the process. The minimum speed is fixed to $0.5\,m/s$ and the maximum speed varies between 3 to $10\,m/s$, being the pause time set to 15 s. These maximum speeds ($3-10\,m/s \equiv 10.8-36\,km/h$) cover the range from pedestrian walk to a moderate vehicle speed.

The number of traffic flows is fixed to 24, each one simulating point to point voice traffic. Several calls per flow are obtained by modelling the pause time between calls (*inter arrival time* or IAT) with a exponential distribution with $\lambda = 7.5\,s$ and the duration of the call (*call holding time* or CHT), modelled as a lognormal with mean, μ, set to 2.5 and standard deviation, σ, set to 0.5 [17]. For each call, one of the legitimate nodes is randomly chosen as destination, being the traffic a Constant Bit Rate (CBR) connection, with 4 packets/second and payload size equal to 512 bytes.

6.2 Detection Results

We now evaluate the global effectiveness of the proposed IDS by means of several test based on simulations. The effectiveness is evaluated by computing two metrics, namely the true positives rate (TPR) and the false positives rate (FPR).

We study the detection efficiency for different mobility conditions, obtaining various operation points to conform the ROC (Relative Operation Characteristic) space by varying the decision threshold θ in (2). It is important to note that the ROC curve is derived by repeating 20 times (with different seeds) every simulation.

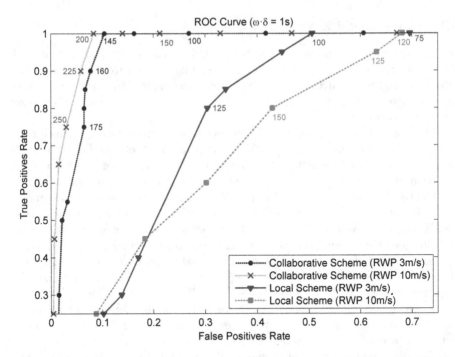

Fig. 4. ROC curve for sinkhole detection, for different values of the decision threshold θ.

Figure 4 depicts the ROC curves obtained by using our collaborative app-roach and those obtained by using an approach that compute a local heuristic only considering the sent and received sequence numbers in the node, as those introduced in [11] or [13]. The curves are obtained under two mobility conditions, by varying the maximum speed of the nodes between 3 m/s and 10 m/s. As it can be seen, by including information from the neighbors, our scheme overcomes the results achieved by the local approach used by some previous schemes.

Besides, it is shown that if the detection threshold is set to a high value, the system is expected to improve FPR, but to achieve worse TPR. On the other hand, the lower the threshold, the better the TPR value, at the expense of increasing the FPR. Thus, the optimal operation point of our system can be achieved empirically, and it depends on the mobility conditions.

As shown, the proposed IDS can achieve excellent results regarding the two metrics considered, TPR and FPR. By selecting the optimal operation point, TPR can achieve 100 % keeping FPR always below 10 %. These results confirm the capabilities of our model.

7   Conclusions and Future Work

In this paper we introduce a new methodology for the detection of sinkhole attacks in MANETs which relies on the existence of contamination zones and border nodes. The scheme is based on a simple heuristic that computes the differences between the sequence numbers on these frontier nodes and those belonging to their neighbors. This heuristic allows to estimate the malicious behavior of the nodes acting as sinkholes.

The use of a simple heuristic overcomes the computational overhead present in more sophisticated approaches based on data mining algorithms. We have confirmed by means of simulation the good performance of our system, where different scenarios have been analyzed. The results obtained clearly highlight the goodness of our IDS approach, which can experience 100 % overall TPR with less than 10 % potential FPR.

As shown, the experimental results obtained are very encouraging. This way, we are going for such direction through the improvement of some aspects of our approach in the near future:

- In distributed IDS for MANETs is highly recommended to reduce the information exchanged and shared. We are working on the development of a communication protocol that takes into account the limited bandwidth resulting from the MANET context, thus involving the lowest possible overhead.
- This way, we are also developing a pre-filtering phase in order to also reduce the overhead introduced by our approach.
- We are planning to extend our approach to include trust-based schemes as response mechanism to face collusions situations carried out to evade the detection process or to accuse legitimate nodes.
- Finally, the inclusion of more realistic mobility models in the experimentation is also of interest.

Acknowledgment. This work has been partially supported by Spanish MICINN through project TEC2011-22579 and by Spanish MECD through the grant "University Professor Training Program" (FPU, Ref.: AP2009-2926).

References

1. Lakhtaria, K.I. (ed.): Technological Advancements and Applications in Mobile Ad-Hoc Networks: Research Trends. IGI Global, Hershey (2012)
2. García-Teodoro, P., Sánchez-Casado, L., Maciá-Fernández, G.: Taxonomy and Holistic Detection of Security Attacks in MANETs, pp. 1–12. CRC Press, April 2014. http://www.crcpress.com/product/isbn/9781466578036
3. Perkins, C., Belding-Royer, E., Das, S.: Ad hoc On-Demand Distance Vector (AODV) Routing. IETF, RFC 3561, July 2003
4. García-Teodoro, P., Díaz-Verdejo, J.E., Maciá-Fernández, G., Vázquez, E.: Anomaly-based network intrusion detection: techniques, systems and challenges. Comput. Secur. **28**(1–2), 18–28 (2009)
5. Zhang, Y., Lee, W., Huang, Y.A.: Intrusion detection techniques for mobile wireless networks. Wirel. Netw. **9**(5), 545–556 (2003)
6. Huang, Y., Fan, W., Lee, W., Yu, P.S.: Cross-feature analysis for detecting Ad-Hoc routing anomalies. In: Proceedings of 23rd IEEE International Conference on Distributed Computing Systems (ICDCS), pp. 478–487, May 2003

7. Alem, Y.F., Xuan, Z.C.: Preventing black hole attack in mobile Ad-Hoc networks using anomaly detection. In: Proceedings of 2nd International Conference on Future Computer and Communication (ICFCC), vol. 3, pp. 672–676, May 2010

8. Kurosawa, S., Nakayama, H., Kato, N., Jamalipour, A., Nemoto, Y.: Detecting blackhole attack on AODV-based mobile Ad Hoc networks by dynamic learning method. Int. J. Netw. Secur. **5**(3), 338–346 (2007)

9. Raj, P.N., Swadas, P.B.: DPRAODV: a dynamic learning system against blackhole attack in AODV based MANET. Int. J. Comput. Sci. Issues **2**, 54–59 (2009)

10. Al-Shurman, M., Yoo, S.M., Park, S.: Black hole attack in mobile Ad Hoc networks. In: Proceedings of 42nd Annual Southeast Regional Conference (ACM-SE), pp. 96–97, April 2004

11. Mistry, N., Jinwala, D.C., Zaveri, M.: Improving AODV protocol against blackhole attacks. In: Proceedings of International MultiConference of Engineers and Computer Scientists (IMECS), pp. 96–97, March 2010

12. Mandhata, S.C., Patro, S.N.: A counter measure to black hole attack on AODV-based mobile Ad-Hoc networks. Int. J. Comput. Commun. Technol. (IJCCT) **2**(VI), 37–42 (2011)

13. Himral, L., Vig, V., Chand, N.: Preventing AODV routing protocol from black hole attack. Int. J. Eng. Sci. Technol. (IJEST) **3**(5), 3927–3932 (2011)

14. Varga, A.: OMNeT++ Discrete Event Simulation System. http://www.omnetpp.org/doc/omnetpp/manual/usman.html. Accessed 14 March 2014

15. Sánchez-Casado, L., Rodríguez-Gómez, R.A., Magán-Carrión, R., Maciá-Fernández, G.: NETA: evaluating the effects of NETwork attacks. MANETs as a case study. In: Awad, A.I., Hassanien, A.E., Baba, K. (eds.) SecNet 2013. CCIS, vol. 381, pp. 1–10. Springer, Heidelberg (2013)

16. Johnson, D., Maltz, D.: Dynamic source routing in Ad Hoc wireless networks. In: Imielinski, T., Korth, H. (eds.) Mobile Computing. The Kluwer International Series in Engineering and Computer Science, vol. 353, pp. 153–181. Springer US, New York (1996)

17. Barceló, F., Jordán, J.: Channel holding time distribution in cellular telephony. Electron. Lett. **34**, 146–147 (1998)

2nd Smart Sensor Protocols and Algorithms – SSPA2014

A Smart M2M Deployment to Control the Agriculture Irrigation

Alberto Reche[1], Sandra Sendra[2], Juan R. Díaz[2], and Jaime Lloret[2(✉)]

[1] Departamento Informática, Servicio Aragonés Salud,
50071 Alcañiz, Zaragoza, Spain
albertoreche@gmail.com
[2] Instituto de Investigación para la Gestión Integrada de zonas Costeras,
Universidad Politécnica de Valencia, 46730 Grao de Gandia, Valencia, Spain
sansenco@posgrado.upv.es, {juadasan,lloret}@dcom.upv.es

Abstract. Wireless sensor networks (WSN) have become in a very powerful infrastructure to manage all kind of services. They provide the mechanism to control a big number of devices distributed around a big geographical space. The implementation of a sensor network is cheap and fast and it allows us to add a smart layer over the physical topology. For these reasons, they have begun to be used in many applications and environments. In this paper, we propose a new smart M2M system based on wireless sensor network to manage and control irrigation sprinklers. Humidity and temperature of soil are used to extract information about soil conditions. The network protocol builds an ad hoc infrastructure to exchange the information over the whole WSN. The proposed algorithm uses the meteorological parameters and characteristics of soil to decide which irrigation sprinklers have to be enabled and when we have to do it. Using our intelligent system we can reduce irrigation water consumption, avoiding activation of sprinklers when they are not needed.

Keywords: Smart algorithm · M2M deployment · WSN · Agriculture irrigation

1 Introduction

There are many reasons to use wireless sensor networks (WSN) in outdoor environments. One of the main advantages of wireless transmission is the significant economical saving in cost of implementation and simplification of wired infrastructure. It has been estimated that the cost of wiring typical in American industrial facilities is about $130–650 per meter and the adoption of wireless technology would eliminate between 20–80 % of this cost [1]. WSNs enable faster deployment and installation of different types of sensors. These networks are able to self-organize, self-configure, self-diagnosis and self-healing. Some of them also allow a flexible extension of the network.

A WSN can be defined as a network of small embedded devices called sensors which communicate wirelessly following an ad hoc configuration [2]. In this new environment, the nodes are involved in decision-making, in maintenance tasks of

© Springer-Verlag Berlin Heidelberg 2015
M. Garcia Pineda et al. (Eds.): ADHOC-NOW Workshops 2014, LNCS 8629, pp. 139–151, 2015.
DOI: 10.1007/978-3-662-46338-3_12

network and taking part in routing algorithms [3]. Ad hoc networks are typically composed of equal nodes that communicate between them over wireless links without any central control. Although primary applications of ad hoc networks were military tactical communication, nowadays, commercial interest in this type of networks continues to grow. Applications such as rescue missions in times of natural disasters, law enforcement operations, human tracking [4], animals monitoring [5], commercial and educational use, and sensor networks are just a few commercial examples [6]. The most common wireless technology used in sensor networks is Bluetooth standardized by the IEEE 802.15.1 protocol. This technology can reach ranges of 30 m and it consumes lower energy than other wireless technologies, like IEEE 802.11. In the recent years there have appeared new sensor-based M2M systems for agriculture monitoring [7].

In this paper, we present the deployment of a smart M2M system where the wireless sensor network, which can be reached from internet, manages and controls the irrigation of the fields. In order to achieve this goal, we have developed an smart algorithms that takes into account the humidity and temperature of soil and weather parameters to decide which sprinklers should be enabled or not. We have also developed an algorithm for network discovering and have performed a set of test to check the correct operation of system.

The rest of this paper is structured as follows. Section 2 presents some previous works about wireless sensor networks used in agriculture. Section 3 describes the outdoor scenario where this research was carried out and the used tools. In Sect. 4 we detail the proposed system and explain the network protocol and the system algorithm. Section 5 describes the test bench results. Finally, conclusion and future works will be presented in Sect. 6.

2 Related Work

The use of new technologies and sensor development platforms for agriculture remote monitoring is being extensively investigated. In fact, we can find some researches uniquely dedicated to the study of sensors and deployments for agriculture and food industry monitoring [8, 9]. In this section, we present some works where new technologies are used to develop WSNs for agriculture monitoring.

Nowadays, there are lots of implementation of WSN and mechanisms for agricultural monitoring. On the one hand, Z. Chen and C. Lu [10] present a review of material and mechanisms for implementing humidity sensors. The most used materials for implementing humidity sensors are ceramic materials, semiconductor materials and polymer materials. To use a humidity sensor is important to know the electrical properties of humidity sensors such as sensitivity, response time, and stability that each sensor can offer. To perform this kind of measures, we can use mirror-based dew-point sensors which are more costly in fabrication with better accuracy, while the Al_2O_3 moisture sensors can be fabricated with low cost and offer better results.

Regarding to WSN implementations, we can find proposals such as presented by F.J. Pierce and T.V. Elliott present in [11] where the system can be used as an agricultural weather network and as on-farm frost monitoring network. This system is based on AWN200 data logger equipped with a 900 MHz, frequency hopping, spread spectrum (FHSS) radio configured into master–repeater–slave network for broad geographic coverage. The network is deployed in a star topology where a strategically placed base radio is in charge of the network synchronization, data collection from remote stations within the network, and re-broadcasting collected data to roamer radio units attached to mobile computers and/or directly to the Internet.

Finally, several wireless technologies can be used to communicate all nodes. The most used technologies are Bluetooth [12] and ZigBee or IEEE 802.15.4 standard [13, 14]. The main reason to use these technologies is their low energy consumption compared to IEEE 802.11 standard. In fact, Y. Kim and R.G. Evans [15] proposed the design of decision support software and its integration with an in-field wireless sensor network (WSN) to implement site specific sprinkler irrigation control via Bluetooth wireless communication. Authors also deployed an user-friendly software that allow growers a simple management of these systems. The software permits a real-time monitoring of irrigation operations via Bluetooth wireless radio communication. An algorithm for nozzle sequencing was developed to minimize hydraulic pressure surges by staggering and uniformly distributing the nozzle-on timeslots during 60-s duty cycle. As results show, the system successfully enabled real-time remote access to the field conditions and feedback control for site-specific irrigation with high correlation near to 1 to water collected by catch cans. Authors proposed to extend the design of this system to adapt automated site-specific fertilizer or chemical applications. Raul Morais et al. [13] and S.-E. Yoo et al. [14] used ZigBee or IEEE 802.15.4 standard as wireless technology. On the one hand, Raul Morais et al. [13] presented a ZigBee-based remote sensing network, intended for precision viticulture in the Demarcated Region of Douro. Authors use the MPWiNodeZ that provides a mesh-type array of acquisition devices for deployment in vineyards. The network nodes are powered by batteries that are recharged with energy harvested from the environment. It also contains a software based method that prevents automatically switch off nodes from being switched-on soon after, as their batteries are charged again. On the other, S.-E. Yoo et al. [14], proposed a system for automated agriculture based on IEEE 802.15.4 wireless nodes. The paper describes the results of a real deployment called A2S which consists of a WSN to monitor and control the environments and a management sub-system to manage the WSN. It also provides various and convenient services to consumers with hand-held devices such as a PDA living a farming village. Although the proposal was used to monitor the growing process of them and control the environment of the greenhouses. Authors conclude that this system could be useful in consumer electronics field such as home network as well as automated agriculture field.

As we can see, there are many WSN applications for agriculture environments. They are developed using different wireless technologies but none of them use mobile devices to control, monitor and provide system maintenance. Our system includes these features and, in addition, we developed a platform to control the field irrigation system, which operation is shown in mobile devices.

3   Scenario Description

In this section, we describe the environment and tools used in this research. The selected cultivation area is shown in Fig. 1. It is a rectangular field without high obstacles; the only exception is the cultivation trees. There are several irrigation sprinklers uniformly distributed over the whole area. This deployment uses sprinklers in a fixed location, but the proposed system could be able to work with mobile sprinklers, if available. The system pursues two main objectives: The first one is reduce the water consumption and the second goal aim to improve productivity and competitiveness of this crop.

Arduino has been selected for implementing the smart wireless node. It is based on an open-source hardware and software platform [16]. The Arduino development environment is particularly designed to offer electronic support to multidisciplinary projects. Arduino system is based on an Atmel AVR microcontroller. The most used are the Atmega168, Atmega328, Atmega1280 and ATmega8 because they are cheaper and easier to manage and because of their flexibility and versatility. Figure 2-a shows the input and output ports of an Arduino board. Moreover, Arduino presents lots of complements, sensors and electronic boards to develop almost any application. Figure 2-b shows the Grove Base Shield board. It is an electronic board that can be connected to the Arduino board in order to connect several sensors. This board avoids having to weld sensors directly to the Arduino board and simplifies the wired part of system.

Fig. 1. Field where the WSN is deployed. Campo de Alcañiz (Spain)

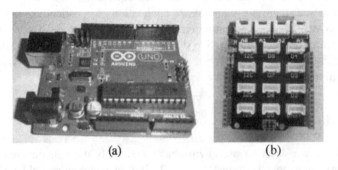

(a)   (b)

Fig. 2. Arduino Board (a) and Grove - Base Shield board (b)

The parameters of soil are provided by a soil moisture sensor (VH400) [17] which provides an accuracy in measurements of 2 % and a soil temperature sensor (THERM200) [18] which provide an accuracy in its measurements of 0.5 °C. The results of temperature sensor will allow us to perform the compensation in temperature for the moisture measurements. Both sensor are connected to the Grove - Base Shield board.

In order to send the information of sensors, a Bluetooth transmitter/receiver module has been installed on the motherboard. The Bluetooth transceiver module allows the device to send and receive data using the TTL Bluetooth technology without connecting any cable. It is easy to use and completely encapsulated.

4 System Description

This section presents the smart algorithm used to make the correct decisions and the network protocol. Clustering of devices is performed by Piconets [12], i.e. groups of up to 8 devices. There is always one master by each Piconet and other devices become slaves that communicate with the master. Each device in the same Piconet has share the frequency channel and use a common frequency hopping pattern. Moreover, they are synchronized by the same clock (the master device provides clock synchronization). The connection between piconets to form a scatternet is performed by a sensor node that is both master and slave of another piconet. The Scatternet increases network scalability up to 255 devices. Figure 3 shows a WSN scatternet with three different piconets in a field of 40 m × 100 m.

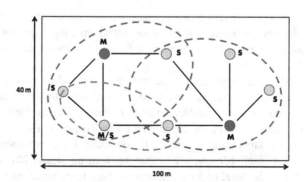

Fig. 3. Wireless sensor network with 3 Piconets (M = Master, S = Slave)

4.1 Data Transfer

Two kinds of data transfer can be used between devices: SCO (Synchronous Connection Oriented) and ACL (Asynchronous Connectionless). A piconet can have up to 3 links. A master can support up to three SCO links with one, two or three slaves. An ACL link can be established between a master and a slave or can be broadcasted from the master to all slaves. A slave can only transmit at the request of the master.

There are two main procedures: Inquire and Paging. The Inquire procedure is used to discover other devices or to be discovered by them. A device sends polling messages and keeps waiting for the answer. Both devices polling and devices answering can belong to other piconets. The Paging Procedure is an asymmetric procedure where a device performs the login procedure while another accepts it. This procedure is point to point and it is performed with a device that had previously answered the polling message. Both devices can previously be connected to other piconets. If the procedure was successful, devices are linked by a logical channel. By default this channel is an ACL logical link.

4.2   System Algorithm

In this section we are going to explain the algorithm procedure. When a new sensor node starts in the network, it will decide its role in the piconet as a function of a set of rules. There are four options: Slave of a single piconet, master of a single piconet, slave of a piconet and master of another piconet or master of two different piconets. The algorithm has three different steps:

(1) If the sensor node starts and it cannot discover any master, it is configured as master and a message is sent to the network to announce it.
(2) If there is a master in the network, then the new sensor node is configured as slave and tries to connect to the master.
(3) If the master node of the piconet has already connected to certain number of customers, for instance 4 slaves, the new sensor node is configured as master of a new piconet.

For the 3rd step, when there are two or more sensor nodes that can become master for a new piconet, the decision will be made based on their capacity. The system assesses the most appropriate sensor node as a function of the higher level of battery and its geographic location.

When the WSN is created, each node runs the smart algorithm responsible for acquiring of the atmospheric parameters and the relative humidity (RH) of soil in order to make decisions on the activation/deactivation of the sprinklers. Firstly, the system takes data of temperature, RH of the air, rain and solar radiation. These data are stored for further processing, labeling and stored in the database. It makes no sense activating a sprinkler if it is raining because we would be wasting water. Nor it is good to watering plants when the level of solar radiation is very high, since the leaves may be damaged. Furthermore, if the temperature is too low, there is the possibility of freezing of plants due to water from the sprinklers. The main factor to consider is the relative humidity of the soil that will have a minimum value as a function of the crop. If this limit is not reached, this signal will be processed along with meteorological parameters to decide whether the sprinkler should be activated or not. The algorithm needs a learning phase that the system will use for the event processing. Figure 4 shows the Smart algorithm for event tagging and sprinkler activation.

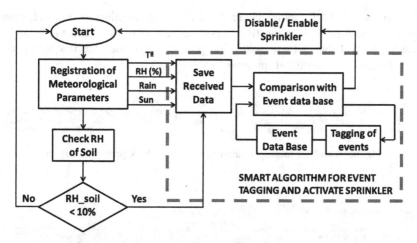

Fig. 4. Smart algorithm for event tagging and sprinkler activation

4.3 Network Protocol

A network protocol has been developed to manage the information exchanged between machines. In Fig. 5 the protocol header is shown. Version field carries the protocol version. Protocol field indicates what technology was selected. It can be IEEE 802.11 or Bluetooth. In this study Bluetooth technology has been selected. NodeID and groupID fields allow sensor nodes identify other sensor nodes and the group they belong to. Number of sequence field is a correlative number to receive messages orderly. Type of message field is the code assigned to the message. The Value of message field has the required information for each Type of message.

Protocol Header						
0 - 3	4 - 10	11 - 16	17 - 32	33 - 40	41 - 45	47 - 47
Version	Protocol	NodeID	GroupID	No. Sequence	Type of message	Value of Message

Fig. 5. Protocol header field

The types of messages and the protocol operation have been designed to be independent from the selected wireless technology.

In Bluetooth technology, the broadcast messages can only be sent by the master, while in IEEE 802.11 all sensor nodes can send broadcast messages to all sensor nodes in the same broadcast domain. In order to send a broadcast from a Bluetooth slave device, it sends the broadcast message to the master and the master forwards it to the remaining slaves in the piconet. If the master is connected with a slave or master of another piconet then the message is forwarded to other sensor nodes from other piconets.

In order to verify that the broadcast messages are working properly, we have performed capture with Wireshark protocol analyzer. Figure 6 shows a capture of some broadcast messages. Protocol messages are encapsulated on UDP transport protocol.

Figure 7 shows this UDP capture. Given that UDP is a connectionless protocol, when a datagram is received, it uses the IP and port address to reply. When the IP address destination is available but the destination port is closed then the destination node notifies the error with an ICMP message.

No.	Time	Source	Destination	Protocol	Length	Info
1	0.000000	Htc_17:e3:eb	Broadcast	ARP	42	192.168.1.129 is at 18:87:96:17:e3:eb
2	51.852783	IntelCor_26:15:e5	Broadcast	ARP	42	Who has 192.168.1.1? Tell 192.168.1.130
3	51.854891	AyecomTe_2b:ea:37	IntelCor_26:15:e5	ARP	42	192.168.1.1 is at 00:1a:2b:2b:ea:37

Fig. 6. Broadcast messages captured with Wireshark protocol analyzer

Filter: ip.addr==192.168.1.133   Expression... Clear Apply

No.	Time	Source	Destination	Protocol	Length	Info
489	450.473903	192.168.1.133	192.168.1.129	UDP	45	Source port: 53833 Destination port: 6000
493	452.765942	192.168.1.133	192.168.1.129	UDP	Protocol	Source port: 38133 Destination port: 6000
496	457.936210	192.168.1.133	192.168.1.129	UDP	45	Source port: 58463 Destination port: 6000
497	457.936360	192.168.1.133	192.168.1.129	UDP	45	Source port: 50632 Destination port: 6000
533	509.184056	192.168.1.133	192.168.1.129	UDP	45	Source port: 32948 Destination port: 6000
536	512.710146	192.168.1.133	192.168.1.129	UDP	45	Source port: 46382 Destination port: 6000
537	513.568739	192.168.1.133	192.168.1.129	UDP	45	Source port: 37553 Destination port: 6000
538	513.569567	192.168.1.133	192.168.1.129	UDP	45	Source port: 46799 Destination port: 6000
543	515.541653	192.168.1.129	192.168.1.133	ICMP	73	Destination unreachable (Port unreachable)
548	520.463332	192.168.1.133	192.168.1.129	UDP	45	Source port: 48761 Destination port: 6000
575	568.675081	192.168.1.133	192.168.1.129	UDP	45	Source port: 60336 Destination port: 6000
576	569.529995	192.168.1.133	192.168.1.129	UDP	45	Source port: 55554 Destination port: 6000
577	569.530452	192.168.1.133	192.168.1.129	UDP	45	Source port: 56118 Destination port: 6000
578	569.530492	192.168.1.129	192.168.1.133	ICMP	73	Destination unreachable (Port unreachable)
601	629.484762	192.168.1.133	192.168.1.129	UDP	45	Source port: 48435 Destination port: 6000
603	642.467664	192.168.1.133	192.168.1.129	UDP	45	Source port: 52632 Destination port: 6000
606	647.646330	192.168.1.133	192.168.1.129	UDP	45	Source port: 58241 Destination port: 6000
607	647.646536	192.168.1.133	192.168.1.129	UDP	45	Source port: 38715 Destination port: 6000
611	659.000126	192.168.1.133	192.168.1.129	UDP	45	Source port: 41336 Destination port: 6000
622	685.669564	192.168.1.133	192.168.1.129	UDP	45	Source port: 42028 Destination port: 6000
624	686.383869	192.168.1.133	192.168.1.129	UDP	45	Source port: 41969 Destination port: 6000
625	686.384031	192.168.1.133	192.168.1.129	UDP	45	Source port: 56053 Destination port: 6000

Fig. 7. UDP Capture and ICMP error notification of unreachable port

In the scope of this research, sprinklers are static or they have very low mobility. Only when a sprinkler does not have any visibility with other nodes, it could be placed in another position to improve the ad hoc network performance. When the sprinkler is moved to the new place it has to start again the discovery process. The neighbor table must be updated so keep-alive messages are sent every 10 s to confirm to the neighbor nodes that it remains active. If a neighbor node does not reply in 30 s, the neighbor table will be updated and that information will be sent to the active neighbor nodes. When neighbor table is updated each node knows the best path to every node. The proposed protocol is also used to notify to a remote node that it can enable the sprinkler.

In order to check the correct operation of system, we used a master node simulated by a computer with a Tomcat application server to check that the slave nodes are sending the relative humidity, temperature and their IP information. The master node runned a Java web application that allowed it to listen all possible slave nodes. It let us to know the management information about the messages sent by the server and the sensors. Figure 8 shows the class Java code used to listen slaves nodes.

5 Test Bench

In order to verify the system performance and its correct operation, we have carried out different tests. The first one is focused on monitoring the number of Bytes generated by each device and the bandwidth consumed by the network when devices are sending information. The second test is focused on checking the operation of the system and how it gathers the data.

5.1 Network Performance Test

In this test, we have used 3 wireless sensor nodes. Two of them are slave sensor nodes that are wirelessly connected to the master node. The master node is connected to a switch that allows us to register all network traffic received and sent by the master node. Figure 9 shows the topology used in this test bench.

```java
public class RecibeMedidas extends HttpServlet {
    private static final long serialVersionUID = 1L;
    final int port=6000;
    DatagramSocket socketSensor;
    DatagramPacket measureSensor;
    byte[] buffer=new byte[3];
    int sensor,temperature,humidity;
    int humidity_min=10;
    int humidity_max=70;
    boolean sprinkler_active=false;
    String ip_client="";
    public RecibeMedidas() {
        super();
    }
    protected void doGet(HttpServletRequest request, HttpServletResponse response) throws
ServletException, IOException {
        PrintWriter out=response.getWriter();
        do{
            try{
                measureSensor=new DatagramPacket(buffer,buffer.length);
                socketSensor=new DatagramSocket(port);
                out.println("Server On");
                socketSensor.receive (measureSensor);
                // Retrieve the client IP
                int puerto = measureSensor.getPort();
                // Internet address from which it was sent
                InetAddress direccion = measureSensor.getAddress();
                ip_client=direccion.getHostAddress();
                socketSensor.close();
            }
            catch(Exception e){  }
                sensor=new Byte(buffer[0]).intValue();
                temperature=new Byte(buffer[1]).intValue();
                humidity=new Byte(buffer[2]).intValue();
                MedidasSensores.inserta(sensor, temperature, humedad, ip_cliente);
        }
        while (true);
    }
}
```

Fig. 8. Class Java code used to listen slaves nodes

The first test is performed during 18 min. Figure 10 shows the Bytes sent from each device as a function of time. As we can see, master node that has more bytes sent than the slave nodes. This happens because it is responsible for announcing the network state and registers existing nodes. The two slave nodes only take care of sending the data results recorded from the physical sensors.

Figure 11 shows the total bandwidth consumed in the network during the test. The major peaks are observed when the master node sends broadcast packets to discover the network.

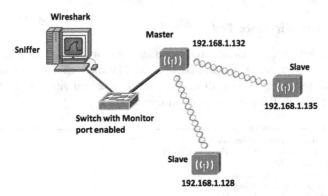

Fig. 9. Topology used in the first test bench.

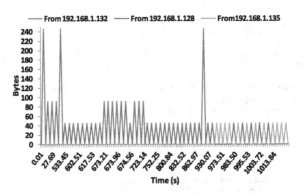

Fig. 10. Bytes sent by each node.

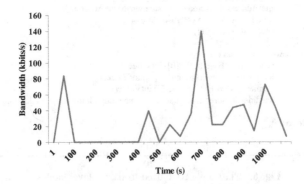

Fig. 11. Bandwidth in kbits/s registered in network.

5.2 Platform Operation Test

In order to show the humidity and temperature data provided by the sensors, we have employed three smartphones connected to the same network. Sensor nodes are placed in a 40 m × 100 m field, as it is shown in Fig. 3. There are not objects affecting to their communication. In this case, the first sensor node has detected a humidity and temperature of soil below the threshold, so it sends that information to the rest of nodes and considering the meteorological parameters the closest sprinkler is notified.

We have developed a Java application for Android Smartphones. Figure 12 shows the main windows of application This application let us switch on and switch off the

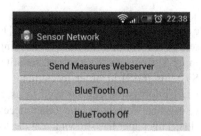

Fig. 12. Main menu of the Android application.

http://localhost:8080/servicio_web_redes_adhoc/PanelDeMedidas

Node	Sensor	Temperatur °C	Humidity %	IP	MAC	Sprinkler on	Sprinkler off
Sensor 1	Slave	53	4	192.168.1.133		Yes	No
Sensor 2	Slave	43	34	192.168.1.130		No	No
Sensor 3	Slave	33	11	192.168.1.134		No	No

Fig. 13. Central node registering measurements in real time.

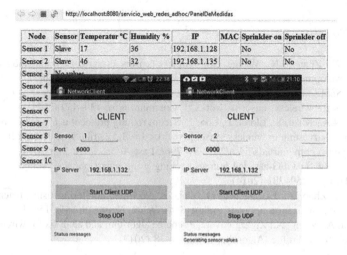

Fig. 14. Temperature and humidity sent by UDP

Bluetooth interface, read the values obtained by the sensor node and send the humidity and temperature values to a master node.

The minimum humidity threshold is 10 % and the maximum 70 %. In the webpage of the master node (see Fig. 13) we can observe gathered measures in real time. If the humidity value is below 10 %, the "sprinkler activation" field changes to green and changes to "Yes".

When any node detects higher humidity value than the maximum humidity threshold, it directly sends a message to stop the sprinkler (see Fig. 14).

6 Conclusion

Most farmers would like to have a precision agriculture [19], i.e. a management of agricultural parcels based on the continuous field monitoring. It requires the use of technology Global Positioning Systems (GPS), sensors, satellite and aerial images with Geographic Information Systems (GIS) in order to estimate evaluate and understand the field variations. Collected information can be used to assess more accurately the optimum planting density, estimate fertilizers and other necessary inputs, and more accurately predict crop yields. This paper has shown a real deployment of a smart M2M system to control the agriculture irrigation. Sensor nodes control the humidity and temperature and as a function of the weather parameters and based on the smart algorithm, the system decides which sprinklers should be enabled or not. Our system allows farmers saving water and makes the watering of their crops more efficiently, even with drip irrigation. The proper operation of system has been shown over mobile devices. As future work, we would like focus our efforts on improving our system by using more specific sensors, such as chemical sensors to measure the soil nutrients. We would like to adapt the systems to other kind of crops with more specific care, such as growing flowers to be exported to other counties.

References

1. Wanga, W., Zhangb, N., Wangc, M.: Wireless sensors in agriculture and food industry - recent development and future perspective. Comput. Electron. Agric. **50**(1), 1–14 (2006)
2. Sendra, S., Lloret, J., García, M., Toledo, J.F.: Power saving and energy optimization techniques for wireless sensor networks. J. Commun. **6**(6), 439–459 (2011)
3. Alrajeh, N.A., Khan, S., Lloret, J., Loo, J.: Secure routing protocol using cross-layer design and energy harvesting in wireless sensor networks. Int. J. Distrib. Sens. Netw. (2013). http://www.hindawi.com/journals/ijdsn/2013/374796/. Last Accessed 18 Mar 2014
4. Mao, Y., Wu, J.: GFG-assisted human tracking using smart phones. Adhoc Sens. Wirel. Netw. **21**(3–4), 259–281 (2014)
5. Zhang, L., Zhao, Z., Li, D., Liu, Q., Cui, Li: Wildlife monitoring using heterogeneous wireless communication network. Adhoc Sens. Wirel. Netw. **18**(3–4), 159–179 (2013)
6. Hawbani, A., Wang, X.: Zigzag coverage scheme algorithm and analysis for wireless sensor networks. Netw. Protoc. Algorithms **5**(4), 19–38 (2013)

7. Karim, L., Anpalagan, A., Nasser, N., Almhana, J.: Sensor-based M2M agriculture monitoring systems for developing countries: state and challenges. Netw. Protoc. Algorithms 5(3), 68–86 (2013)
8. Lloret, J., Bosch, I., Sendra, S., Serrano, A.: A wireless sensor network for vineyard monitoring that uses image processing. Sensors 11(6), 6165–6196 (2011)
9. Ruiz-Garcia, L., Lunadei, L., Barreiro, P., Robla, J.I.: A review of wireless sensor technologies and applications in agriculture and food industry: state of the art and current trends. Sensors 9(6), 4728–4750 (2009)
10. Chen, Z., Lu, C.: Humidity sensors: a review of materials and mechanisms. Sens. Lett. 3(4), 274–295 (2005)
11. Pierce, F.J., Elliott, T.V.: Regional and on-farm wireless sensor networks for agricultural systems in Eastern Washington. Comput. Electron. Agric. 61(1), 32–43 (2008)
12. IEEE Std 802.15.1-2002 – IEEE Standard for Information technology – Telecommunications and information exchange between systems – Local and metropolitan area networks – Specific requirements Part 15.1: Wireless Medium Access Control (MAC) and Physical Layer (PHY) Specifications for Wireless Personal Area Networks (WPANs)
13. Moraisa, R., Fernandes, M.A., Matos, S.G., Serôdio, C., Ferreira, P.J.S.G., Reis, M.J.C.S.: A ZigBee multi-powered wireless acquisition device for remote sensing applications in precision viticulture. Comput. Electron. Agric. 62(2), 94–106 (2008)
14. Yoo, S.E., Kim, J.E., Kim, T., Ahn, S., Sung, J., Kim, D.: A 2S: automated agriculture system based on WSN. In: Proceedings of IEEE International Symposium on Consumer Electronics, (ISCE 2007), Dallas, Texas, USA, 20–23 June 2007
15. Kim, Y., Evans, R.G.: Software design for wireless sensor-based site-specific irrigation. Comput. Electron. Agric. 66(2), 159–165 (2009)
16. Arduino web site. http://www.arduino.cc/es/. Last Accessed 18 Mar 2014
17. VH400 Soil Moisture Sensor features. http://www.vegetronix.com/Products/VG400/. Last Accessed 18 Mar 2014
18. THERM200 Soil Temperature Sensor features. http://www.vegetronix.com/Products/THERM200/. Last Accessed 18 Mar 2014
19. López, A., Soto, F., Suardíaz, J., Sánchez, P., Iborra, A., Vera, J.A.: Wireless sensor networks for precision horticulture in Southern Spain. Comput. Electron. Agric. 68(1), 25–35 (2009)

A Location Prediction Based Data Gathering Protocol for Wireless Sensor Networks Using a Mobile Sink

Chuan Zhu[1,2](✉), Yao Wang[1], Guangjie Han[1,2,3],
Joel J.P.C. Rodrigues[4], and Hui Guo[1]

[1] College of Internet of Things Engineering, Hohai University,
Changzhou, China
{dr.river.zhu,wangyao.hhuc,hanguangjie,
guohuiqz}@gmail.com
[2] Guangdong Provincial Key Laboratory of Petrochemical Equipment Fault
Diagnosis, Guangdong University of Petrochemical Technology,
Maoming, China
[3] Changzhou Key Laboratory of Photovoltaic System Integration and Production
Equipment Technology, Changzhou, China
[4] Instituto de Telecomunicações, University of Beira Interior, Covilhã, Portugal
joeljr@ieee.org

Abstract. Traditional data gathering protocols in wireless sensor networks are mainly based on static sink, and data are routed in a multi-hop manner towards sink. In this paper, we proposed a location predictable data gathering protocol with a mobile sink. A sink's location prediction principle based on loose time synchronization is introduced. By calculating the mobile sink location information, every source node in the network is able to route data packets timely to the mobile sink through multi-hop relay. This study also suggests a dwelling time dynamic adjustment method, which takes the situation that different areas may generate different amount of data into account, resulting in a balanced energy consumption among nodes. Simulation results show that our data gathering protocol enables data routing with less data transmitting time delay and balance energy consumption among nodes.

Keywords: Location prediction · Data gathering · Mobile sink · Wireless sensor networks

1 Introduction

A wireless sensor network (WSN) is composed of hundreds or thousands of battery-powered tiny sensors that monitoring their interesting surroundings and reporting the sensed data to the base station or sink through multi-hop message relay. Typical applications of WSNs include environment monitoring, military surveillance, target tracking, health monitoring, natural disasters monitoring and so on [1–3]. In these applications, manual replacement of sensor batteries is often infeasible due to operational factors. As a result, it is expected to minimize and balance energy consumption among sensor nodes. It has been proved that in static networks, the sensors deployed

© Springer-Verlag Berlin Heidelberg 2015
M. Garcia Pineda et al. (Eds.): ADHOC-NOW Workshops 2014, LNCS 8629, pp. 152–164, 2015.
DOI: 10.1007/978-3-662-46338-3_13

near the sink exhaust their battery power faster than those far apart due to their heavy overhead by relaying messages for the nodes far from the sink, and this is the so called "hot-spot" problem [4]. In addition, in the case of node failure or malfunction, the network connectivity and coverage around the sink may not be guaranteed [5]. Unbalanced energy consumption causes network performance degraded and network lifetime shortened. Recently, various new strategies that using mobile attributes of elements in WSNs have been introduced to reduce and balance energy expenditure among sensors. The usage of mobile sink is favored by many researchers. When the sink moving, the role of the "hot-spot" rotates among sensors [6], resulting in balanced energy consumption. The effectiveness has been demonstrated both by theoretical analysis and by experimental study [7–10].

Many data gathering protocols and schemes for mobile sinks have been proposed and can be classified into five categories, full flooding-based [11, 12], local flooding-based [13, 14], grid-based [15, 16], location predicting-based [17], and rendezvous point based [18] solutions. A. Kinalis et al. [15] proposed a biased, adaptive sink mobility scheme (*Adaptive*). The regions called "pocket" are deployed specifically with higher node density than the rest of network area. In order to achieve accelerated coverage of the network and fairness of service time of each region, the sink moves probabilistically, favoring less visited areas and adaptively staying longer in network regions that tend to produce more data. Because the mobile sink has to traverse all vertexes in the graph, it may cause a rigorous time delay problem in large scale networks. Based on time synchronization, K. Shin and S. Kim [17] proposed a predictive routing for mobile sinks in WSNs. A concept of milestone node is introduced, which plays a role of spreading the estimated sink's future location information to the nodes located in the vicinity of the recent trail of the sink by multi relaying a beacon packet. During the process of relaying beacon packet, the neighbors of these relay nodes can update their own "routing information" by overhearing the beacon packet, as a result, all local nodes can acquire the latest location information of the mobile sink. Although this protocol improves energy consumption, milestone based approach still needs substantial overhead for transmitting the location information of mobile sinks, especially when sinks change their moving direction frequently.

In this article, we propose a location prediction based data gathering protocol for wireless sensor networks using a mobile sink. The trajectory of mobile sink is a predefined circle, and the moving velocity of the sink is a constant. These two strategies make the mobile sink location predictable, and reduce the energy overhead for broadcasting location update messages of the mobile sink while maintaining low data transmitting delay. When reporting or forwarding data to the mobile sink, sensors calculate the location of the mobile sink based on a loose time synchronization mechanism among sensor nodes and the mobile sink. The sink collects data from sensors only when it is dwelling at sojourn points. Different from [17], mobile sink needs no location updating message to inform nodes the latest location of the mobile sink, which saves a lot of control overhead. The sink dwelling time at sojourn points is dynamically adjustable, but different from the criterion proposed in [15] that the dwelling time is determined by the density of local nodes, the time adjustment method in our protocol is based on the amount of historical data generated in each quadrant, which is more applicable to real environments.

The rest of this paper is organized as follows: The network model of our protocol is given in Sect. 2. Moving strategy of the mobile sink is presented in Sect. 3 and the data reporting process of nodes towards the mobile sink is described in Sect. 4. The performance of our protocol is compared with that of the adaptive sink mobility scheme (*Adaptive*) in terms of data transmitting latency and energy consumption through simulations in Sect. 5. And finally, the conclusion is given in Sect. 6.

2 Network Model

The network model is shown in Fig. 1. Sensor nodes are deployed randomly in a rectangle area. The network consists of N nodes and one mobile sink that gathering data from the whole network. All sensor nodes are quasi-stationary and location-aware (i.e. equipped with GPS-capable antennae). The mobile sink is not constrained by energy and can move at uniform velocity along the predefined trajectory in the two-dimensional area. It can be a vehicle or an aerocraft. The whole network area is a $W*L$ rectangle, and for the simple of time-location calculation of the sink node, the mobile sink moves along a predefined circle trajectory with radius R. The deployed area is divided into four quadrants and its center is denoted as origin point O. The mobile sink turns off its radio transceiver while moving between two sojourn points and collects data from sensors only when it is dwelling at sojourn points. The number of sojourn points n is a multiple of four, and evenly distributed on the trajectory. An anticlockwise rule is used to determine which quadrant a sojourn point belongs to, when it locates exactly on a coordinate axis (e.g., point A belongs to quadrant I as shown in Fig. 1.).

Sensor nodes are able to communicate with mobile sink by multi-hop relay. The nodes can communicate directly with the sink within their communication radius r, and they are one-hop neighbors of the mobile sink. For the sake of convenience, main symbols used in this paper are listed in Table 1.

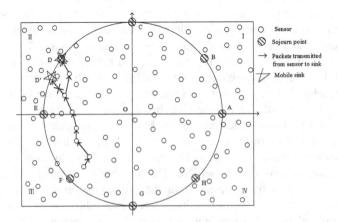

Fig. 1. Network Model

3 Moving Strategy

In this section, we explain integrated data gathering process and introduce the dwelling time adjustment method. The data gathering process of the mobile sink can be divided into three phases, they are loose time synchronization, regular data gathering, and ending declaration.

Table 1. Notations

$T_s(i, k)$	The dwelling time of mobile sink at each sojourn point in quadrant i during the kth circle, $i \in \{1, 2, 3, 4\}$
T_{syn}	The time needed for the network to achieve time synchronization
$P_{data}(i, k)$	The proportion of the collected data from quadrant i to the whole network during the kth circle
T_p	The time since the beginning of present circle
T_{bl}	The time before the mobile sink leaving the network in present round
r	The communication radius of nodes and the mobile sink
R	The radius of the sink's moving trajectory
V	The speed of the mobile sink
n	The number of sojourn points
N	The number of sensor nodes

After the network achieved time synchronization, the mobile sink starts to collect data packets from the network. And the time for time synchronization can be ignored because it is quite small compared with the time for one round data gathering.

3.1 Loose Time Synchronization

In our protocol, there are two important concepts, round and circle. A round is defined as the process from the beginning of the sink starting to gather data packets to its leaving the network, while a circle refers to the process of the sink moving along the trajectory, and backing to the initial point. For example, the sink starts from point A, moves along the trajectory, passes sojourn point from B to H, and comes back to A again, as shown in Fig. 1. This process is a circle.

The mobile sink and every node in the network have their own clock. At the beginning of one round data gathering, the mobile sink broadcasts a time synchronization message, HELLO, as a result, all nodes in network achieving loose time synchronization.

The loose time synchronization phase is the first phase during one round data gathering. When entering into the network, the mobile sink broadcasts a HELLO message to the whole network, it consists of the starting location information $S(x, y)$, current time t_0, the moving velocity V of the mobile sink, the number of sojourn points n during one circle data gathering, and the dwelling time at each sojourn point in quadrant i ($i \in \{1, 2, 3, 4\}$) during the first circle $T_s(i, 1)$. Every node changes its clock to t_0 when it receives the HELLO message for the first time, and then re-transmits this message. Note that the parameter $T_s(i, 1)$ in the HELLO message is equal to each other, that is $T_s(1, 1) = T_s(2, 1) = T_s(3, 1) = T_s(4, 1) = T_s$.

3.2   Regular Data Gathering with Dwelling Time Adjustment

During the regular data gathering phase, the dwelling time is adjusted dynamically. According to the degree of the variation of $P_{data}(i, k-1)$ and $P_{data}(i, k)$, the dwelling time in the $(k+1)$th circle, $T_s(i, k+1)$, at sojourn points in each quadrant is adjusted dynamically. In this way, the energy consumption of entire network nodes can be further balanced.

As the randomness of data packets generation in each quadrant, intuitively, the routing path for the data packets generated in quadrant i will be longer than quadrant j ($i \neq j$) where the mobile sink locates. As a result, the former will consume much more energy than the latter. To reduce the energy consumption caused by long distance data packets routing, after finishing each circle of data gathering, the mobile sink statistics the amount of packets generated from each quadrant and then calculates the proportion of these packets to the entire network data packets P_{data1}, P_{data2}, P_{data3} and P_{data4}, accordingly. Depending on these proportions, the dwelling time at sojourn points in each quadrant is adjusted dynamically, which makes the energy consumption in the network more balanced, and extends the network lifetime.

The method of adjusting the dwelling time $T_s(i, k+1)$ in the $(k+1)$th circle is described in detail as follows:

Source sensors report the sensed data to its next hop and add to 2-bit quadrant information at the head of the data packet. The quadrant information can be calculated based on their position information $loc(x_i, y_i)$ relative to the origin point O's location information. Note that only the source nodes add their own quadrant information to the head of the data packet.

During one round data gathering, the mobile sink calculates the $(k+1)$th circle dwelling time in each quadrant according to the proportions P_{data1}, P_{data2}, P_{data3} and P_{data4} in the $(k-1)$th circle and the proportions in the kth circle. When the value $P_{change}(k, k-1)$ is greater than the threshold value T_h, the dwelling time in corresponding quadrant will be adjusted as $T_s(i, k+1) = 4P_{datai}T_s$. The value of $P_{change}(k, k-1)$ is calculated as the following formula:

$$P_{change}(k, k-1) = \sqrt{\sum_{i=1}^{4} (P_{data}(i, k) - P_{data}(i, k-1))^2} \qquad (1)$$

The value $P_{change}(k, k-1)$ represents the degree of the variation that the proportion of data gathering between two adjacent circles in the network. When this value is greater than the threshold T_h, it means that the amount of data packets generated in each quadrant has changed significantly, and the dwelling time needs adjusted. Under this circumstance, the mobile sink will broadcast a UPDATE message to all nodes in the network, which includes the adjusted sojourn time in each quadrant $T_s(i, k+1)$. Otherwise, there is no necessary to modify the dwelling time, and the mobile sink maintains the dwelling time in each quadrant the same as the previous circle.

3.3 Ending Declaration

Ending declaration phase is the last circle data gathering in one round. The mobile sink broadcasts a BYE message to inform all nodes in the network at the beginning of this circle, which means this round data gathering is coming to an ending. The BYE message consists of the time T_{bl}, which is the time interval between current time and the mobile sink finishing current round data gathering. Instead of routing the data to the mobile sink, when receiving BYE messages, all nodes will buffer the data sensed from the surroundings after time T_{bl}.

$$T_{bl} = nT_s + 2\pi R/V - 2T_{syn} \tag{2}$$

T_{syn} is the time needed for the network to achieve loose time synchronization, and as to T_{syn}, there is $T_{syn} \ll nT_s + 2\pi R/V$, therefore T_{syn} has little effect on T_{bl} and can be ignored in practical applications.

4 Data Reporting Process

In the network, source nodes transmit data packets to the mobile sink by multiple hops. The principle of selecting next hop is to make the path between a source node and the mobile sink approximately shortest. Nodes need to calculate the mobile sink current location based on their own clocks, which are loose time synchronization to the mobile sink, and then choose one of its neighbor nodes as next hop.

The time step T_{step} is the time interval for moving between two adjacent sojourn points. It is calculated by the following formula:

$$T_{step} = \frac{2\pi R}{nv} \tag{3}$$

During the loose time synchronization phase, the mobile sink broadcast a HELLO message to achieve the loose time synchronization among all nodes. The parameter $T_s(i, 1)$ in the HELLO message is equal to each other, that is $T_s(1, 1) = T_s(2, 1) = T_s(3, 1) = T_s(4, 1) = T_s$. T_s is a constant value, and keeps unchanging during a round data gathering.

To determine the location of the mobile sink at time t, in this paper, we have the moving trajectory of the mobile sink map into a polar coordinate system. Assuming the mobile sink starts its data gathering process from point A at time t_0 and reaches point B at time t, the arc length of AB is $R\theta$ ($\theta < 2\pi$). When the sink moves at speed V in the network, there is $V(t-t_0) = R\theta$. The corresponding B polar coordinate is (R, θ).

The moving time of the mobile sink in current circle is denoted as T_p, which is calculated by the following formula:

$$T_p = \left[\frac{t - t_0}{n * (T_{step} + T_s)} \right] \tag{4}$$

Based on our loose time synchronization, the location of sink can be calculated. For example, when the mobile sink locates at the first quadrant, T_p meets $0 \le T_p < n/4T_s(1, k) + 2\pi R/4\ V$, the polar angle of the sink is calculated by the following formula:

$$
\theta = \begin{cases}
\dfrac{2\pi}{n} * \left\lceil \dfrac{Tp}{Tstep+Ts(1,k)} \right\rceil, & if \left\lceil \dfrac{T_p - \left\lfloor \frac{T_p}{Tstep+T_s(1,k)} \right\rfloor * \left(T_{step}+T_s(1,k)\right)}{T_s(1,k)} \right\rceil = 0 \\[4ex]
\pi V * \dfrac{Tp - \left\lfloor \frac{Tp}{Tstep+Ts(1,k)} \right\rfloor * Ts(1,k)}{\pi R}, & if \left\lceil \dfrac{T_p - \left\lfloor \frac{T_p}{Tstep+T_s(1,k)} \right\rfloor * \left(T_{step}+T_s(1,k)\right)}{T_s(1,k)} \right\rceil \ne 0
\end{cases}
\tag{5}
$$

The according polar coordinate is (R, θ). We define $\lfloor x \rfloor$ as the largest integer of no more than x, $[x1/x2]$ as the reminder of $x1$ divided by $x2$, and define $\lceil x \rceil$ as the smallest integer no less than x. When the sink locates at other quadrants, the corresponding location information of it can be obtained in a similar manner.

To keep the formula as simple as possible, we require the starting point of data gathering must be on the intersection of x axis or y axis. Without loss of generality, we choose the location A as shown in Fig. 1 as the starting point and deduce a series of formulas above.

While events occur in the monitoring area, the sensors outside the communication range of the mobile sink route the data packets to its next hop directly, without the necessary of judging the state of the mobile sink, i.e., moving between sojourn points or gathering data packets at a sojourn point. Only the neighbor nodes of the mobile sink need to judge the state of the mobile sink. If the mobile sink is moving between sojourn points, the neighbor nodes have to wait for a period of time T_{wl}, otherwise, they transmit the data packet to the mobile sink directly. For instance, as shown in Fig. 1, we assume the current location of the mobile sink is D, if the events occurring in quadrant *III*, then data packets can be routed along the shortest routing path to D. When the data packets reach the neighbor node of the mobile sink, it will judge the state of mobile sink according to its time clock. If the time of the node meets $t_0 + k*(T_{step} + T_s(i, k)) < t < t_0 + k*(T_{step} + T_s(i, k)) + T_s(i, k)$, which means the mobile sink is still gathering data at the sojourn point D, then this one-hop neighbor node of the sink transmits the data packets directly to the mobile sink; otherwise, e.g., the sink is now located at D', it needs to wait time T_{wl} and then transmit the data packet to the mobile sink. The time T_{wl} is calculated by the following formula:

$$
T_{w1} = t_0 + (k + 1) * (T_{step} + T_s(i, k)) - t
\tag{6}
$$

During routing data packets to the mobile sink, hop-by-hop acknowledgement mechanism is applied to ensure the data transmission rate, i.e., if the receiver *Node2* gets the data packets from sender *Node1*, it will replies an ACK message to *Node1*. If *Node1* does not receive the ACK message from its next hop node *Node2* within time T, *Node1* considers that the packet transmission is failed, and will cache the data packets and wait for a random time, and then re-transmit the packets to its next hop again. We assume that T is equal to the propagation time of a packet between two farthest nodes of the network.

5   Simulation and Performance Evaluation

In this section, we evaluate our protocol through extensive simulations. Two performance metrics, energy consumption and data delivery latency, are investigated. Energy consumption is the average energy that consumed by nodes during one round data gathering. Data delivery latency is the time interval between a message creation and the mobile sink receiving it.

5.1   Simulation Environment

We implement the proposed protocol in MatLab. In our simulation, the deployment area length L equals to its width W, that is, the sensor network has a 500 m*500 m square sensing field and sensor nodes are randomly deployed. The communication range of the nodes and the mobile sink is set to 60 m. The mobile sink moves along the predefined trajectory for 10 circles every round, and in every circle, 5 % of sensor nodes act as source nodes, which send message toward the mobile sink continually when the sink is dwelling at sojourn points.

Different simulation environment with varying number of nodes N, mobile sink moving speed V and sink's moving trajectory are studied. We varied N from 800 to 1200, V from 4 m/s to 20 m/s, which is the same as that in [15], and set the radius of trajectory as $L/2$, $L/4$, $3L/8$ and $L/8$. Additionally, several groups of simulation experiments are carried out. The threshold of adjusting dwelling time T_h is varying among 0, 0.25, 0.5, 0.75 and 1. $T_h = 0$ means the dwelling time of the mobile sink needs to be changed if the proportion of packets amount generated from every quadrant is not exactly same as previous circle, while $T_h = 1$ means the dwelling time keeps unchanging during one round data gathering.

5.2   Simulation Results with Varying Number of Sensor Nodes

Now we discuss the performance of our protocol by setting the number of sensor nodes N varying from 800 to 1000. The simulation results are shown in Figs. 2 and 3.

As illustrated in Fig. 3, the energy consumption of nodes decreases first with increased number of nodes and then increases when N is more than 1000. It is because, as N increasing, the amount of selectable next hop neighbors increases, as a result, the hop distance and the routing path between the mobile sink and source nodes are improved, which results in less energy consumption for relaying the same amount of data packets. When the number of sensor nodes is more than 1000, the role of the amount of sources nodes has more influence than hop distance. Energy consumption of nodes increases when N increases. The performance is outstanding than the others when T_h is 0.75, this is because the dynamic adjustment of dwelling time is beneficial to the performance of network. Besides, when $T_h = 0.75$, as shown in Fig. 2, the energy consumption of control message is very low. There is an appropriate tradeoff between the control overhead and the balanced energy consumption among different quadrants.

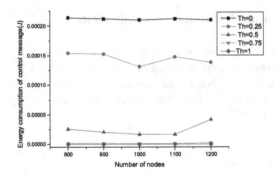

Fig. 2. Energy consumption of control message under number of nodes

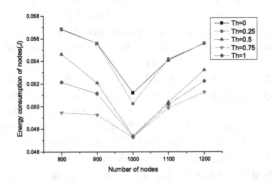

Fig. 3. Energy Consumption under varying number of nodes

In contrast, the dwelling time adjustment frequency is too high when $T_h = 0$, which results in much energy overhead. When $T_h = 1$, the energy consumption of control message equals to 0, which means there is no dwelling time adjustment, the energy consumption among nodes is not well balanced.

5.3   Simulation Results with Varying Radius of Moving Trajectory

Now we turn to study the influence of moving trajectory by setting moving trajectory radius R as $L/2$, $L/4$, $3L/8$ and $L/8$. The simulation results are shown in Figs. 4 and 5.

As shown in Fig. 5, the energy consumption is the lowest while mobile sink moves along the track with value of R is $L/4$. It is different from the theory proposed in [10] that peripheral movement is the best strategy, because the ideal load-balanced routing is hard to satisfy. Under the condition of a certain trajectory, there is an outstanding performance when $T_h = 0.75$. The reason is the same as explained in Sect. 5.2.

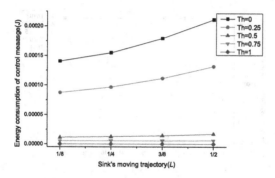

Fig. 4. Energy consumption of control message under varying trajectory of sink

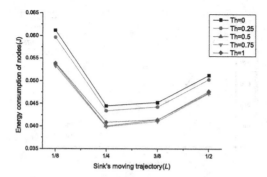

Fig. 5. Energy Consumption under varying trajectory of sink

5.4 Simulation Results with Varying Sink's Moving Speed

Now we evaluate the network performance when sink's moving speed V varying from 4 m/s to 20 m/s. The results are shown in Figs. 6 and 7.

It is noticed that as the mobile sink speed goes up, the energy consumption decreases. This is because with the increasing of sink movement speed, the time the mobile sink spends for moving between two adjacent sojourn points decreases, and the mobile sink can receive the data packets timely from sensor nodes. This result in less energy consumed.

5.5 Simulation Results of Data Transmitting Delay and Energy Consumption

We simulated our proposed algorithm, as well as the *Adaptive* algorithm and *Constant* algorithm described in the adaptive sink mobility scheme proposed by A. Kinalis et al. [15] to evaluate the performance of data delivery latency and energy consumption by varying sink movement speed.

The results are shown in Fig. 8. Our *Predictive* algorithm outperforms in attribute of latency compared with *Adaptive* scheme. The reason is that in our algorithm, latency

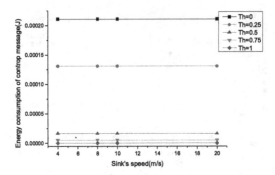

Fig. 6. Energy consumption of control message under varying sink speed

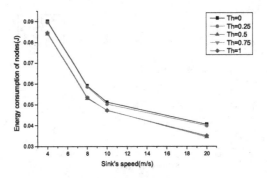

Fig. 7. Energy consumption of nodes under varying sink speed

is mainly caused due to the mobile sink turning off its communication model when moving between two adjacent sojourn points. But in *Adaptive* and *Constant*, the sink has to traverse all vertexes, which results in large time delay. Besides, the increase of speed is beneficial for our *Predictive* algorithm. It is because the time needed decreases for moving the same distance with higher movement speed so the data transmission delay significantly reduced with the increase of sink speed.

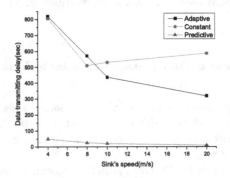

Fig. 8. Data transmitting latency

Figure 9 shows the performance of energy consumption with the change in velocity. When mobility speed is relatively small, *Adaptive* algorithm performance is almost same to our *Predictive* algorithm. However, with the increasing of sink's moving speed, the energy consumption of *Adaptive* and *Constant* are much more than our algorithm. It is because with the increasing of sink's speed, the time for mobile sink moving between two sojourn points decreases. As a result, the mobile sink can receive the data packets timely from sensor nodes, and the energy consumption decreases accordingly.

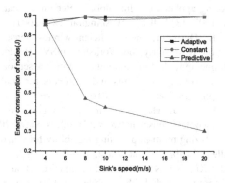

Fig. 9. Energy consumption under varying sink speed

6   Conclusion

In this paper, we propose an energy-balanced location predictable data gathering protocol for data communication among sensors and one mobile sink. Based on loose time synchronization, the latest location information of the mobile sink can be predicted, and by using this information, source nodes are able to route data packets timely by multi-hop relay to the mobile sink. As a result, the energy overhead for updating sink's location is largely reduced. Along with the predictive algorithm, this study also introduces a time adjustment method for the mobile sink to efficiently balance the energy consumption among nodes. Simulation results show that the proposed protocol achieves some improvements on time latency and energy consumption. For future research, we plan to research into the location prediction data gathering protocol with multi sinks to further improve the performance of data gathering algorithm.

Acknowledgement. The work is supported by the Science & Technology Pillar Program of Changzhou (Social Development), NO.CE20135052 and Jiangsu Province Ordinary University Graduate Innovation Project, NO.CXLX13_227. Part of this work is supported by the Instituto de Telecomunicações, Next Generation Networks and Applications Group (NetGNA), Portugal, and by National Funding from the FCT – Fundação para a Ciência e a Tecnologia through the PEst-OE/EEI/LA0008/2013 Project.

References

1. Han, G., Xu, H., Jiang, J., Shu, L., Hara, T., Nishio, S.: Path planning using a mobile anchor node based on trilateration in wireless sensor networks. Wireless Commun. Mobile Comput. **13**(14), 1324–1336 (2013)
2. Zhu, C., Zheng, Č., Shu, L., Han, G.: A survey on coverage and connectivity issues in wireless sensor networks. J. Netw. Comput. Appl. **35**(2), 619–632 (2012)
3. Han, G., Xu, H., Duong, T.Q., Jiang, J., Hara, T.: Localization algorithms of wireless sensor networks: a survey. Telecommun. Syst. **52**(4), 2419–2436 (2013)
4. Wang, G., Cao, J., Wang, H., Guo, M.: Polynomial regression for data gathering in environmental monitoring applications. In: Global Telecommunications Conference, 2007, GLOBECOM'07, pp. 1307–1311. IEEE, Washington, DC (2007)
5. Chen, C., Ma, J., Yu, K.: Designing energy efficient wireless sensor networks with mobile sinks. In: Proceedings of ACM Sensys'06 Workshop WSW, Boulder, CO, pp. 1–9 (2006)
6. Li, X., Nayak, A., Stojmenovic, I.: Sink mobility in wireless sensor networks. In: Wireless Sensor and Actuator Networks: Algorithms and Protocols for Scalable Coordination and Data Communication, pp. 153–184. Wiley (2010)
7. Lee, K., Kim, Y.H., Kim, H.J., Han, S.: A myopic mobile sink migration strategy for maximizing lifetime of wireless sensor networks. Wireless Netw. **20**(2), 303–318 (2014)
8. Rao, J., Biswas, S.: Analyzing multi-hop routing feasibility for sensor data harvesting using mobile sinks. J. Parallel Distrib. Comput. **72**(6), 764–777 (2012)
9. Liang, W.F., Luo, J., Xu, X.: Network lifetime maximization for time-sensitive data gathering in wireless sensor networks with a mobile sink. Commun. Mobile Comput. **13**(14), 1263–1280 (2013)
10. Luo, J., Hubaux, J.-P.: Joint mobility and routing for lifetime elongation in wireless sensor networks. In: 24th Annual Joint Conference of the IEEE Computer and Communications Societies, vol. 3, pp. 1735–1746 (2005)
11. Shah, D., Shakkottai, S.: Oblivious routing with mobile fusion centers over a sensor network. In: Proceedings of 26th IEEE International Conference on Computer Communications, Anchorage, AK, pp. 1541–1549 (2007)
12. Ye, F., Zhong, G., Lu, S., Zhang, L.: Gradient broadcast: a robust data delivery protocol for large scale sensor networks. Wireless Netw. **11**(3), 285–298 (2005)
13. Wang, G., Wang, T., Jia, W., Guo, M., Li, J.: Adaptive location updates for mobile sinks in wireless sensor networks. J. Supercomput. **47**(2), 127–145 (2009)
14. Lin, P.-L., Ko, R.-S.: An efficient data-gathering scheme for heterogeneous sensor networks via mobile sinks. Int. J. Distrib. Sensor Netw. **2012**, 1–14 (2012)
15. Kinalis, A., Nikoletseas, S., Patroumpa, D., Rolim, J.: Biased sink mobility with adaptive stop times for low latency data collection in sensor networks. Inf. Fusion **15**(SI), 56–63 (2014)
16. Ye, F., Luo, H., Cheng, J., Lu, S., Zhang, L.: A two-tier data dissemination model for large-scale wireless sensor networks. In: Proceedings of 8th International Conference on Mobile Computing and Networking, Atlanta, GA, pp. 148–159 (2002)
17. Shin, K., Kim, S.: Predictive routing for mobile sinks in wireless sensor based on milestone-node. J. Supercomput. **62**(3), 1519–1536 (2012)
18. Lee, J., Yu, W., Fu, X.: Energy-efficient target detection in sensor networks using line proxies. Int. J. Commun. Syst. **21**(3), 251–275 (2008)

Deployment and Performance Study of an Ad Hoc Network Protocol for Intelligent Video Sensing in Precision Agriculture

Carlos Cambra[1], Juan R. Díaz[2], and Jaime Lloret[2(✉)]

[1] University of San Jorge, Saragossa, Spain
alu.22962@usj.es
[2] University Polithecnic of Valencia, Valencia, Spain
{juandasan, jlloret}@dcom.upv.es

Abstract. Recent advances in technology applied to agriculture have made possible the Precision Agriculture (PA). It has been widely demonstrated that precision agriculture provides higher productivity with lower costs. The goal of this paper is to show the deployment of a real-time precision sprayer which uses video sensing captured by lightweight UAVs (unmanned aerial vehicles) forming ad hoc network. It is based on a geo-reference system that takes into account weeds inside of a mapped area. The ad hoc network includes devices such as AR Drones, a laptop and a sprayer in a tractor. The experiment was carried out in a corn field with different locations selected to represent the diverse densities of weeds that can be found in the field. The deployed system allows saving high percentage of herbicide, reducing the cost spent in fertilizers and increasing the quality of the product.

Keywords: Precision Agriculture · UAV · Video sensing · Geo-references · Weeds · Ad Hoc Network Protocol

1 Introduction

AR Drones are becoming the last revolution in technology, not only on military technologies but also in agricultural industry and electrical companies. This revolution is based on the low price of ARM processors and the need of professionals who want efficiency solutions in their works. Currently, there is an ongoing research in the field of Mobile Ad hoc Networks (MANET), mainly in Wireless Sensor and Actor Networks [1, 2]. A large interest is arising in ad hoc applications for vehicular traffic scenarios, mobile phone systems, sensor networks and low-cost networking [3]. Up to now, research has been focused on topology-related challenges such as node organization, routing mechanisms or addressing systems [4], as well as security issues like traceability of radio communication, attack prevention or encryption [5]. The distribution and location of the nodes and the energy efficiency are the key issues that to maximize the lifetime of the whole network [6] and enlarge the coverage area [7]. Some of the Most of this research aims either general approach to wireless networks in a broad setting (and so operate on a more abstract level) or it focus on an extremely

© Springer-Verlag Berlin Heidelberg 2015
M. Garcia Pineda et al. (Eds.): ADHOC-NOW Workshops 2014, LNCS 8629, pp. 165–175, 2015.
DOI: 10.1007/978-3-662-46338-3_14

special issue that bundles software and hardware challenges into one tailored problem. In general, unmanned aerial vehicles (UAVs), and unmanned aerial systems (UAS) need wireless systems to communicate. Current UAS are very flexible and allow a wide spectrum of mission profiles by using different UAVs.

Precision Agriculture (PA) is a new concept that is focused on monitoring the field in order to gather a lot of accurate data [8]. The interpretation of the data is what will lead us to make changes in the management practices. An example of a wireless sensor network for precision agriculture, for the case of vineyards, can be seen in [9]. Caution is advised when interpreting values obtained from grid sampling as different labs or different sampling techniques can also yield different results.

Our motivation to perform this research comes because farmers demand a video sensing application to monitor their agricultural land, especially for fertilizer tasks. Common Agricultural Policy (CAP) 2014/2020 implements new points inside of the new term "Greening" [10]. This policy aims the energy efficiency and the reduction of fertilizers in productions. It can be achieved by applying new technologies in farms with strong traditional thinkings in grow methods, obtaining a sustainable and healthy way of productions destined to human consumptions, while decreasing around 70 % the use of chemical herbicides.

The paper is structured as follows. Section 2 includes some works related with aerial vehicles used for agriculture monitoring. Section 3 describes our proposed system. The video processing system and weeds mapping is shown in Sect. 4. Finally, Sect. 5 draws the conclusion and future work.

2 Related Work

The work presented in this paper is based on a previous work [11]. We implemented an Ad Hoc network using AR Drones. The captured video was used to control de irrigation system and the block of sprayers, which may produce problems and crop death if they are not irrigated properly. In this paper we continue an analysis of the ad hoc network infrastructure, using UAVs and agricultural precision machines to demonstrate the potential of the proposed system in terms of energy efficiency in agricultural productions. We studied the potential of these technologies (drones, GPS and Ad Hoc network) to reduce the quantity of herbicides used and to obtain a more efficient production with environment and human health.

In [12] Koger Clifford et al. presented an analysis of the potential of multispectral imagery for late-season discrimination of weed-infested and weed-free soybean. Weed infestations were discriminated from weedfree soybean with at least 90 % accuracy. The discriminant analysis model used in one image obtained from 78 % to 90 % of accuracy discriminating weed infestations for different images obtained from the same and other experiments.

Recently, C. Zhang and J. Kovacs [13] presented a study of Unmanned Aerial Vehicle (UAV) images focused in image-acquisition dedicated to PA. Their proposal combines several contextual and image-adquisition features that discriminate corn rows to the weeds. This algorithm creates a weed infestation map in a grid structure that gives the opportunity to reduce the use of fertilizers and decrease the environment pollution.

Authors of [14] define site-specific weed control technologies as the machinery or equipment embedded with technologies that detect weeds growing in a crop, taking into account predefined factors such as economics, in order to maximise the chances of successfully controlling them. They define the basic parts and review the state-of-the-art. They also discuss some limitations and barriers.

Robotic weed control systems [15] allow the automation of the agriculture operations and provide a means of reducing herbicide use. Despite of it, a few robotic weed control systems have demonstrated the potential of this technology in the field. Because of it, additional research is needed to fully realize this potential.

The system presented in this paper has three main pillars. The communication system is inspired on the work presented by Maria Canales et al. in [16]. They presented a system that allows aerial vehicles to acquire visual maps of large environments using comparable setup with an inertial sensor and low-quality camera pointing downward. In order to include QoS, in the ad hoc network we took into account this research. The system video analyser is based on the work presented in [17], but we adapted it to colour lines recognition. In this work, the experimental results showed the highly accurate classifications of green lines and asynchronous green points sign patterns with complex background images. Furthermore, they provide the computational cost of the proposed method. The object detector function included in our system was initially proposed by P. Felzenszwalb et al. in [18]. It is based on a Dalal-Triggs detector that uses a single filter on histogram of oriented gradients (HOG) features to represent an object category.

3 System Description and Operation

3.1 System Overview

The communication ad hoc protocol used in this system was presented in [11]. This paper presents the real time analysis of the images on corn crops taken from the video captures of the AR Drones. We have included a weeds geoposition video processing system that creates a map allowing the fertilizer sprayer system to work only in areas with weeds. This process is performed thanks to the rules we have included in the GPS Autopilot device, which is included in Sprayer System. The system scheme is shown in Fig. 1.

There is an ad hoc communication between several devices. The devices were: several AR Drones, which use ARM processors, a computer for video processing, which have an i386 processor, and the sprayer tractor, which uses an ARM processor. All devices had full mobility, but they can only move to places keeping the wireless link. The mobility feature makes mandatory the use of ad hoc networks. It is needed an discovery scheme and autonomous handshakes selecting the best route for the integration of robotic devices, UAV and sensors.

Fig. 1. Scheme of the ad Hoc network for video-sensing used on agricultural precision spreayers.

3.2   Q-Ground AutoPilot Configuration for the Autonomous AR Drone Flight

QgroundControl [19] is an Open Source application destined to control by coordinate points (waypoints) planes, helicopters and drones. It implements a framework for the autonomous pilotage of AR Drones with GPS. Its main features are:

- Open Source Protocol.
- Aerial maps 2D/3D with drag and drop waypoint.
- Change of control parameters in flight time of AR Drone.
- Real time monitoring of video, data sensors and telemetry.
- It works in Windows, Linux and MacOS platforms.

MAVlink protocol [20] is used to implement flexible and open source libraries. It allows working with different air vehicles and radio-control devices using C programming language.

The design of AR Drone flight map is shown in Fig. 2. The tool map module uses Google Maps framework for the flight map creation and allows drawing the flight traces on a satellite image according to the section of the corn crop. Concretely, Fig. 2 an image of a field that has 18 m width and 15 m length.

Fig. 2. Flight map of the AR Drone created by Q-Ground Control

3.3 Communication Between AR Drones and a Video Receiver

OLSR protocol has been included in the devices in order to route the information between devices [11]. OLSR is a dynamic routing protocol which uses the status of the links (gathering their data) and dynamically measures the best routes to transfer the data in the ad hoc network [21]. It is currently one of the most employed routing algorithms for ad hoc networks [1]. The routing table is estimated whenever there is a change in the neighborhood or the topology information changes. OLSR protocol allows increasing the number of AR Drones in the ad hoc network dedicated to video recording and video processing.

The system must determine the distance between devices (sprayer system, AR Drones and computer). If distance between devices is too large, the transfer rate is lower than video transfer, so the radio coverage distance is not a limitation, but the maximum distance for the proper data rate to transmit video. Thanks to the GPS system included in all our devices, we can update the flight map values during the mission and estimate the distances between devices in the ad hoc network.

3.4 Media Player and Video Codecs

In order to watch the received images, we used the ffmpeg multimedia player [22]. It is shown in Fig. 3. Ffmpeg provides the best technical solution for developers and end users. In order to achieve this, we have combined the best available free software options. We kept low the dependencies to other libraries in order to maximize code sharing between parts of ffmpeg.

In order to watch the real time video in the computer, we installed Ubuntu 10.04 and ffmpeg. Ffmpeg uses SDL 2.0 for displaying and decoding video concurrently in separate threads and synchronizing them.

Fig. 3. Ffmpeg multimedia player receiving video from an AR Drone

3.5 Geo-references in Video Frames (Metadata)

The waypoint protocol describes how waypoints are sent to a MAV (Micro Air Vehicle) and read from it. The goal is to ensure a consistent state between sender and receiver. QGroundControl has an implementation of the Groundcontrol side of the protocol. Every waypointplanner on a MAV implementing this protocol using MAVLINK can communicate with QGroundControl and exchange and update its waypoints. The GPS file format is shown in Fig. 4.

```
Format

QGC WPL <VERSION>
<INDEX> <CURRENT WP> <COORD FRAME> <COMMAND> <PARAM1> <PARAM2> <PARAM3> <PARAM4> <PARAM5/X/LONGITUDE>
```

```
Example

QGC WPL 110
0      1      0      16      0.149999999999999994      0      0      0      8.54800000000000004
1      0      0      16      0.149999999999999994      0      0      0      8.54800000000000004
2      0      0      16      0.149999999999999994      0      0      0      8.54800000000000004
```

Fig. 4. GPS coordinates file format

4   Video Processing and Weeds Mapping

4.1   Video Processing Using OpenCV

The free- and non-commercial Intel® Open Source Computer Vision Library (OpenCV) [23] has C++, C, Python and Java interfaces and supports several operative systems such as Windows, Linux, Mac OS, iOS and Android. OpenCV was designed with a strong focus on real-time applications, trying to have high computational efficiency. It is able to use multi-core processing and can take advantage of the hardware acceleration of the underlying heterogeneous computer platform. The object detector function is based on a Dalal-Triggs detector. It uses a single filter on Histogram of Oriented Gradients features to represent an object category. It has a sliding window approach that allows applying the filter to all positions and scales of an image.

In our deployment, we only took into account green and brown color lines because these shapes are generally present in many types of maize crops. We used a Multilayer Perceptron [24] as a learning algorithm to recognize groups of weeds. The scheme to process an image is shown in Fig. 5.

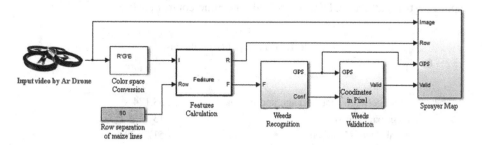

Fig. 5. Video Processing scheme.

4.2 Data File with Geoposition Weeds and Video Frames

The deployed function calculates an average motion direction in the selected region, and returns the angle between 0 and 360 degrees. The average direction is computed from the weighted orientation histogram, where a recent motion has a larger weight and the older motion has a smaller weight. In OpenCV, images are represented by matrices. Thus, we used the same convention for both cases, the 0-based row index (or y-coordinate) goes first and the 0-based column index (or x-coordinate) follows it. Figure 6 shows an example of the output of the motion direction estimation function.

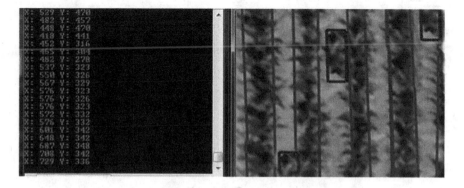

Fig. 6. Adding geo-positions to video pixels coordinates of video frames.

5 Performance Study

In order to make a performance study of our system, we set up a scenario with 2 AR Drones sending video streaming with one hop or two hops. We used Wireshark network analyzer in a laptop in order to gather QoS data in several points of the scenario. It gathered all packets of the ad hoc network corresponding to the ffmepeg and video processing, and for the network control (mainly from the routing protocol). The system used RTSP & RTP protocols control the quality of service and the route tables of the neighbors to optimize the use of OLSR protocol on communication.

Table 1 shows the summary of the information obtained by Wireshark during the capture test. It gathered UDP, OLSR and Ar Drone control packets.

Table 1. Data captured in the test

Traffic	Quantity
Captures	37691
Time between first package and last package	38.174 s
Avg. package size	1373.58 bytes
Total Bytes Captured	51778297
Avg. bytes/sec	1356747.891
Avg. Mbits/sec	10.859

Figure 7 shows the graph obtained when we gathered data during 10 s. We have distinguished the video traffic from the control frames. This test let us know the bandwidth required for video delivery from an AR Drone.

In order to compare the deployed system with and without drones (only laptops), we tested video transfers between Laptops, checking the dynamic route tables, and then we took measurements between an AR Drone and a Laptop when transferring video. Figure 8 shows the obtained results.

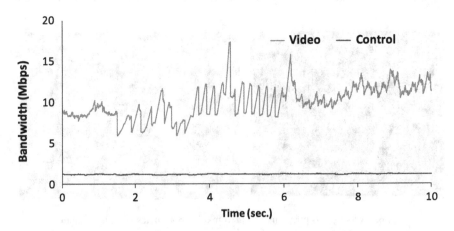

Fig. 7. Data traffic during the test (RTP and AR Drone control protocol)

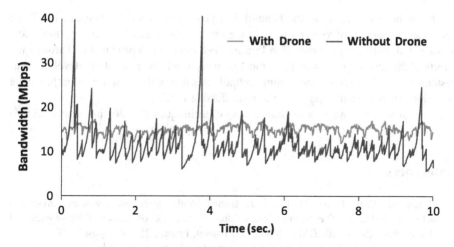

Fig. 8. Total bandwidth obtained with and without AR Drones.

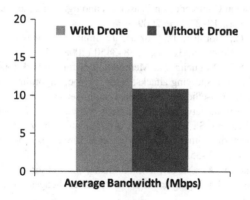

Fig. 9. Average Bandwidth

Figure 9 shows the Average Bandwidth (in Mbps) when we are using AR Drones (an average value of 15.4 Mbps) and without AR Drones (an average value of 10.91 Mbps).

6 Conclusion

In this paper, we have presented the deployment of a real-time precision sprayer which uses video sensing captured by lightweight UAVs (unmanned aerial vehicles) forming ad hoc network. It uses OLSR routing protocol to route video efficiently in the dynamic network, which is formed by AR Drones, a video processing computer and a sprayer system. It is based on a geo-reference system that takes into account weeds inside of a mapped area. This technology offers the potential to improve production efficiency in the precision agriculture and, at the same time, reduce agriculture impact on the environment. To achieve all the benefits and potentials of these advances, it will be necessary to open new research lines in the agriculture research. These machines and services will help to apply precision farming techniques crop production. Through the efforts of farmers, engineers and chemical industry it will be possible to develop smart systems for production and supporting technologies that will be required to improve the precision agriculture turning it into a cost-effective model.

In a future work we will include new QoS techniques [25, 26] in the system.

References

1. Garcia, M., Coll, H., Bri, D., Lloret, J.: Using MANET protocols in wireless sensor and actor networks. In: The Second International Conference on Sensor Technologies and Applications (SENSORCOMM 2008), Cap Esterel, France, 25–31 August 2008
2. Garcia, M., Bri, D., Sendra, S., Lloret, J.: Practical deployments of wireless sensor networks: a survey. J. Adv. Netw. Serv. 3(1&2), 170–185 (2010)
3. Bri, D., Garcia, M., Lloret, J., Dini, P.: Real deployments of wireless sensor networks. In: The Third International Conference on Sensor Technologies and Applications (Sensorcomm 2009), Atenas (Grecia), 18–23 June 2009
4. Lloret, J., Palau, C., Boronat, F., Tomas, J.: Improving networks using group-based topologies. Comput. Commun. 31(14), 3438–3450 (2008)
5. Lopes, P., Salvador, P., Nogueira, A.: Methodologies for network topology discovery and detection of MAC and IP spoofing attacks. Netw. Protoc. Algorithms 5(3), 153–197 (2013)
6. Liu, Y., Xu, B.: Energy-efficient distributed multi-sensor scheduling based on energy balance in wireless sensor networks. Adhoc Sens. Wirel. Netw. 20(3–4), 307–328 (2014)
7. Liao, Z., Wang, J., Zhang, S., Zhang, X.: A deterministic sensor placement scheme for full coverage and connectivity without boundary effect in wireless sensor networks. Adhoc Sens. Wirel. Netw. 19(3–4), 327–351 (2013)
8. Karim, L., Anpalagan, A., Nasser, N., Almhana, J.: Sensor-based M2 M agriculture monitoring systems for developing countries: state and challenges. Netw. Protoc. Algorithms 5(3), 68–86 (2013)
9. Lloret, J., Bosch, I., Sendra, S., Serrano, A.: A wireless sensor network for vineyard monitoring that uses image processing. Sensors 11(6), 6165–6196 (2011)
10. European Commission, Overview of Common Agricultural Policy (CAP) Reform 2014–2020. http://ec.europa.eu/agriculture/policy-perspectives/policy-briefs/05_en.pdf (December 2013). Accessed 18 March 2014
11. Cambra Baseca, C., Diaz, J. R., Lloret, J.: Communication Ad Hoc protocol for intelligent video sensing using AR drones. In: IEEE Ninth International Conference on Mobile Ad-hoc and Sensor Networks (MSN 2013), Dalian (China), 11–25 December 2013
12. Koger, C.H., Shaw, D.R., Watson, C.E., Reddy, K.N.: Detecting late-season weed infestations in soybean (*Glycine max*). Weed Technol 17, 696–704 (2003)
13. Zhang, C., Kovacs, J.: The application of small unmanned aerial systems for precision agriculture: a review. Prec Agric 13, 693–712 (2012)
14. Christensen, S., Sogaard, H.T., Kudsk, P., Norremark, M., Lund, I., Nadimi, E.S.: Site-specific weed control technologies. Weed Res. 49(3), 233–241 (2009)
15. Slaughter, D.C., Giles, D.K., Downey, D.: Autonomous robotic weed control systems: a review. Comput. Electron. Agric. 61, 63–78 (2008)

16. Canales Compes, M., Gallego Martinez, J.R., Hernández Solana, A., Valdovinos Bardaji, A.: QoS provision in mobile Ad Hoc Networks with an adaptive cross-layer architecture. ACM/Springer Wirel. Netw. **15**(8), 1165–1187 (2009)

17. Kang, D.S., Griswold, N.C., Kehtarnavaz, N.: An invariant traffic sign recognition system based on sequential color processing and geometrical transformation. In: IEEE Southwest Symposium on Image Analysis and Interpretation, pp. 88–93, 21–24 April 1994

18. Felzenszwalb, P.F., Girshick, R.B., McAllester, D., Ramanan, D.: Object detection with discriminatively trained part-based models. IEEE Trans. Pattern Anal. Mach. Intell. **32**(9), 1627–1645 (2009)

19. Krajník, T., Vonásek, V., Fišer, D., Faigl, J.: AR-drone as a platform for robotic research and education. In: Obdržálek, D., Gottscheber, A. (eds.) EUROBOT 2011. CCIS, vol. 161, pp. 172–186. Springer, Heidelberg (2011)

20. MAVLink Micro Air Vehicle Communication Protocol. http://qgroundcontrol.org/mavlink/start

21. Rosati, S., Krizélecki, K., Traynard, L., Rimoldi, B.: Speed-aware routing for UAV Ad-Hoc networks. mobile communications laboratory. In: École Polytechnique Féderale de Laussane (EPFL), Laussane, Switzerland

22. FFmpeg multimedia framework. http://www.ffmpeg.org/ (2014). Accessed 18 March 2014

23. Open Source Computer Vision Library. http://opencv.org/ (2014). Accessed 18 March 2014

24. Bradski, G., Kaehler, A.: Learning OpenCV: Computer Vision With the OpenCV Library, p. 580. O'Reilly Media, Inc., Cambridge, September 2008

25. Suganthe, R.C., Balasubramanie, P.: Improving QoS in delay tolerant mobile Ad Hoc network using multiple message ferries. Netw. Protoc. Algorithms **3**(4), 32–53 (2011)

26. Chalouf, M.A., Mbarek, N., Krief, F.: Quality of Service and security negotiation for autonomous management of next generation networks. Netw. Protoc. Algorithms **3**(2), 54–86 (2011)

8th International Workshop on Wireless Sensor, Actuator and Robot Networks – WiSARN 2014

Virtual Localization for Robust Geographic Routing in Wireless Sensor Networks

Tony Grubman[1], Y. Ahmet Şekercioğlu[1,2][(✉)], and Nick Moore[1]

[1] Wireless Sensor and Robot Networks Laboratory,
Monash University, Melbourne, Australia
[2] Université de Technologie de Compiègne, Compiègne, France
asekerci@ieee.org

Abstract. Geographic routing protocols are well suited to wireless sensor networks because of their modest resource requirements. A major limiting factor in their implementation is the requirement of location information. The *virtual localization* algorithm provides the functionality of geographic routing without any knowledge of node locations by constructing a virtual coordinate system. It differs from similar algorithms by improving efficiency – greedy routing performs significantly better over virtual locations than over physical locations. The algorithm was tested and evaluated in a real network environment.

Keywords: Wireless sensor networks · Geographic routing · Localization

1 Introduction

Efficiency of routing protocols is very important in wireless sensor networks [2], where nodes are cheap, resource-limited devices, and power consumption and use of the constrained wireless channel are key issues. The most *scalable* routing scheme is geographic routing [13], which requires the use of local information only (1-hop neighbourhood).

The simplest geographic routing protocol is greedy routing [13], where routing decisions are made to locally optimise the *progress* of a packet (usually measured as the distance to the destination). This generally finds efficient paths, especially in dense, uniform networks. Packet delivery, however, is not guaranteed, as packets can get stuck in local minima (called *voids*). The success rate of greedy routing is heavily dependent on the network's *topology* and *geometry*.

The main drawback of geographic routing is that it requires the knowledge of node locations. The most common solution is to equip each node with a GPS receiver, but this adds to the cost and power consumption of the nodes. Also, GPS signals may not always be available. Alternatively, some nodes may be

This work has been partially supported by the Labex MS2T, which is funded by the French Government, through the program "Investments for the Future", managed by the National Agency for Research (Reference ANR-11-IDEX-0004-02).

© Springer-Verlag Berlin Heidelberg 2015
M. Garcia Pineda et al. (Eds.): ADHOC-NOW Workshops 2014, LNCS 8629, pp. 179–186, 2015.
DOI: 10.1007/978-3-662-46338-3_15

anchored, with known locations, while others run a localization algorithm to find their coordinates relative to the anchored nodes. A comprehensive analysis of three such algorithms (APS [8], Robust positioning [11], and N-hop multilateration [12]) is provided in [6]. Another localization algorithm is LASM [4].

For a completely self-organising wireless mesh network, there should be no requirement for any nodes to know their physical location. To achieve this, there are algorithms that construct *virtual* locations purely for routing purposes. These algorithms attempt to reproduce the functionality of geographic routing without using location information. This was first done in [9], where the algorithm relies on finding 'perimeter' nodes on the edges of the network. These nodes then exchange information to determine their virtual locations, after which they become anchor nodes (the other nodes perform a localization algorithm). The first two stages of the algorithm require many packets to be *flooded* through the network, and the resource requirements at the perimeter nodes are linear with respect to the network size.

In [5], a more scalable approach is used, where distances to some fixed anchors are used as the virtual coordinates directly (the anchors do not require physical locations). VCap [3] adopts a similar technique for constructing coordinates, but also defines a method to determine distances (in hops) to anchors. This involves packet flooding, but only to choose the anchors; the anchors do not need to flood messages to all other anchors. Discrete Ricci flows are used in [10] to construct virtual coordinates from a triangular mesh (which can be created without location information). The locations generated provide guaranteed delivery for greedy routing.

While the operation of these algorithms is more scalable, the performance of the routing algorithm itself (in terms of *reachability* and *path length*) is not considered in detail when constructing the virtual coordinate systems. The performance of greedy routing in [5] and [3] is comparable to (and sometimes worse than) using the physical coordinates. In [10], the reachability is always 100 %, but the average path length is considerably higher than the case with physical coordinates. Thus all of these methods necessarily sacrifice performance to provide geographic routing capabilities to networks without location information.

Our virtual localization algorithm is explained in Sect. 2, an overview of the test network, we set up in Monash Wireless Sensor and Robot Networks Laboratory (WSRNLab) [14], for collecting the experimental data can be found in Sect. 3, and the results of the experiments are presented in Sect. 4. Finally, in Sect. 5, we offer our concluding remarks.

2   Virtual Localization

The virtual localization algorithm [7] constructs virtual coordinates using only local connectivity information (topology). Each node stores the virtual locations of its 2-hop neighbourhood and uses this information to calculate its own coordinates. Consistency is achieved with periodic broadcast packets containing the locations of the sender's 1-neighbours.

Nodes determine an optimal location to "place" themselves in by minimising an energy function. This corresponds to virtual forces acting on the node. The forces are based on a simple model: nodes are attracted to their 1-neighbours with a spring-like force, and experience a repulsive electrostatic-like force from their 2-neighbours. The energy can be minimised using any optimisation technique, but as the nodes are assumed to have limited computational capabilities, the stochastic hill climbing method is chosen for its simplicity. Figure 1 summarises the operation of the algorithm.

As updated locations of neighbours are received, the energy function (and hence the optimum virtual location) changes. In fact, the energy minimisation algorithm can be continuously iterating as the neighbours' locations are being updated. This is especially useful in mobile ad-hoc networks, where the network topology changes constantly.

This algorithm is arbitrarily scalable because it uses only local information. All storage, computational, and network overhead requirements depend only on the node degree (i.e. network density), and not on the size of the network.

Require: Virtual locations of 1-neighbours, N
and 2-neighbours, M
Ensure: Own virtual location, ℓ
Parameters: Constants k_a, k_r, N_ITERATIONS

```
 1: function ENERGY(a)                    ▷ Calculates energy at location a
 2:     U ← 0
 3:     for b ∈ N do                      ▷ Energy from 1-neighbours
 4:         U ← U + k_a ‖a − b‖²
 5:     end for
 6:     for b ∈ M do                      ▷ Energy from 2-neighbours
 7:         U ← U + k_r / (1 + ‖a − b‖)
 8:     end for
 9:     return U
10: end function

11: procedure LOCALIZATION                ▷ Finds optimal location
12:     ℓ ← RANDOM                        ▷ Initialise location
13:     for i ← 1 to N_ITERATIONS do
14:         t ← ℓ + RANDOM                ▷ Perturb location
15:         if ENERGY(t) < ENERGY(ℓ) then
16:             ℓ ← t                     ▷ Update location
17:         end if
18:     end for
19:     return ℓ
20: end procedure
```

Fig. 1. Virtual Localization Algorithm

3   Packet Radio Network

A wireless mesh network testbed (Fig. 2(a)) was created using 33 packet radio modules (Fig. 2(b)). These modules are cheap devices operating in the 433 MHz ISM band, with a serial connection (over USB) to a computer. The network was set up in a computer lab, with each module controlled by a desktop computer. The lab contains many obstacles between the nodes, including chairs, tables and other computers.

The virtual localization algorithm was implemented in Python, along with a basic wireless medium access control at the data link layer (ALOHA [1]). The program was run on the lab computers for each node, and the connectivity and location information were recorded and analysed.

(a) Laboratory network testbed   (b) Packet radio module

Fig. 2. Monash WSRNLab's [14] experimental wireless mesh network.

4   Results

4.1   Virtual Locations

Virtual localization can generate coordinates in almost any metric space, but three dimensional Euclidean space was chosen. Even though the actual geometry of the network is planar (the nodes were placed at the same height), the extra dimension allows more complex *topologies* to be represented. Figures 3(a) and (b) show a typical topology of the lab network, and the corresponding virtual configuration achieved by the algorithm. The virtual locations vaguely resemble the actual locations, but the denser parts of the network tend to spread out more, as the algorithm cannot distinguish between 'long' and 'short' links.

The network topology in Fig. 3(a) would not usually be considered when conducting simulations. This is because simulations frequently use the unit-disk graph (UDG) model, or some variant of it (usually quasi-UDG). In actual wireless networks such as the one in Fig. 2(a), slight differences between different nodes (such as antenna length/transceiver sensitivity) have a dramatic impact on the quality of links between nodes. Some very long links are stable and reliable, while some shorter links are very noisy and cannot be used reliably. Obstacles and

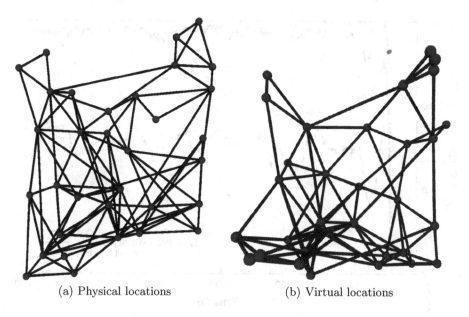

(a) Physical locations   (b) Virtual locations

Fig. 3. Network topology and geometries

the environment the network operates in also affect the topology significantly. No current simulation tools can accurately model such complex conditions, so a realistic evaluation of the algorithms can only be obtained on real networks.

4.2   Performance

The performance of the algorithm was assessed using the *reachability* and *path length* metrics. These were calculated at each time step by considering each pair of nodes (total of $33 \times 32 = 1056$) and applying the greedy routing algorithm. Packets are dropped when a void is encountered. The statistics were calculated using the actual (physical) locations and the virtual locations, and were compared to the optimal solution (packet flooding for reachability and shortest routes for path length). Figures 4 and 5 show that using the virtual locations significantly improves the performance of greedy routing. This is because the virtual locations more accurately describe the topology of the network than the actual locations, thus reducing the number of voids.

The topology of the network was observed to change dramatically over the duration of each one hour run, with some links being made and others broken at seemingly random times. This is likely due to the probabilistic nature of correctly receiving packets, and reflects a typical real-world scenario where, for example, the weather may influence the quality of wireless links. The success of the algorithm in such dynamic network conditions suggest that it may also be suitable for mobile ad-hoc networks.

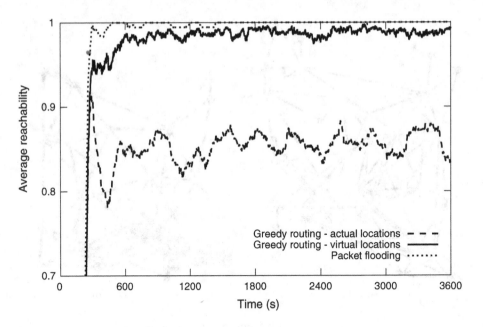

Fig. 4. Reachability of network

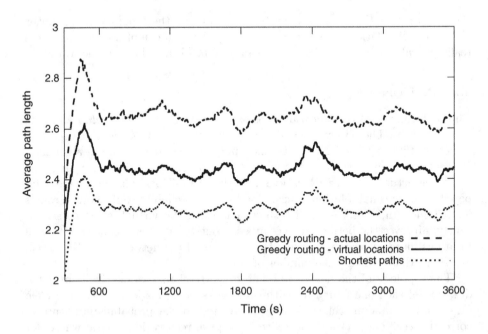

Fig. 5. Length of paths in network

5   Conclusion

A wireless mesh network of packet radio modules was created. The virtual localization algorithm [7], which generates virtual coordinates for networks of arbitrary size scalably, was implemented in this network. The virtual coordinates of the nodes represent the topology of the network in three dimensions better than the two dimensional coordinates in physical space. Greedy routing over the virtual coordinates delivers packets much more reliably than over the physical locations, and results in paths with a lower average hop count. Virtual localization not only improves performance of greedy routing, but also removes the requirement of external localization hardware for the nodes.

References

1. Abramson, N.: The ALOHA system - another alternative for computer communications. In: Proceedings of the Fall Joint Computer Conference. November 17–19, 1970, pp. 281–285. ACM, New York (1970)
2. Akyildiz, I.F., Su, W., Sankarasubramaniam, Y., Cayirci, E.: Wireless sensor networks: a survey. Comput. Netw. **38**, 393–422 (2002)
3. Caruso, A., Chessa, S., De, S., Urpi, A.: GPS free coordinate assignment and routing in wireless sensor networks. In: INFOCOM 2005. Proceedings of the 24th Annual Joint Conference of the IEEE Computer and Communications Societies, vol. 1, pp. 150–160. IEEE (2005)
4. Chen, W., Mei, T., Meng, M.Q.H., Liang, H., Liu, Y., Li, Y., Li, S.: Localization algorithm based on a spring model (LASM) for large scale wireless sensor networks. Sensors **8**(3), 1797–1818 (2008)
5. Huc, Florian, Jarry, Aubin, Leone, Pierre, Rolim, José: Virtual raw anchor coordinates: a new localization paradigm. In: Scheideler, Christian (ed.) ALGOSENSORS 2010. LNCS, vol. 6451, pp. 161–175. Springer, Heidelberg (2010)
6. Langendoen, K., Reijers, N.: Distributed localization in wireless sensor networks: a quantitative comparison. Comput. Netw. **43**(4), 499–518 (2003)
7. Moore, N., Şekercioğlu, Y.A., Egan, G.: Virtual localization for mesh network routing. In: Proceedings of IASTED International Conference on Networks and Communication Systems (NCS2005), ACTA Press (2005)
8. Niculescu, D., Nath, B.: Ad hoc positioning system (APS) using AOA. In: INFOCOM 2003. Proceedings of the 22nd Annual Joint Conference of the IEEE Computer and Communications Societies, vol. 3, pp. 1734–1743 (2003)
9. Rao, A., Ratnasamy, S., Papadimitriou, C., Shenker, S., Stoica, I.: Geographic routing without location information. In: Proceedings of the 9th Annual International Conference on Mobile Computing and Networking, MobiCom '03, pp. 96–108. ACM, New York (2003)
10. Sarkar, R., Yin, X., Gao, J., Luo, F., Gu, X.D.: Greedy routing with guaranteed delivery using Ricci flows. In: Proceedings of the 2009 International Conference on Information Processing in Sensor Networks, IPSN '09, pp. 121–132. IEEE Computer Society, Washington, DC (2009)
11. Savarese, C., Rabaey, J.M., Langendoen, K.: Robust positioning algorithms for distributed ad-hoc wireless sensor networks. In: Proceedings of the USENIX Technical Annual Conference, pp. 317–327. USENIX Association, Berkeley, CA, USA (2002)

12. Savvides, A., Park, H., Srivastava, M.B.: The bits and flops of the N-hop multilateration primitive for node localization problems. In: Proceedings of the 1st ACM International Workshop on Wireless Sensor Networks and Applications, WSNA '02, pp. 112–121. ACM, New York (2002)
13. Stojmenovic, I.: Position-based routing in ad hoc networks. IEEE Comm. Mag. **40**(7), 128–134 (2002)
14. Wireless Sensor and Robot Networks Laboratory (WSRNLab), Monash University, Melbourne, Australia. http://wsrnlab.ecse.monash.edu.au

Micro Robots for Dynamic Sensor Networks

Boaz Benmoshe[1]($\boxtimes$), Kobi Gozlan[2], Nir Shvalb[3], and Tal Raskin[2]

[1] Department of Computer Science, Ariel University, Ariel, Israel
benmo@g.ariel.ac.il
[2] Department of Physics and Department of Electrical Engineering,
Ariel University, Ariel, Israel
[3] Department of Industrial Engineering, Ariel University, Ariel, Israel

Abstract. In a network of micro sensors, the network capabilities can be greatly enhanced if participant nodes are able to fine-tune their positions. Even when a node is optimally located, it could benefit from subtle maneuvers that optimize the functionality of the node's directional sensors. In particular, the ability to aim a directional networking interface (e.g., antenna) is essential for low-power networking that is enforced by the tiny form-factor of the nodes. This paper presents a prototype design for a multi-terrain sensor-carrying micro robot, which excels in subtle movements. The robot has an egg-shaped shell, which provides protection, recover-ability and unique maneuvering capabilities in versatile terrains. We demonstrate the advantages of the suggested design in the context of dynamic sensor networks.

Keywords: Dynamic sensor network · Micro-robot · Bio-inspired robot · Egg-shaped robot · Directional data link

1 Introduction

This paper presents a framework for a network of micro-sensors with improved sensing and wireless networking. These improvements are the result of micro-robotic movement capabilities, which are added to each network node. Such node can rotate its camera, use directional antenna or aim its solar panel to the sun. This dynamic sensor network is based on a new multi-terrain micro robot, which can perform orientation changes and minor movements with little energy-consumption overhead. We introduce movement mechanisms that allow versatile mobility in different types of terrain: A vibration mechanism is used to move the micro robot across mostly plane and negative slopes terrain. For positive slopes, a novel crawling-pushing mechanism is used to move the robot. An orientation adjustment mechanism is being used to stabilize the robot to a specific orientation. The micro-robot is equipped with both orientation and movement sensors (i.e., 9 DoF: accelerometer, gyroscope, magnetic-field and an optical flow sensor). Using these sensors, the robot can either move autonomously, or follow movement instructions from external components of the system, such as an airplane, which flies above a swarm of micro-robot and functions as the network's sink.

© Springer-Verlag Berlin Heidelberg 2015
M. Garcia Pineda et al. (Eds.): ADHOC-NOW Workshops 2014, LNCS 8629, pp. 187–202, 2015.
DOI: 10.1007/978-3-662-46338-3_16

1.1   Motivation

The motivation for our research stems from sensor networks problems, in which a large number of very small sensors should be spread in a given region in a way that the region is well-covered by the sensors. The sensors should form a robust network with good sensing and networking capabilities, and the network should remain operational for a long period of time. For example, consider the use case of a sensor-network, which tracks wild-life in a forest. To fulfill its role, each sensor in this use case may need to rotate its camera, move to a place with better visibility, direct its solar panel to the sun, or aim its directional-antenna in order to establish a reliable data-link with a network sink. For such use cases, we need to have a simple yet robust method of aiming and moving the sensors on a wide range of surfaces with minimal power and weight overhead. Figure 1 illustrates a simplified use case that is not fully supported by current configurations of sensor networks. Specifically, the sensors in the figure must be able to direct their networking interface (e.g., antenna) in order to efficiently transmit data to the airplane above. This requires subtle maneuvering capabilities, which are not yet integrated in most sensors.

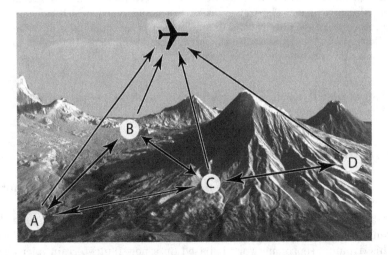

Fig. 1. An illustration of a typical scenario of a dynamic sensor network: the edges represent a line of sight relation between the sensor nodes, while the plane represents the network sink to whom the sensors need to transmit data.

1.2   Related Work

Theoretical and applied research on sensor networks often emphasize topics such as node collaboration, load-balancing and wireless routing of data. For example, using wireless multihoping, a large number of sensors can cope with the problem of limited wireless transmission range by relaying data down the multihop chain

[4,14]. A precondition for such routing and optimization problems is the ability to deploy the sensor network in the target area. This raises the need for sensor mobility, thus links the topic of sensor networks to the fields of micro-robotics [1,16] and swarms of micro-robots [3,13]. Provided the apparent relation between the topics, we can revise the definition of the stated sensor-spreading problem as follows: Given a set of (n) micro robots at a point on a terrain (T), design the micro-robot such that after a short given time (dt), the overall entropy of the micro-robots' positions will be maximized.

Striving for maximal spreading of micro robots in versatile terrains is a scarcely documented topic. Nevertheless, a promising research field that can assist in this task is "bio-inspired" micro robots [5,6,11,15,19]. Designing a micro robot that mimics an ant or a caterpillar can solve many issues of movement in difficult terrains. In some use cases, such abilities to move and spread may be the only roles of the micro-robot. However, future networks of sensor-carrying micro robots will likely require more than that: Specifically, a key topic in advance networks of micro robots is how to fine-tune the positions of directional sensors that the micro-robot is carrying. This topic is important due to multiple reasons. For example, a typical difficult problem for sensor-carrying micro robots is high energy consumption and limited range of wireless communication. A possible solution is to use a well-directed transmitter that is generally more energy-efficient and long-range, in comparison to an omnidirectional antenna. In rough terrains, however, the micro robot's exact posture on the ground is unpredictable. Therefore, performing tasks such as pointing an antenna, requires subtle in-place maneuvers and fine-tuning capabilities. We were not able to locate prior studies in the fields of micro-robots and sensor networks, which address such subtle in-place maneuvers.

1.3 Our Contribution

The focus of this research is in-place maneuvers that allow a micro robot to fine-tune its posture for the purpose of optimizing directional sensors (e.g., directional RF antenna, directional light data-link, camera, solar panel). We present several realistic use cases, which illustrate the potential of fine-tuning directional sensors in the context of micro-robot sensor networks. For each such use case, we explain the movement mechanisms within the suggested design, which enable the use case.

1.4 Paper Structure

The rest of this paper is organized as follows: In Sect. 2 we elaborate on the suggested new model for dynamic sensor network. In Sect. 3 we present the robot's mechanical design and movement types. In Sect. 4 we discuss the mechanisms for controlling the robot's movement. In Sect. 5 we present our experimental results using both: simulation and field experiments. In Sect. 6 we conclude the paper and suggest some future work.

2   Networking and Directional Sensors

In this section we present our model of dynamic sensor network. The focus here is on scenarios in which the sensors are assumed to be in-place and mainly perform orientation changes and minor movements. We consider an outdoor network of very small sensors that were manually-placed or simply spread (e.g., dropped from a drone flying above the region of interest). Each sensor-node has a low-power micro robotic mechanism allowing local positioning optimization for each sensor.

2.1   The Benefit of Orientation Change Capabilities

To illustrate the benefits associated with orientation change capabilities in the context of sensor networks, consider an imaging sensor such as a camera, with a len's angle of $50 * 40$ degrees (horizontal, vertical). Assume we throw such sensors in a region, where the positions of the sensors are uniformly random. Even in a perfect visibility case, the probability for such sensor to "see the moon" is approximately $\frac{50*40}{360*360} = \frac{1}{64.8}$. In case of 5 degrees directional antenna, the probability of aiming it to the desired direction (i.e., the network sink) is approximately $\frac{1}{5184}$, where in the case of 3 mili-radian laser communication module, the probability goes down to less than $\frac{1}{4000000}$.

On the other hand, directional sensors and antennas allow longer-range sensing and communication with significantly lower power consumption. Therefore, if we could rotate the directional antenna or the imaging sensor (of each sensor node) towards the proper direction, the over sensor-network should have significantly improved sensing and communication capabilities. Another major benefit of orientation change movement is the ability of a sensor to change its direction as the scene-interest changes, or to perform directional communication with several other nodes in different locations. In order to allow such orientation change, we have designed a mechanism that rotates the sensor-node itself instead of just rotating the camera/directional antenna. The detailed design of the sensor-node is presented in Sect. 3.

2.2   Improved Networking Model

In a network of sensor-carrying micro robots, where the micro robots are already positioned, efficient system performance requires the following capabilities from each sensor node:

1. Ability to perform subtle in-place maneuvers in order to adjust positions and angles of various directional sensors (e.g., camera, ground temperature sensor, solar panel).
2. Low-power and long-range directional wireless transmission interface.
3. Low-power omni-directional interface for capturing wireless radio frequency signals.

To address these requirements, a possible approach would be to design a micro-robot, such that each directional sensor and networking interface can be moved independently. However, a simpler approach is to design a micro robot that changes the direction of sensors by moving the robot's body. Our suggested egg-shaped robot (see Figs. 3, 4, 5, 6 and 7) follows this latter approach.

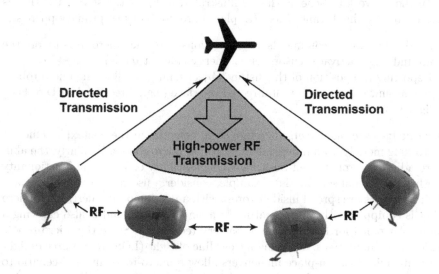

Fig. 2. The networking model: In-place maneuvers of the micro robots allow long-range and low-power directed transmission. Without the ability to aim a transmission interface, the sensors would need the assistance of capable agents in order to communicate with distant system components

A key aspect in many types of sensor networks relates to acquisition and real-time transmission of sensory data. When the sensors are carried by micro robots, the transmission must be very lightweight in order to conserve battery. When the network is rather dense, nodes can use short-range and low-power RF protocols such as Bluetooth Low Energy (BLE) or ZigBee/802.15.4 [17] to form an ad-hoc network, then exchange data with other system components (e.g., an airplane) through capable agents or gateways located at selected locations. However, when the network is less dense, or when the nodes operate as an independent swarm (with no gateways) the nodes cannot fully rely on short-range RF transmission using omnidirectional antennas. Moreover, when located on the ground, the range of an RF transmitter tends to drop significantly, and may cause many cases of *hidden node* [10,12].

An independent swarm of micro robots must therefore comply with the following:

1. At least some nodes must be able to transmit to a distant sink, such as an airplane or a distant antenna.
2. The transmission to the distant sink must be done using a module that is both small in size as well as energy efficient.

Note that the problem of capturing data from a distant object is not as dire as the problem of transmission to a distant object. That is because a distant object such as an airplane is typically not restricted by energy consumption and size of antenna, and can therefore transmit strong RF signals that the nodes can capture from distance.

In the networking scheme that is illustrated in Fig. 2 the strong RF signals transmitted by the distant object (airplane) are used for two primary purposes:

1. Sending various "commands"/"instructions" to the micro robots on the ground (e.g., activate sensors, enter energy-efficient mode).
2. Reporting the position of the distant object (sink), to allow the micro robots to transmit data using a well-directed interface (e.g., directional LED or laser light source).

Versatile in-place maneuvers therefore seem essential in the context of efficient networking model for an independent swarm of micro robots. Similarly, the ability to adjust the robot's self-position can significantly enhance the functionality of other directional sensors. For example, consider a use case in which a swarm of micro robots are spread inside a forest with a mission of fire alert. Each micro robot is equipped with a temperature or a smoke sensor, and is also carrying a small solar panel for the purpose of battery recharging. Since the micro robots are located among trees, detecting a clear line of sight (LOS) to a source of light is a non-trivial task. In-place maneuvers allow a micro-robot in this scenario to adjust itself, such that the solar panel is directed to the sun. Moreover, using a camera and self-orientation the sensor-node can construct a visibility map [9] which presents the set of directions in which it has no obstacles and therefore can use its directed transmission unit, as discussed above.

2.3   Network Initialization and Maintenance

The general construction of the suggested dynamic sensor network is composed of the following steps:

1. **Spreading the sensors:** all the sensors are being spread - i.e., thrown from an airplane above the region of interest.
2. **Sensor init:** after a sensor was located (stopped moving), the sensor approximates its position using a short (hot-start) cycle of its GNSS[1] sensor (should take 3–30 s). After acquiring a position the GNSS device is turned off. Next, the 9DoF (9 Degrees of Freedom) sensor is read to compute the current orientation and the barometric sensor data is fused with the elevation computed by the GNSS to compute an improved high-precision approximation.
3. **Forming an ad-hoc network:** each sensor transmits a periodical (RF) beacon and listens to other beacons to acquire a list of close neighbors. Then each sensor broadcasts its list-of-neighbors, which allows constructing and maintaining the ad-hoc sensor network [18].

---

[1] Global Navigation Satellite System.

4. **Communicate with the sink:** the network sink (e.g., airplane) broadcasts its accurate location. Each sensor can then aim itself to the sink's direction (by computing the position difference) and then start transmitting with the directional-long range transmitter.

In some cases the sensor-nodes may not have GNSS units due to energy and space limitations. In such cases the sink can use high power LED-lights in order to mark itself. The sensor-node can then use its camera in order to detect the sink's lights. Then, the sensor correlates the direction to the sink with the sink's transmitted position and forms an accurate temporal ray to the sink. This allows the robot to: (1) aim its directional light with high accuracy; (2) compute its 3D position.

3 Micro Robot Design

To construct the sensor network envisioned in the previous section, a node in the network should have sensing capabilities typical to nodes in static sensor networks, as well as specific micro-robotic movement capabilities. A sensor in a standard sensor-network is typically composed of the following components:

- A logic unit - Micro controller allowing ultra-low power computations.
- Sensors: such as camera, microphone, thermometer and many others.
- Communication module: An omni directional RF micro transceiver for short range communication. communication (possible with no line of sight NLOS).
- Energy source: A battery (e.g., Lipo), with some energy harvesting mechanism such as solar panel.

The micro-robot (sensor-network mobile node) is also equipped with the following components:

- Position sensors: 9DOF-orientation (3-axis gyroscope,accelerometer and magnetic field), optical flow, barometric sensor, and in some cases a micro GNSS sensor.
- Directional Communication module: a directional micro-transceiver based on directional RF antenna or LED light source for long range communication (line of sight - LOS).
- Actuators: Vibration motors, Linear servo, center of mass actuators.
- Frame: a special outer shell, for protection and mobility.

Figure 3 presents the conceptual extra components that are added to a standard sensor-node. The maneuverability of the micro-robot is accomplished by four types of movement mechanisms: Vibration, Motion adjustment, Orientation adjustment and Crawling-Pushing mechanism. We next discuss each of these movement mechanisms.

Figures 3, 4, 7 show the various components of the robot. The battery (Lipo) is used for changing the center of mass. Two linear motors are used as "legs". The vibration motors allow a smooth rotation. $A-C$ present the standard sensor components: micro-controller, standard sensors, and RF communication module (respectively). $D - G$ present the extra robot components: movement sensor (optic flow and 9DoF), the robot shell, the directional communication module and vibration motors (respectively).

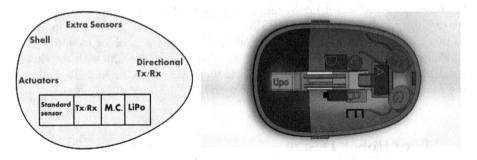

Fig. 3. Left: the additional concepts of a mobile sensor node. Right: the general design of the micro robot includes movement mechanisms that are based on linear motors, vibration actuators and dynamic center of mass.

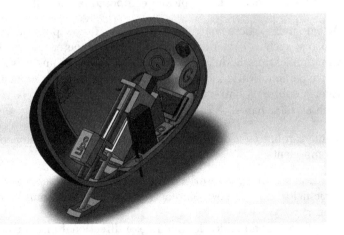

Fig. 4. The general robot construction: F represents the directional communication module. It can be directed using the circular vibration motors (G).

3.1   Vibration

The electrical eccentric vibration motor consists of a central mass, which when in operation creates vibration to the X and Z axis. The vibrations are created as a result of the centrifugal force of the angular movement of the mass. The motor operates at a high speed, thus we can assume the mass to be a solid mass that creates moment at the Y axis.

As can be seen in Fig. 5, the mass produces internal forces so the mass momentum is conserved and thus creates circular motion.

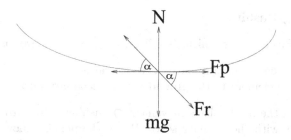

Fig. 5. Physical forces applied on the egg-shaped element

One Motor. Using one vibration motor we achieved two distinct patterns of movement:

1. While $\theta < 40°$ the element is yawing.
2. While $\theta > 50°$ the element performed Random Walk. The element moved a lot, but remained relatively in close proximity to the starting point.

Two Motors. When using two vibration motors, the forces on the X axis are canceled by each other, and the moment created on the Y axis is also canceled. The cancellation of the motors' moments nulls the yawing motion around the central mass pole, and the vibration movement on the Z axis is the only main active force. The Z axis vibration is sinusoidal (because of the circular motion). In the negative part of the wave the motion created is pushing against the ground and in the positive part of the wave the motion created is pulling from the ground up. This type of movement enables micro jumps of the element on the Z axis of the motor and the Y axis of the element. By placing a single vibrating motor the element can be rotated by its central axis of rotation.

3.2   Changing the Center of Mass

A vibrating robot tends to rotate itself according to its center of mass. In particular, if the robot has a round shape a smooth rotation can be achieve using minor vibration which cancels the robot static friction with the ground.

We have developed two mechanisms for changing the center of mass:

1. Placing two linear motors (one for X axis and one for Y axis) attached to the central mass of the element.
2. By constructing a conveyor on the inner shell of the element in 2 axis (X and Y).

These methods that are illustrated in Fig. 6 allow us control over the central mass position inside the element, thus control over the element's orientation.

Both methods allow orientation change yet their location inside the robot's shell limited the free space. We have therefore used a somewhat simpler design, which mainly moves the mass in a single dimension, while the main rotation is performed using the vibration motors. As shown in Fig. 7 the back leg is used for accurate orientation modifications and for fixing the orientation.

3.3 Crawling-Pushing

The Crawling-Pushing mechanism (see Figs. 4, 8) works at two distinct modes:

1. Pushing the element with a linear servo forwards.
2. Anchoring the element to the ground with a secondary linear servo.

The first step of the mechanisms movement is pushing it forward by extending a leg backwards with the primary servo. When the pushing movement is complete the element anchors it self to the ground using the secondary servo. When anchored the primary servo retracts to the point of no contact with the ground and the secondary servo retracts to the elements body. The whole process creates a motion of Crawling-Pushing.

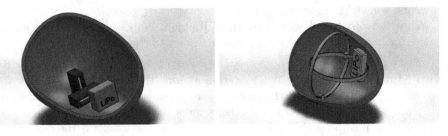

Fig. 6. Mechanisms for changing the center of mass. Left: simplified mechanism based on 2–3 linear motors. Right: advance mechanism allowing to locate the battery mass practically anywhere on the shell.

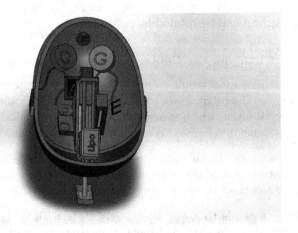

Fig. 7. Orientation change: the vibration motors (G) are used for the main rotation, while the Lipo position affects the center of mass and therefore allows controlled vertical rotation.

4 Motion Control

In this section we present the general framework for controlling the micro robot. The controlling mechanism is mainly designed for efficient orientation change, yet, it also allows the robot to have controlled movements.

The following components are taking part in the control loop:

1. **Vibration motor**
2. **Orientation sensor (mems 9DoF):** This sensor allows the micro robot to compute self orientation. The accuracy level of the sensor affects the ability of the robot to accurately aim directional devices, such as lasers, antennas, cameras and solar panels. While in case of a camera or a solar panel an accuracy of 2–5 degrees is mostly sufficient, the needed accuracy for laser communication can be very high - often refer is 1 mili-radian which is smaller than 1/17 degree. We have tested many orientation sensors and found that the YEI sensor is very small, accurate (with its inner implementation of Kalman Filter) and has a relatively fair energy consumption. This sensor allows the micro robot to have a minor angular drift of less than 0.2 degrees per minuet.
3. **Center of mass actuator**
4. **Micro Controller**

The element is designed to work on various types of terrains. For successful operation over different types of terrain, we implemented a control loop over the operation of the movement mechanisms to control their operation, and ideally keep them in the most effective operation mode. When addressing the vibration motors operation the control loop moderates and adjusts the voltage of the motor to control the frequency of the motor. While the frequency of the motor reaches the self-resonance region its at its peak operation (the amplitude is the highest).

The self-resonance region of a vibrating motor is very narrow. This is due to the physical size of the motor, its construction and materials used to assemble it with. We aim to adjust the motors operation so the whole element will operate at the self-resonance region while adjusting each motor separately. We use an accelerometer sensor to measure the frequency of the element and adjust it according to its readings.

The control loop logic, which is implemented using a Arduino Nano Micro-Controller [7], consists of a PD controller with a negative feedback. In most cases, a control loop uses positive feedback, since its purpose is to minimize noise and interference, but since we want to amplify the interference we use the negative feedback.

By adding Optical Flow sensor to the element, we can track its movement and measure exact distance its traveling while operating. These readings can indicate whether the element is advancing forward, or stuck in-place and rotating in the vicinity of the same spot.

5   Experimental Results

Since we only have a small number of prototype robots, it may not be convincing to attach quantitative and statistical data to our conducted field experiments. Some impressional results of the robot's movement experiments are shown in Fig. 8.

In the current results section we mainly address non-robotic aspects of the proposed model. Specifically, we empirically test the aspects of: (1) Energy consumption, (2) Performance and range of an optic data link and (3) LOS/NLOS[2] simulation which tests the feasibility of the networking model proposed in Sect. 2.

Fig. 8. Movement experiments: Left: vibration based crawling. Right: 1 cm up-hill climbing step.

5.1   Energy Consumption

The energy consumption associated with the movement mechanisms is roughly as follows:

1. A vibration motor consumes 40–70 mA Current.
2. A linear motor consumes up to 100–200 mA Current.
3. An Arduino Nano [7] consumes 9–15 mA, according to operation load.
4. A 9DOF sensor consumes about 5–15 mA.
5. A directional light source unit 10–50 mA (Omni directional, e.g., BLE low energy < mA).

An operation cycle of the Crawling-Pushing which consists of 4 operations takes about 8 s (2 s for each operation) and consumes about 250 mA. With a common Li-Po 1 Cell battery of 350 mAh we can produce about 700 such operation cycles. If we estimate that in each operation cycle the Micro-Robot can move 1–2 cm that means that the range of operation with 1 Cell Li-Po battery 7–14 m, or hundreds of orientation changes. When adding more motors, as the mass of the element increases, the resulting operation is decreased. Although more motor power is added, the operation effectiveness does not increase accordingly. This is mainly due to the "insect" behavior of the micro robot which requires it to be very light weight (preferably below 20 g).

---

[2] Line of Sight vs Non Line of Sight.

5.2 Long Range Light Source Data Link

To examine the micro robot's ability to communicate with remote network sink
(such as an airplane), we have tested the following light source data link:
Tx: directional LED light source (5–20 degrees angle).
Rx: Camera with optical zoom with a corresponding narrow-band light filter.

Below are several configurations which were able to support a long range (LOS)
data link:

– 20 degrees LED, No optical zoom, no IR filter: 200 m.
– 10 degrees LED, x10 optical zoom, 100 nano-meter filter: 2500 m.
– 10 degrees LED, x30 optical zoom, 30 nano-meter filter: 7000 m.
– 5 degrees LED, x100 optical zoom, 10 nano-meter filter: 13000 m.

The above results assume good visibility. In case of limited visibility (e.g., dust
or fog), the corresponding ranges will be significantly reduced.

As stated above, it is hard to provide statistical data that demonstrates the
robot's performance in the task of aiming its LED to a distant sink. We can say
by impression that in most cases, the prototype robot succeeded to aim itself
to the needed direction with an angular error of less then 3.5 degrees, in less
than 20 s.

5.3 LOS/NLOS Simulation

The purpose of the following stimulative analysis is to examine the suggested
networking model with respect to LOS/NLOS probabilities. As explained in
Sect. 2.1, the initial probability of a randomly-placed directional sensor to be
towards to a distant sink is low (e.g., in case of a standard 45 degrees camera
the probability is less than 2 %). The various in-place maneuvers proposed in this
paper are designed to enable the micro-robot to aim its networking interface to
the position of the sink, thus enabling the entire swarm to establish a networking
link with the distant sink. This networking model is highly-dependent on the
LOS/NLOS conditions in a given outdoor environment. In order to test the
suggested communication model, we have implemented a computer simulation
that allow us to test the performance of dynamic sensor network over a wide
range of terrains.

We define a **LOS-ratio**(p) to be the angular area of the visible sky from a
point p divided by angular area of half a ball. In practice we compute LOS-ratio
from a point using an efficient radar-like heuristic as described in [2].

The simulation uses Delaunay triangulation [8] in order to present natural surfaces. Each terrain point has two elevation values (z,s) representing
the height and the type of the surface in the current location. The close-by-
environment of each sensor-node was modeled in high resolution using a probabilistic LOS/NLOS map which is suitable to the local type of the terrain
(e.g., rocky mountains, dunes, hills with low vegetation, 3–12 m trees in a
forest, etc.).

Various terrains representing several types of surfaces were used (each with 5,000–50,000 vertices). Each part of each terrain was assigned with surface type. Then, we have randomly located 1000 points on each terrain and computed for each point its LOS-ratio

In general, the *LOS-ratio* of ground sensors is significantly lower than observer which are few meters above the ground, as presented in Fig. 9. Yet, most cases the ground sensors have a LOS-ratio larger than 20%. In other words: a plane positioned over a swarm of sensors has a high probability to see at least 20% of the sensors. Moreover, while flying over such swarm, at each point different sensors may have a LOS to the plane, while other sensors with NLOS may use neighbour-sensors (with LOS) as relays.

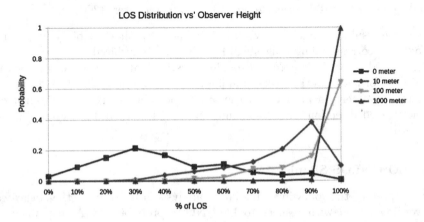

Fig. 9. The distribution of LOS with respect to the height of the observer above the ground. Observers on the ground are much more sensitive to local obstacles.

6   Conclusion

We have presented an egg-shaped micro-robot that can perform minor movements and robust orientation change in versatile terrains. The proposed design introduces several unique movement mechanisms, such as crawling-pushing and adjusting orientation by controlling the central mass position inside the egg. The paper demonstrates that these movement mechanisms enable the micro robot to perform subtle in-place maneuvers, thus controlling directional sensors and networking interfaces. Most notably, the lower energy-consumption and longer transmission-range associated with well-directed wireless networking forms the basis for a next-generation sensor network, in which sensor-carrying micro robots operate in distant locations as an independent swarm. Finally We have also addressed the validity of the suggested network model with respect to limited visibility of ground based sensors, and found via simulations that although few randomly spread sensors may have no LOS, most members of a swarm of such sensors will be able to have a reasonable LOS-ratio which as a swarm is sufficient for communicating with flying-sink such as an airplane.

In future work, we intend to develop robot models that integrate the advantages of the proposed design with more efficient methods for longer-range movement. The main challenge is to mimic insect-like movement (that usually requires legs), and still maintain the apparent advantages associated with the robot's egg shape.

References

1. Akyildiz, I.F., Su, W., Sankarasubramaniam, Y., Cayirci, E.: Wireless sensor networks: a survey. Comput. netw. **38**(4), 393–422 (2002)
2. Ben-Moshe, B., Carmi, P., Katz, M.J.: Approximating the visible region of a point on a terrain. GeoInformatica **12**(1), 21–36 (2008)
3. Dorigo, M., Gambardella, L.M., Birattari, M., Martinoli, A., Poli, R., Stützle, T. (eds.): ANTS 2006. LNCS, vol. 4150. Springer, Heidelberg (2006)
4. El-Hoiydi, A., Decotignie, J.-D.: WiseMAC: An ultra low power MAC protocol for multi-hop wireless sensor networks. In: Nikoletseas, S.E., Rolim, J.D.P. (eds.) ALGOSENSORS 2004. LNCS, vol. 3121, pp. 18–31. Springer, Heidelberg (2004)
5. Floreano, D., Mattiussi, C.: Bio-inspired Artificial Intelligence: Theories, Methods, and Technologies. The MIT Press, Cambridge (2008)
6. Franceschini, N., Ruffier, F., Serres, J.: A bio-inspired flying robot sheds light on insect piloting abilities. Curr. Biol. **17**(4), 329–335 (2007)
7. Gibb, A.M.: New media art, design, and the arduino microcontroller: a malleable tool. Ph.D. thesis, Pratt Institute (2010)
8. Lee, D.T., Schachter, B.J.: Two algorithms for constructing a delaunay triangulation. Int. J. Comput. Inf. Sci. **9**(3), 219–242 (1980)
9. Luo, J., Etz, S.P.: A physical model-based approach to detecting sky in photographic images. IEEE Trans. Image Process. **11**(3), 201–212 (2002)
10. Ng, P.C., Liew, S.C., Sha, K.C., To, W.T.: Experimental study of hidden node problem in IEEE 802.11 wireless networks. Sigcomm Poster, p. 26 (2005)
11. Pfeifer, R., Lungarella, M., Iida, F.: Self-organization, embodiment, and biologically inspired robotics. Science **318**(5853), 1088–1093 (2007)
12. Rahman, A., Gburzynski, P.: Hidden problems with the hidden node problem. In: Proceedings of 23rd Biennial Symposium on Communications, pp. 270–273 (2006)
13. Şahin, E.: Swarm robotics: from sources of inspiration to domains of application. In: Şahin, E., Spears, W.M. (eds.) Swarm Robotics WS 2004. LNCS, vol. 3342, pp. 10–20. Springer, Heidelberg (2005)
14. Scaglione, A., Servetto, S.: On the interdependence of routing and data compression in multi-hop sensor networks. Wirel. Netw. **11**(1–2), 149–160 (2005)
15. Scarfogliero, U., Stefanini, C., Dario, P.: The use of compliant joints and elastic energy storage in bio-inspired legged robots. Mech. Mach. Theory **44**(3), 580–590 (2009)
16. Sibley, G.T., Rahimi, M.H., Sukhatme, G.S.: Robomote: a tiny mobile robot platform for large-scale ad-hoc sensor networks. In: Proceedings of IEEE International Conference on Robotics and Automation, ICRA 2002, vol. 2, pp. 1143–1148. IEEE (2002)

17. Siekkinen, M., Hiienkari, M., Nurminen, J.K., Nieminen, J.: How low energy is blue-tooth low energy? comparative measurements with zigbee/802.15. 4. In: Wireless Communications and Networking Conference Workshops (WCNCW), 2012 IEEE, pp. 232–237. IEEE (2012)
18. Tubaishat, M., Madria, S.K.: Sensor networks: an overview. IEEE Potentials **22**(2), 20–23 (2003)
19. Wood, R.J.: The first takeoff of a biologically inspired at-scale robotic insect. IEEE Trans. Robot. **24**(2), 341–347 (2008)

A Pragmatic Approach for Effective Indoor Localization Using IEEE 802.11n

P. Shanmugaapriyan, H. Chitra, E. Aiswarya,
Vidhya Balasubramanian[(⊠)], and S. Ashok Kumar

Department of Computer Science and Engineering, Amrita School of Engineering,
Amrita Vishwa Vidyapeetham (University), Coimbatore, India
b_vidhya@cb.amrita.edu

Abstract. Wi-Fi based Indoor Localization is commonly used in pervasive systems due to its ease of use and relatively low cost. In recent times, IEEE 802.11n is gaining more attention due to the operation of devices in dual band (2.4 GHz and 5 GHz) simultaneously. However the utility of dual band in Wi-Fi indoor localization is still a subject of study and has not been widely implemented. The focus of this paper is to evaluate the feasibility of using both these bands by comparing their indoor localization performance using fingerprinting techniques in a real indoor environment. The effects of interference and localization accuracy are the subject of the experimental study. Based on the study, we propose intelligent policies which effectively utilize the advantages of both the bands. Our experiments and analysis have demonstrated the effectiveness of our policies in improving the accuracy of indoor localization.

1 Introduction

Indoor pervasive systems are becoming widespread, therefore the need for effective indoor localization techniques has grown. Indoor localization refers to tracking people or devices within any indoor environment and Wi-Fi is the most widely employed technology for this purpose. The reason for widespread use of Wi-Fi is that, unlike other localization technologies like RFID, ZigBee and UWB (Ultra Wide Band) [1], it does not demand additional infrastructure for indoor localization and is relatively inexpensive [2].

Among the different Wireless Local Area Network (WLAN) standards, IEEE 802.11b which operates in 2.4 GHz band has been extensively deployed for locating devices in indoor environments. In Wi-Fi based indoor localization, the most commonly used techniques are Fingerprinting and Trilateration. In IEEE 802.11b, it has been generally observed that fingerprinting outperforms trilateration [3]. However, fingerprinting is expensive and involves a lot of effort for mapping and updating [4]. This has forced researchers to look at other Wi-Fi technologies for the purpose of localization. In recent times, IEEE 802.11n is

This work has been funded in part by DST(India) grant DyNo. 100/IFD/2764/2012-2013.

M. Garcia Pineda et al. (Eds.): ADHOC-NOW Workshops 2014, LNCS 8629, pp. 203–216, 2015.
DOI: 10.1007/978-3-662-46338-3_17

gaining more attention since devices operate simultaneously in the 2.4 GHz and 5 GHz bands. In this paper we explore the practical usage of both these bands in IEEE 802.11n for indoor localization.

Existing studies have shown that the 5 GHz band has higher localization accuracy than the 2.4 GHz band [5,6]. This is widely attributed to the stability exhibited by the 5 GHz band and its propagation effect. However most such studies are focused on either demonstrating signal level properties or analyzing the performance of fingerprinting, generally in ideal environments. However, there is a need to study the relative performance of fingerprinting based localization in actual environments, which are affected by interference from obstacles and regular human activities. Hence in this paper, we implement fingerprinting based localization using 2.4 GHz and 5 GHz bands in an indoor environment with regular human activity. Based on the results of this analysis we propose policies, which help utilize the specific advantages of both the bands. The specific contributions of our work are as follows:

- A detailed analysis of the relative performance of 2.4 GHz and 5 GHz band in fingerprinting based solutions by experimenting in indoor environments, both during the presence and relative absence of human activity.
- Analyze when 5 GHz band performs better than 2.4 GHz band and vice versa, based on which, we map the regions of the environment where each band performs better than the other.
- Design policies that use the above map to effectively combine information from both the bands for the purpose of localization, so that the overall accuracy of localization can be improved. These policies can be practically and easily implemented in any environment.

In the next section we will provide an overview of the existing body of work, and motivate further the need for this work. The comparative study is highlighted in Sects. 3 and 4.

2   Related Work

As mentioned previously, Wi-Fi based solutions are the most commonly used for Indoor Localization. In this section, we review the current state of the art Indoor Localization techniques using Wi-Fi and motivate the necessity of our contributions. Several technologies are being used for indoor localization such as IEEE 802.11a, 802.11b, 802.11g, 802.11n. IEEE 802.11b WLAN standard, operating in the 2.4 GHz, has become very popular in public and enterprise locations during the last few years. The reason for this, is the wide availability of routers and almost pervasive operation of devices at 2.4 GHz frequency band.

The most popularly used methods in the process of localization using 2.4 GHz are fingerprinting and trilateration [7]. In fingerprinting a reference radio map is generated based on the strength of the radio signal received from the access point. This map is used to locate the position of the unknown devices, provided the position of the APs are fixed [8]. Lateration which is the process of determining the location/position of the device, is based on the simultaneous range measurement from the nearby stations [9]. Both fingerprinting and trilateration need ranging techniques to measure the distance between nodes such as Time of Arrival (TOA), Time Difference of Arrival (TDOA), Angle of Arrival (AOA) and RSSI [10]. RSSI is most commonly used because it does not require any additional infrastructure [11]. Trilateration is easy to implement, but its accuracy is very low. Fingerprinting while being more accurate, has the drawback of requiring generation of radio map which is time-consuming, labor intensive and vulnerable to environmental changes. In both the methods there have been attempts to improve accuracy like the one done in [12]. However they fall short due to the poor stability of 2.4 GHz band which results in accuracy degradation.

To overcome the inaccuracies in Wi-Fi based localization using 2.4 GHz we look at other technologies like UWB and IEEE 802.11a, which operates at the 5 GHz band. UWB is very accurate, but it is very expensive and not yet ready for widespread implementation. IEEE 802.11a is now becoming common and the routers are getting cheaper. The stability of RSSI in 5 GHz is relatively higher than that of 2.4 GHz band [6,13,14]. Thus the 5 GHz band has the potential to improve the accuracy in Wi-Fi based indoor localization. Low signal penetration capability and low coverage area in the 5 GHz band are the two major deficiencies of using this band [13]. Due to the low penetration and higher propagation loss of the signals in the 5 GHz band, the RSSI value decreases more rapidly in 5 GHz than the 2.4 GHz band as the distances between the access point and the users increases [15]. As a result, for larger spaces, more routers are needed when localizing using the 5 GHz band. Study of fingerprinting based localization using 5 GHz has been done in [6]. This study indicates that 5 GHz provides more accuracy due to its relative stability. However it has not been studied, if this performance holds true in the presence of human activity and other environmental factors.

In order to utilize the advantages of both the bands, when they coexist, we look at the approaches that combine them. Till date this approach is used to maximize throughput in data transfer and not used in localization. For improving accuracy in localization, different combination of technologies involving Bluetooth [16], RFID [17], Infrared and Wi-Fi have been used. In [16], a method for merging Bluetooth and WLAN technologies for indoor localization has been proposed. Initially, trilateration is applied on RSSI value of Bluetooth and then an approximate region of location is determined. Based on the confined zone found using Bluetooth, they have performed Wi-Fi fingerprinting to locate the position accurately. Combining two different technologies as in this approach, however, complicates the infrastructure requirements of the localization system. Our goal is therefore, to implement the localization effectively using minimal infrastructure. So we devise an approach which uses both 2.4 GHz and 5 GHz for indoor localization using just dual-band routers. In [18] they utilize both 2.4 GHz and 5 GHz bands during localization by combining the top k neighbors in each band and directly use these neighbors in the k-NN algorithm. To our knowledge this is the only work combining both the bands for the purpose of localization. Hence there is scope for designing better policies that can utilize both these bands more effectively, and this is one of the goals of this paper.

3   Analysis of 2.4 GHz and 5 GHz Bands in Indoor Environments

Before looking into the concept of band switching, a comprehensive knowledge about relative performance of 2.4 GHz and 5 GHz is vital. For this study, we have to analyse the properties of the signals, and their impact on localization, in both the bands in different kinds of environments (such as presence and absence of human activity, reflection or penetration of signals because of different obstacles) so that a common experimental setup can be established which helps in exploiting both the bands. For this purpose we have selected two environments, one an office location (see Fig. 1), and a smaller lab space as our experimental environments, such that they are representative of most indoor environments. The Belkin N600 DB Wireless Dual-Band N+ Router has been used with IEEE 802.11n for localization. In this experimental environment no routers other than dual band N+ is present for ensuring that both the bands' signals are not affected and adequate safeguards are taken to keep the load on both the bands uniform. RSSI is chosen as the primary metric for range determination since it is the only parameter that can be retrieved without any additional hardware. To get better accuracy in indoor localization the following considerations must be addressed.

Number of routers: Based on the guidelines prescribed in [19], and the difference in coverage area of the two bands, we identify a maximum of 4 routers as the optimal number of routers for supporting localization in both bands.

Placement of router: For ensuring accuracy, optimal placement of routers is essential. The routers should be placed in positions such that the variance between the routers is maximum [20]. In addition, the height at which the router is placed also plays an important factor in the localization accuracy. Positioning of routers is influenced by the RSSI variance between two routers. We analyze variance of RSSI for both bands to determine an optimal common positioning. Based on our experiments we have analyzed the RSSI variations for both the bands. From our analysis, we have observed that unique mapping between RSSI and distance is present upto a distance of 80 ft and 60 ft between the access point and device in 2.4 GHz and 5 GHz bands respectively. This unique correspondence is necessary for fingerprinting algorithms to distinguish various positions accurately. In addition, for utilizing both the bands, the correlation between RSSI from 2.4 GHz and 5 GHz must be high. It is observed that beyond the distance of 18 meter, RSSI fluctuates with respect to distance and therefore correlation is reduced. So for maintaining accuracy, we ensure that, for any localization point, there are three routers within a distance of 18 meters each, and the router positions are chosen accordingly. In an indoor environment there may be signal loss

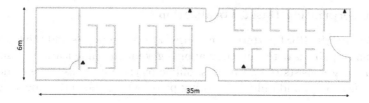

Fig. 1. Office environment

in both the bands due to penetration through obstacles existing in the environment. Placing the routers at an optimal height can help reduce the impact of the obstacles and this also aids to have a proper line of sight with minimum path loss [20]. In general the 5 GHz band has less penetration capability than the 2.4 GHz band and therefore the penetration loss in 5 GHz will be higher [5,21]. For the office environment chosen, due to the presence of large number of cubicles, a height of 2.43 m has been chosen after careful consideration. Our analysis has shown that this height is suitable for our fingerprinting based localization in both the bands. In order to cover the whole area and also to have distinct RSSI vectors at each point within the localization area we placed our routers in an asymmetric fashion instead of having symmetric placement. Thus the routers are placed as shown in the Fig. 1. With this experimental environment setup as the base, we explain the fingerprinting process and our observations in the next section.

4   Fingerprinting Using IEEE 802.11a/b

As discussed earlier, there are not enough studies to demonstrate the relative performance of 2.4 GHz and 5 GHz bands in fingerprinting based localization. The limited studies are inconclusive about which band is better for localization and hence we first analyze the relative performance of these two bands in the context of indoor fingerprinting based localization in our environment. Specifically the goal of the study is to analyze.

1. If 2.4 GHz or 5 GHz band outperforms the other during localization in general.
2. If the answer is true to the previous question, then analyze the criteria and reasons for the same.
3. Based on the analysis determine the specific contexts where one might be better than the other.

Hence we conduct various experiments analyzing the performance of fingerprinting based localization using 2.4 GHz and 5 GHz bands. The experiments use the k-NN based technique [7] for localization due to its relative simplicity. We believe that the observations would hold true irrespective of the technique (k-NN or probabilistic methods) chosen. The next paragraphs will explain in detail the fingerprinting process.

4.1   Fingerprinting Based Localization

In this process the RSSI pattern is measured at certain points and the location of the unknown point is determined by applying some machine learning algorithm. Fingerprinting consists of two phases: offline or training phase where the radio map is constructed and online or testing phase where the actual localization is performed.

Firstly a reference radio map is generated for the office environment, where the considered area is split into several cells or grids. For our environment (considering that it is rectangular and narrow), we have arrived at a grid size of $2\,m \times 1\,m$ for ensuring maximum variance between adjacent cells. This is also based on the recommendations given in [19]. The center of each grid cell is defined as the reference point and its RSSI is considered representative of the entire cell. The construction of the radio map is done by taking sample RSSI values at each reference point. At every reference point, a sample set of RSSI values are retrieved from the different routers for about $2\,min$. The constructed map is thereby used to locate the position of an unknown point by obtaining a sample of the RSSI values at this point from the mobile device. k-NN is used to determine the position of the unknown point. In k-NN, different distance metrics are used to find the best match among fingerprint vectors and localization vectors, and in this paper we choose the following metrics for our studies after finding that other metrics like Cosine similarity give lesser accuracy.

1. Euclidean distance:
   Euclidean distance is the most commonly used metric to find the nearest neighbors. It is computed using the sum of squared distance between the two positions x and y with Cartesian coordinates for x as $(x_1, x_2, x_3, ...x_n)$ and y as $(y_1, y_2, y_3, ...y_n)$.

$$d\left(x, y\right) = \sqrt{\sum_{i=1}^{i=N} \left(x_i - y_i\right)^2}$$

2. Manhattan distance:
   Manhattan distance is calculated as sum of absolute distance between two positions x and y.

$$d\left(x, y\right) = \sum_{i=1}^{i=N} |x_i - y_i|$$

The localization samples have been collected in two iterations, once when human activity is low, and once when human activity is high. About 85 unknown points have been collected in both these cases, and they have been localized using each of these metrics and the average accuracy is shown for the different metrics in both $2.4\,GHz$ and $5\,GHz$ band. Accuracy is defined as the Euclidean distance between the estimated position using the localization algorithm and original position. To analyze the impact of choice of k in the k-NN algorithm, we run it for different values of k. The Tables 1 and 2 show the localization accuracy for data taken when human activity is low, and human activity is high respectively.

Table 1. Accuracy of 2.4 GHz (left) and 5 GHz bands using k-NN in a non-busy office environment

k	Euclidean	Manhattan
2	3.12	3.03
3	2.87	2.87
4	2.77	2.86
5	2.73	2.80

k	Euclidean	Manhattan
2	2.54	2.53
3	2.50	2.41
4	2.45	2.40
5	2.42	2.38

Table 2. Accuracy of 2.4 GHz (left) and 5 GHz bands using k-NN in a busy office environment

k	Euclidean	Manhattan
2	3.49	3.52
3	3.35	3.27
4	3.18	3.26
5	3.19	3.22

k	Euclidean	Manhattan
2	2.52	2.57
3	2.44	2.38
4	2.39	2.37
5	2.42	2.34

From both the tables we see that irrespective of the distance metric chosen, human activity and choice of k, the localization accuracy is higher in the 5 GHz band than the 2.4 GHz band. This improvement is due to the stability in RSSI values exhibited by the 5 GHz. While existing studies have shown that the degradation in signal strength due to obstacles and human activity, is higher in 5 GHz band, our results show that the localization accuracy in the 5 GHz range is consistently better than that using the 2.4 GHz band, due to the absence of major obstacles like concrete walls.

From the tables we also observe that the Euclidean distance metric performs better in general and mostly for k = 4. While the environment has many obstacles, Manhattan distance usually works best [22], we believe that the lower height of the obstacles (here cubicles walls) allows for a straight line distance metric to work better in most cases. However for the 5 GHz band the Manhattan metric performs marginally better, accounting for the poor performance of 5 GHz in the presence of obstacles.

To determine if these results hold good for all kinds of environment, we chose a smaller lab space, where a major wall separates two rooms. The same experiments were replicated for this space, and the results when human activity is low is shown in Table 3, and when human activity is high as shown in Table 4. The base accuracy is much lower since the space is smaller, and the influence of obstacles is more pronounced. In this space we can see that the accuracy of the 5 GHz band is much higher when there is lesser human activity. This is similar to what we have observed in the other environment. Hence we can reasonably conclude that the 5 GHz band provides better accuracy when the human activity is lower, in any environment. Both Euclidean and Manhattan distance metrics work well in the 5 GHz band. However when activity is much

Table 3. Accuracy of 2.4 GHz (left) and 5 GHz bands using k-NN in a non-busy lab environment

k	Euclidean	Manhattan
2	3.94	3.71
3	3.74	3.62
4	3.76	3.52
5	3.71	3.38

k	Euclidean	Manhattan
2	3.01	2.87
3	2.91	2.76
4	2.85	2.72
5	2.69	2.61

Table 4. Accuracy of 2.4 GHz (left) and 5 GHz bands using k-NN in a busy lab environment

k	Euclidean	Manhattan
2	5.32	5.14
3	5.01	4.92
4	4.99	4.91
5	5.05	5.00

k	Euclidean	Manhattan
2	5.78	5.77
3	5.71	5.78
4	5.69	5.65
5	5.57	5.72

higher, the performance of the 5 GHz band degrades. While in the previous environment the absence of influential obstacles helped overcome the signal decay due to human activity, in this space, combination of obstacles like major walls, and high human activity contributes to higher propagation loss, and hence lower accuracy in 5 GHz band. The performance of 2.4 GHz band is just marginally better in such cases.

We have observed that the 5 GHz band works better as seen in Environment1, despite human activity or obstacles, unless there are major obstacles. This contradicts the penetrating property of the frequency bands. The reason for the 5 GHz overcoming 2.4 GHz is the stability of 5 GHz. In our experiment, the fingerprint database and the unknown database are not taken at the same time, but the stability of RSSI values in the 5 GHz band contributes to the accuracy of localization. To understand the relative stability, we observed the variation in RSSI of both bands over an entire day both in a busy and non busy environment. The results are plotted as shown in Fig. 4. We observe that with limited human activity, the 5 GHz band displays almost no change over time. During busy hours the variation is still limited in the 5 GHz band as compared to that exhibited by the 2.4 GHz. This stability most likely takes precedence over the penetrating property thereby resulting in better accuracy in 5 GHz over 2.4 GHz.

From our experimentation we have observed that localization is more accurate in the 5 GHz band. However this cannot be generalized to all environments as we can see from our experimental results in the second environment with more human activity and larger influence of obstacles. This shows that there is scope for improving the overall accuracy of localization, by considering both the 2.4 GHz and 5 GHz band. In order to do that we analyze the performance of both these bands in specific locations in the environment. For each reference point in the map, we determine the difference between the localization accuracy using 2.4 GHz and 5 GHz. We call this map the "accuracy influence map".

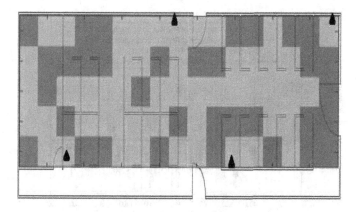

Fig. 2. Location where 2.4 GHz and 5 GHz works best in office environment

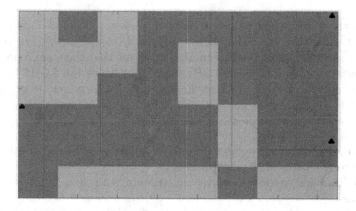

Fig. 3. Location where 2.4 GHz and 5 GHz works best in lab environment

For computing this, we take unknown point readings immediately after finger-printing is taken. We study which frequency band provides better accuracy in the different reference points of the radio map and a plot of the frequency which works better in the two environments is shown in Figs. 2 and 3. In the figures the regions whose accuracy is higher in 5 GHz is shaded in light gray, while those whose accuracy is higher in 2.4 GHz band is shaded in dark gray.

By observing the regions where 5 GHz gives better accuracy from the accuracy influence map, it is clear that it gives better results in regions that lie in the line of sight between majority of the routers and receiver. In the presence of obstacles 2.4 GHz gives better accuracy. This observation is in concordance with existing studies on the penetrating property of 2.4 GHz and 5 GHz. As per this property, 2.4 GHz has high penetrating capacity than 5 GHz thereby yield-ing better results in the areas which are located behind an obstacle. There are certain points which can be accurately localized using both 2.4 GHz and 5 GHz. Since the area where 5 GHz works better is larger than that of 2.4 GHz (it may

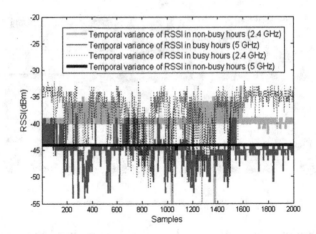

Fig. 4. Stability plots of both the bands in both busy and non busy hours

be due to the lower height of obstacles in this environment) in Environment1, our results show that the 5 GHz band provides better accuracy. On the other hand the accuracy plot of the Environment2, shows that there are more regions having better accuracy in the 2.4 GHz band, and hence the 5 GHz band fares poorly specially in presence of high human activity. However, it is also clear that the 2.4 GHz band does not fare well either. These accuracy influence maps demonstrate some key points about the two bands:

- the 5 GHz band provides better accuracy even if the influence of this band is limited to few parts of the environment.
- the 5 GHz band performs better in regions with less interference from obstacles.
- the 2.4 GHz band performs better in the central regions and regions surrounded by obstacles.

This indicates that by using intelligent policies, we can combine the two bands to obtain better localization accuracy. We will describe a few policies for intelligently using this information and evaluate the accuracy of the same.

5   Combining 2.4 GHz and 5 GHz Bands for Improving Localization Accuracy

In the previous section we analyzed the relative performance of 2.4 GHz and 5 GHz bands in indoor localization using fingerprinting. From the extensive studies we understand that there is scope to improve the accuracy of the localization by utilizing the advantages of 2.4 GHz and 5 GHz bands. In this section we propose two policies which use the information in the accuracy influence map for improving localization accuracy. We assume that we have fingerprint data for both 2.4 GHz and 5 GHz band, and accuracy influence map of these bands in the chosen environment. Since the user's location is unknown, we cannot directly use the accuracy influence map. We propose two greedy policies that use the accuracy influence map to improve the location accuracy of k-NN. The greedy strategy is in the neighborhood selection process and the strategies are as follows:

1. Greedy Band Selection: which chooses either 2.4 GHz or 5 GHz as the medium for the current localization, by selecting the best of either 2.4 GHz neighbors or 5 GHz neighbors.
2. Greedy Neighborhood Selection: which combines the advantages of both bands by selecting the best 4 neighbors both in the 2.4 GHz and 5 GHz fingerprint map.

For both the policies we first measure the RSSI vector of both the bands from the unknown point's device. Next we calculate the Euclidean distance (or Manhattan) from this unknown point's RSSI value to all the points in the reference radio map and find the distance values, separately in both bands. Let $P1 = P1_1, P1_2, P1_3, P1_4$ be the first four coordinates based on the sorted distance values in 2.4 GHz band and $P2 = P2_1, P2_2, P2_3, P2_4$ be the first four coordinates in 5 GHz band. By using these 8 neighbourhood points as the base, we explain the two policies as follows.

Greedy Band Selection. This strategy is a greedy strategy which chooses either the neighbors recommended by the 2.4 GHz map, or neighbors in the 5 GHz map. We take the neighborhood points in sets $P1$ and $P2$ and determine the number of points in each set falling in the corresponding accuracy influence map position. That is, $P1_i$ is a candidate point if it is a position where 2.4 GHz provides higher accuracy than 5 GHz in the accuracy influence map. $n1$ gives the number of candidate points in $P1$ and similarly $n2$ gives the corresponding number of candidate points in $P2$. If $n2 >= n1$, 5 GHz points are chosen as the k neighbors, else, 2.4 GHz points are chosen. In the case of a tie, preference is given to 5 GHz due to its observed accuracy.

Greedy Neighborhood Selection. In this method, instead of choosing a particular frequency band neighborhood for localization, we choose the neighbors from both bands depending on the information in the accuracy influence map. Similar to the previous strategy for each $P1_i$ and $P2_i$, we determine if it is a candidate point. Let the total number of candidate points in both $P1_i$ and $P2_i$ be n_c. We choose the final k neighbors to be provided to the k-NN algorithm based on the following:

- If $n_c = 4$, choose them as is.
- If $n_c > 4$, choose the 4 points closest to the unknown point based on distance between RSSI vectors.
- If $n_c < 4$, find the closest $4 - n_c$ points to the unknown point based on distance between RSSI vectors.

In addition to these two strategies we also evaluate the strategy suggested by [18] which provides all the points in both $P1$ and $P2$ as input to the k-NN

Table 5. Average accuracy of proposed algorithms in office environment

Greedy band	Greedy neighborhood	Multiband
1.61	2.56	2.48

Table 6. Average accuracy of proposed algorithms in lab environment

Greedy band	Greedy neighborhood	Multiband
3.25	2.16	5.50

algorithm. We refer to this strategy as "Multiband". To the best of our knowledge this is the only other suggested solution for combining both the bands for the purpose of localization. We evaluate all these strategies in both Environment1 and Environment2. All algorithms are tested for unknown points taken when human activity is high.

The results of these three approaches which combine 2.4 GHz and 5 GHz bands in different ways for localization are shown in Tables 5 and 6. The results demonstrate that in general all approaches perform better than using only one of the bands, hence reiterating the need to combine both bands.

On closer observation we see that in Environment1, the "Greedy Band Selection" policy works best. It outperforms the other two approaches. The standard deviation of this approach is also low i.e. 0.34. This is because about half the regions are influenced by 5 GHz, and there is a good clustering of each of these bands in the region. This provides an opportunity to select either 5 GHz neighbors or 2.4 GHz neighbors in their best positions. Hence we see a drastic improvement in accuracy. The performance of the second policy "Greedy Neighborhood Selection" is low, demonstrating that when the influence map shows equal distribution for both bands, it is better to select either of the two bands for localization based on the region, rather than combining them. This is also reiterated by the poor performance of the "Multiband" approach.

On the other hand in Environment2, the influence of 2.4 GHz is most predominant. Very few points are influenced by the 5 GHz band. As a result, in the "Greedy Band Selection" policy, the 2.4 GHz band is chosen most of the times, which is indicated by the poor result. The accuracy is close to 2.4 GHz accuracy; however as observed in previous section, the 5 GHz band provides better accuracy, so even a small contribution by this band, can help improve accuracy. Hence in this environment the "Greedy Neighborhood Selection" policy improves the accuracy the most. Even if the influence of 2.4 GHz is high, it has the option of choosing a 5 GHz band neighbor whenever possible, thereby improving accuracy.

From the above results we believe that the accuracy of indoor localization can be improved by using simple greedy approaches that intelligently combine both 2.4 GHz and 5 GHz bands. This is true in all the kinds of environments as we have evaluated the policies in environments with different kinds of obstacles under different circumstances, which leads to different behaviors of the signals.

By just performing an additional iteration after fingerprinting, we can generate an accuracy influence map which helps in efficiently using both the bands, consequently improving the accuracy.

6 Conclusion

In this paper we have analyzed the relative performance of the IEEE 802.11a (5 GHz) and IEEE 802.11b (2.4 GHz) bands in indoor localization. Specifically we have analyzed the performance of these bands, using fingerprinting where a k-NN based approach is used for the localization. Extensive experimentation comparing the two bands have been done in different environments with real-world conditions. Our experiments have demonstrated that the 5 GHz band performs better than 2.4 GHz in most situations, except in small busy environments. In general where environments are relatively larger, and influences of major obstacles like walls is limited, the 5 GHz band performs very well. We also proposed novel approaches that use a accuracy influence map to intelligently use the benefits of both these bands. Our experiments show that there is a large improvement in accuracy by either intelligently choosing the right band, or by combining them. The paper has demonstrated that these simple yet effective techniques using the IEEE 802.11n can improve indoor localization without need for extra infrastructure. Our initial studies have shown that the 5 GHz band maintains accuracy even if the fingerprinting map or accuracy influence map is not updated. Impact of freshness of fingerprinting map, and its updation is potential future work. Also there is scope for utilizing the advantages of both these bands in lateration techniques.

References

1. Eltaher, A., Kaiser, T.: A novel approach based on UWB beamforming for indoor positioning in non-line-of-sight environments. In: Proceedings of RadioTeCc 2005 (2005)
2. Liu, H., Darabi, H., Banerjee, P., Liu, J.: Survey of wireless indoor positioning techniques and systems. IEEE Trans. Syst. Man Cybern. Part C Appl. Rev. **37**(6), 1067–1080 (2007)
3. Del Mundo, L.B., Ansay, R.L.D., Festin, C.A.M., Ocampo, R.M.: A comparison of wireless fidelity (wi-fi) fingerprinting techniques. In: 2011 International Conference on ICT Convergence (ICTC), pp. 20–25, Sept 2011
4. Yin, J., Yang, Q., Ni, L.M.: Learning adaptive temporal radio maps for signal-strength-based location estimation. IEEE Trans. Mobile Comput. **7**(7), 869–883 (2008)
5. Abu-Sharkh, O.M.F., Al-hamad, A.M., Abdelrahim, T.M., Akour, M.H.: Dynamic multi-band allocation scheme for a stand-alone wireless access point. In: 2012 26th Biennial Symposium on Communications (QBSC), pp. 168–173, May 2012
6. Marina, M.K., Farshad, A., Li, J., Garcia, F.J.: A microscopic look at wifi fingerprinting for indoor mobile phone localization in diverse environments. In: 2013 International Conference on Indoor Positioning and Indoor Navigation (2013)

7. Bahl, P., Padmanabhan, V.N.: Radar: an in-building rf-based user location and tracking system. In: INFOCOM 2000, Nineteenth Annual Joint Conference of the IEEE Computer and Communications Societies, Proceedings, vol. 2, pp. 775–784. IEEE (2000)

8. Brunato, M., Battiti, R.: Statistical learning theory for location fingerprinting in wireless LANs. Comput. Netw. **47**(6), 825–845 (2005)

9. Farid, Z., Nordin, R., Ismail, M.: Recent advances in wireless indoor localization techniques and system. J. Comput. Netw. Commun. **2013**, 12 p. (2013)

10. Roxin, A., Gaber, J., Wack, M., Nait-Sidi-Moh, A.: Survey of wireless geolocation techniques. In: 2007 IEEE Globecom Workshops, pp. 1–9 (2007)

11. Daiya, V., Ebenezer, J., Murty, S.A.V.S., Raj, B.: Experimental analysis of rssi for distance and position estimation. In: 2011 International Conference on Recent Trends in Information Technology (ICRTIT), pp. 1093–1098, June 2011

12. Jiang, J.-A., Zheng, X.-Y., Chen, Y.F., Wang, C.-H., Chen, P.-T., Chuang, C.-L., Chen, C.-P.: A distributed rss-based localization using a dynamic circle expanding mechanism. IEEE Sensors J. **13**(10), 3754–3766 (2013)

13. Lui, G., Gallagher, T., Li, B., Dempster, A.G., Rizos, C.: Differences in rssi readings made by different wi-fi chipsets: A limitation of wlan localization. In: 2011 International Conference on Localization and GNSS (ICL-GNSS), pp. 53–57, June 2011

14. Khun-Jush, J., Schramm, P., Malmgren, G., Torsner, J.: Hiperlan2: Broadband wireless communications at 5 ghz. IEEE Commun. Mag. **40**(6), 130–136 (2002)

15. Sendra, S., Lloret, J., Turro, C., Aguiar, J.: IEEE 802.11a/b/g/n short-scale indoor wireless sensor placement. IJAHUC **15**(1/2/3), 68–82 (2014)

16. Aparicio, S., Perez, J., Bernardos, A.M., Casar, J.R.: A fusion method based on bluetooth and wlan technologies for indoor location. In: IEEE International Conference on Multisensor Fusion and Integration for Intelligent Systems, 2008, MFI 2008, pp. 487–491, Aug 2008

17. Papapostolou, A., Chaouchi, H.: Exploiting multi-modality and diversity for localization enhancement: Wifi & rfid usecase. In: 2009 IEEE 20th International Symposium on Personal, Indoor and Mobile Radio Communications, pp. 1903–1907, Sept 2009

18. Sertthin, C., Fujii, T., Nakagawa, M.: Multiband received signal strength fingerprint based location system. In: 2009 IEEE 20th International Symposium on Personal, Indoor and Mobile Radio Communications, pp. 1893–1897, Sept 2009

19. Kaemarungsi, K., Krishnamurthy, P.: Analysis of wlans received signal strength indication for indoor location fingerprinting. Pervasive Mobile Comput. **8**(2), 292–316 (2012). Special Issue: Wide-Scale Vehicular Sensor Networks and Mobile Sensing

20. Wi-fi location-based services 4.1 design guide. White Paper: Cisco Systems Inc (2008)

21. Bahl, P., Adya, A., Padhye, J., Walman, A.: Reconsidering wireless systems with multiple radios. ACM SIGCOMM Comput. Commun. Rev. **34**(5), 39–46 (2004)

22. Marques, N., Meneses, F., Moreira, A.: Combining similarity functions and majority rules for multi-building, multi-floor, wifi positioning. In: 2012 International Conference on Indoor Positioning and Indoor Navigation (IPIN), pp. 1–9, Nov 2012

Use of Time-Dependent Spatial Maps
of Communication Quality for Multi-robot
Path Planning

Gianni A. Di Caro, Eduardo Feo Flushing[✉], and Luca M. Gambardella

Dalle Molle Institute for Artificial Intelligence (IDSIA), Lugano, Switzerland
{gianni,eduardo,luca}@idsia.ch

Abstract. We consider the path planning problem of mobile networked
agents (e.g., robots) that have to travel towards assigned target locations.
Robots' path planners have to optimally balance potentially conflicting goals: keep the traveled distance within an assigned maximum value
while, at the same time, let the robot reliably and effectively communicate with other robots in the multi-robot network, and reduce the risk
of collisions. We propose a solution approach based on the integration of
two components: a link quality predictor based on supervised learning,
and a path optimizer, based on a mathematical programming formulation. The predictor is built offline and yields spatial predictions of the
expected communication quality of the wireless links in terms of packet
reception rate. Exploiting shared information about planned trajectories,
these spatial predictions are used online by the robots to build time-
dependent spatial maps of communication quality, to iteratively assess
the best path to follow considering both local and prospective links, and
to plan paths accordingly. To deal robustly with dynamic environments,
path planning is implemented as a multi-stage scheme using a receding
horizon strategy. The framework is evaluated in realistic simulation scenarios, showing the effectiveness of using the spatial predictor for the
effective online planning of network-aware trajectories.

Keywords: Network-aware path planning · Mobile ad hoc networks ·
Spatial link quality prediction

1 Introduction

When a team of mobile physical agents (e.g., robots and/or humans) perform a
joint mission, it is often the case that the agents need to concurrently address
multiple *communication* and *navigation* issues. In general, navigation and communication requirements both play a crucial role determining the success or the
failure of the mission and, as such, they should be both taken into account and
optimized when *planning the movements of the agents*. Hereafter, we commonly
refer to an agent as a mobile *robot*, to emphasize both its physical embedding
and its decisional autonomy.

© Springer-Verlag Berlin Heidelberg 2015
M. Garcia Pineda et al. (Eds.): ADHOC-NOW Workshops 2014, LNCS 8629, pp. 217–231, 2015.
DOI: 10.1007/978-3-662-46338-3_18

In this context, we consider the general scenario in which the task assigned to a mobile robot consists in traveling towards a given, known target destination e. The objective is reaching e as soon as possible (e.g., to perform there some additional location-specific tasks) while at the same time reliably exchanging data with the other robots in the wireless ad hoc network formed by the team. Based on available knowledge of the environment, the best path to reach e can be computed by the path planner on-board of the robot. If only the traveling distance would be taken into account, the computed path would be the *shortest path* to reach e from the current robot's location. However, also optimization regarding *networking* must be included during path calculation. Therefore, the problem becomes finding the path to e that provides the optimal balance between robot's traveled distance and the ability to effectively communicate along the path with the other robots in the ad hoc network. In order to select such a path, the core issue becomes how to evaluate the *communication quality* of each one of the possible feasible paths that the robot could follow to reach its destination.

To tackle this problem, we propose the use of *supervised machine learning* to construct *spatial maps of communication quality*. A map provides *predictions* of the quality of wireless links that would exist when the robot moves to certain locations in the space. Prediction values are derived based on the framework that we developed in our previous works [1–3], where the quality of a wireless link is evaluated in terms of the *expected packet reception ratio (PRR)*. The expected quality of a link is expressed as a function of its local network configuration, which is defined in compact way through a set of network features. As a result, we can endow the robot with the ability to answer the question: "What will be the quality of the wireless links with my neighbors when I move to that given point in space?". That is, it can build a spatial map that associates to each point in the space a value of expected link quality. In this way, the robot can predict the expected quality of communication when traveling along different paths, and select the path that offers the best balance between distance and quality of communication.

In practice, the spatial map is used to derive the expected *network reward* associated to points in space that could be reached by the robot in the near future. With network reward we mean the local provisioning of network capability in terms of both connectivity and bandwidth, expressed through the prediction of the PRR associated to each wireless link along the traversed path. However, in order to robustly predict the quality of a prospective link and the reward associated to the related points in space, the *mobility* of the other robots needs to be taken into account: the local network configuration near the link can change over time because of it. To address this issue, we let the robots *locally exchanging information about their planned paths*. In this way, explicitly taking into account the dynamics of the others, each robot can build a *time-dependent* spatial map of link quality and network rewards.

Given that each point in space provides a finite amount of network reward, the *total reward* associated to a full path is computed in an additive way, as the cumulative network reward that can be collected along the path. Different paths will have different values of cumulative rewards, which we use to perform *network-aware path planning*: given a maximum distance the robot can travel over, the planner computes the path that allows the robot to gather the maximum cumulative network reward while satisfying the bounded distance constraint. This is equivalent to solve a single objective constrained optimization problem, which we model through a *mixed integer, linear program* (MIP). The solution consists of a path, expressed as a finite sequence of waypoints, from the current location of the robot to its final destination point. Each robot in the team then plans its path based on the solution of its individual MIP, in a fully distributed way. Moreover, since when we speak in terms of multiple mobile robots, *collisions* among the robots are an issue that has to be addressed. To this end, we include in the MIP formulation a term to avoid them, which penalizes the crossing of trajectories. This is made possible by the availability of the information about the paths planned by the other robots.

In order to deal robustly with *dynamic environments* where all robots are potentially mobile, each robot calculates its best network-aware path as a *multi-stage* scheme using a *receding horizon* [4]. Online, while advancing towards its destination, the robot iteratively *replans* its path based on the newly gathered information about positions, traffic loads, and planned paths of the other robots, that allows to issue new and up-to-date time-dependent spatial predictions.

It is important to remark that without the spatial predictor of link quality, a network-aware path to the destination (i.e., over a relatively long distance horizon) could not be computed. Only purely greedy policies could be iteratively applied to control robot's trajectory in its local neighborhood. The main contribution of this work is precisely to show the advantage of *learning and using time-dependent spatial maps of communication quality for explicitly connecting path planning with network optimization in the same control model of a mobile robot*. Our settings are quite general and address a large number of potential real-world applications.

2   Related Work

A crucial requirement in multi-robot systems is that of maintaining or providing ad hoc communications [5]. Thus, the problem of planning and coordinating robot actions has to account for two potentially conflicting objectives. First, since application-related tasks have to be carried out at well defined locations in the environment (e.g., performing sensing tasks), a robot has to compute paths and navigate to the specified locations. However, in order to enable the formation of local network topologies that permit the required flows of information, the robot also needs to support creation, maintenance, and improvement of wireless links in order to enable data exchange. In practice, supporting wireless networking imposes constraints to the way robots can move throughout the environment. The challenges arising by the interplay between communication and mobility have been addressed in different domains such as search [6], task allocation and planning [7,8], surveillance [9], pursuit and evasion [10].

A common way to address the problem is through the dedicated use of a group of robots as *communication providers*, whose only objective is to enable communication [11–13]. In other works the robots simultaneously play the role of communication providers and task executors, and the provisioning of communication and task planning are usually considered as integrated issues (e.g., [14]). In this paper, we deal with the problem of planning (and continually replanning) the path of each single mobile robot moving from one location in the environment to another, respecting imposed limits on the maximum traveling distance, and aiming to support at the same time wireless communications. Since we are looking for the best balance between traveling for some extra distance and ensuring effective communications, the problem that we tackle is fundamentally different (and complementary) from those addressed in the previously mentioned applications. In fact, neither we want to remove resources from the main task to support communications, nor we want to blur planning and communications together. The most innovative aspect of our approach is however the way we deal with the problem, based on time-dependent spatial maps for predicting networking quality, and on the cooperative exchange of planned paths to support it.

Recently, some works started to highlight the importance of considering realistic communication models to control the trajectory of a mobile robot, such as probabilistic channel prediction [15], and online estimators of wireless link capacity [16]. However, so far these models have been limited to the spatial predictions of the single wireless link between the mobile robot and a stationary base station. Moreover, in these models the interference caused by the concurrent transmissions of nearby nodes is usually neglected. In this work, we use a machine learning approach for spatial predictions of link qualities, and we use the local exchange of planned paths to build time-dependent maps of spatial predictions, that take into account for the mobility of all nodes. In this way, the predictions we make available can relate to wireless links between any pair of nodes, and can account for the dynamic aspects of the scenarios, including the time-varying effects of interference. The fact that we use the iterative replanning of the paths over time provides additional robustness to the approach.

3   Spatial Link Quality Prediction and Network Rewards

The quality of a wireless link depends on the complex interplay of a number of factors related to *hardware, software, traffic generation, environmental* aspects. In previous works [1,2], we proposed a supervised learning framework to learn the mapping between some of these factors and the expected link quality. We consider the expected *packet reception ratio* (PRR) (i.e. $[0, 1]$ ratio between received and sent data packets) as a measure of *link quality metric*. In practice, the framework predicts the quality of a wireless link on the basis of its local network configuration. Based on the literature and our experience, we represent this configuration by a vector of features that are relatively easy to measure and that play a major role determining the quality of a link. We consider the following set of features: the *distance* between the two end-points of the link, and the number, relative positioning, and traffic characteristics of the *neighbor robots* (a robot is a neighbor of one of the end-points of the link if it lies within a distance less or

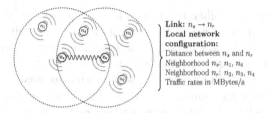

Fig. 1. Local network configuration of a link ($n_s \rightarrow n_r$). The neighborhood of the two end-points n_s, n_r of the link are described in terms of the relative positions of surrounding robots and are depicted as dotted circles.

equal to the transmission range of the network). Figure 1 illustrates the concept of local network configurations.

In this work we employ the aforementioned framework to issue time-dependent spatial predictions of link quality. From an operational point of view, we need to follow three steps in order to enable the robots to issue spatial predictions: (1) collect the set of data to learn from; (2) build a link quality prediction model using a supervised learning approach; and (3) deploy the learned model to robots.

In the initial step, we must collect a set of labeled link quality samples (i.e., pairs composed of a feature vector describing the local network configuration of a link and the corresponding PRR value). To this end, we may employ one of two different procedures: offline or, online data gathering. In the offline procedure [1], a group of mobile robots is deployed in the field, prior the operation of the network. Robots move in a controlled way, trying to maximize the number and the diversity of the observed local network topologies. At the same time, robots generate probing messages at variable rates, and measure the reception rate of the probing packets together with the values of the corresponding features describing the local network configuration of all links to its neighbors. On the other hand, in online data gathering [2], all nodes passively monitor incoming and outgoing network traffic, and exchange minimal amount of information that is required to assemble feature vectors and compute their corresponding PRR. Over time, each node incrementally records a set of link quality samples. In this work we consider the offline procedure, assuming that an initial data gathering phase is executed before starting the system, or that the samples are already available from past operations. Note that the samples can be collected using any number of devices, at any place and time. Therefore, in practice, the same set of samples can be used in different situations and with a different number and type of robots, as long as the hardware/software parameters of the network interfaces remain the same (i.e., PHY-MAC protocols, transmission range, bandwidth).

After collected, the samples are used to learn a link quality model in the form of a regression mapping from the space of the network features to the PRR values. The effectiveness of the selected features and of the learning process has been validated in extensive experiments in simulation [1], sensor networks [17] and mobile robots [3] in various open space and cluttered environments, showing excellent accuracy and ability to automatically capture the effects of complex radio propagation phenomena in the environment. Once trained, the model is installed on each robot and can be used to issue predictions about the expected PRR of a link estimating its related local network configuration. By exploiting its generalization capabilities, the regressor is able to predict the PRR for a wide range of input configurations, including previously unobserved ones. This condition is particularly useful to build spatial prediction maps.

We use the predictor to issue time-dependent spatial predictions of the quality of *prospective* links, that will possibly materialize when a robot reaches certain location in space at certain time. A robot uses the learned model to answer the question: "What will be the quality of wireless links with my neighbors, when I pass through point (x, y) at a certain time t?". At first, in order to answer this question, the robot needs to estimate the positioning, at time t, of other robots whose transmissions might affect the quality of wireless links at point (x, y) (i.e., the local network configuration). This can be accomplished if the information about current positions, data rates, and individual planned trajectories of nearby robots is available. Following a fully distributed approach, we let the robots *locally exchange* this information with each other. Periodically each robot publishes its status, including the above information, by local broadcast. The information is then locally propagated through a simple controlled flooding mechanism, which does not involve a significant overhead. In this work we consider 2-hop neighbor information, and updates each 1 s. As a result, each robot is aware of the planned actions of a subset of robots located in its vicinity, and able to estimate their positions at a future time t. As a second step, the robot calculates the feature vector for each grid point (x, y) of the environment (see Sect. 4) and at the estimated location of each robot, taking into account an upper bound on the transmission range of the network tx_r. Finally, using the feature vectors, the robot uses the predictor to compute the quality of all (incoming and outgoing) links at (x, y).

Figure 2 shows an example of this procedure. A robot a (red square) wants to compute the network reward at point (x, y) (green circle) at a future time t. The robot uses the current information gathered from its neighbors to estimate the future positions of other robots (triangles) at time t, as shown in Fig. 2a. Using the estimated positions, a is able to construct the network topology that will be formed at time t and to identify all wireless links that form if it positions itself at (x, y) at time t. In Fig. 2b, eight links (four incoming and four outgoing) are established at (x, y). Finally, the local network configuration of each link is computed, and given as input to the predictor, which in turn provides the quality (in terms of PRR) of each of these links.

The network reward R_i^t at point i is a value that represents the estimated *attractiveness* of that point in terms of communication. It is derived as a function of the number and quality of prospective links at that point. In practice, the definition of R_i^t should be related to the communication performance goals of a specific scenario. In this work, we seek for a balance between the total amount of data gathered along a path, and the diversity of the data. In other words, we aim at collecting a large amount of information, for a large number of sources.

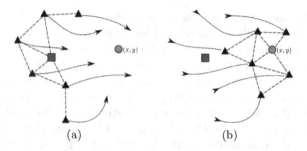

(a)   (b)

Fig. 2. Computing the network reward at a point (green circle) considering the trajectories of surrounding robots. In the left, a robot (red square) gets information about the planned trajectories of its neighbors (triangles). Using these trajectories, it estimates the future network topology, shown in the right (Color figure online).

To this end, we define the network reward R_i^t as follows:

$$R_i^t = \alpha LQ_i^t + (1 - \alpha)Conn_i^t(l), \tag{1}$$

where LQ_i^t denotes a link quality component (the average of predicted PRR values of all links at point i at given time t) and $Conn_i^t$ denotes the connectivity component, related to the number l of links at point i. Given a parameter max_l that controls the value of each additional connection, $Conn_i^t(l)$ equals to 1 if $l > max_l$. On the contrary, $Conn_i^t(l) = l/max_l$. Both LQ_i^t and $Conn_i^t$ are in $[0, 1]$. The parameter $\alpha \in [0, 1]$ balances both components. The setting of these parameters is a strategic choice that depends upon specific application goals. In the experiments, we set $max_l = 6$ and $\alpha = 0.5$.

4   Network-Aware Path Planning Model

Given a multi-robot scenario where each robot is equipped with a wireless network interface, and moves with constant speed, we consider the problem of defining the path for a specific robot a that has to travel from its current location s to an ending location e. The objective is to find a path, expressed through a discrete sequence of waypoints, that maximizes a selected measure of *network performance*, while keeping the traveled distance $\mathcal{D}$ within an assigned maximum limit value $\mathcal{D}_{max}$, and reduces the risk of *collisions*. $\mathcal{D}_{max}$ is strategically defined as an extra percentage of the shortest distance $\mathcal{D}_{min}$ for traveling from s to e (e.g., $\mathcal{D}_{max} = 1.2\mathcal{D}_{min}$). Let $\mathcal{P}$ be the set of feasible paths from s to e with total distance $\leq \mathcal{D}_{max}$. A general formulation of the problem is as follows:

$$\max_{p \in \mathcal{P}} Q(p)$$

where $Q : \mathcal{P} \mapsto \mathbb{R}$ is a metric that quantifies the quality of a path.

In practice, the value of $Q(p)$ depends upon the planned routes of the nearby robots. As the robots move, the network topology changes, which in turn affects the quality of communication links between the robot a and its neighbors. In addition, the risk of collisions between a and nearby robots is also changing if the trajectories of the robots intersect the path p of robot a. Given the time dependency of Q, and for the reason of reducing computational complexity, we assume a time discretization into uniform steps $t_1, t_2, \ldots$, with $t_{i+1} - t_i = \Delta_t$. For a given time step t, we define $\bar{Q}_t(p)$ as the instant quality measure of path p at time t. The value of $\bar{Q}_t(p)$ depends on the estimated positions of surrounding robots at time t, and is related to the quality of the network topology at time t, and the risk of collisions during the interval $[t, t+1]$. Under the given time discretization, we define $Q(p)$ as the cumulative score over time:

$$Q(p) = \sum_t \bar{Q}_t(p) \tag{2}$$

In this work, we assume that motion happens on a discretized 2D plane, and feasible paths between are defined in terms of a numerable set $\mathcal{N}$ of candidate *waypoints* placed on a 2D grid. Furthermore, the movement of the robot between waypoints is restricted to a numerable set of *feasible transitions* $\mathcal{E} \subseteq \mathcal{N} \times \mathcal{N}$ (directed arcs). The union of $\mathcal{N}$ and $\mathcal{E}$ constitute a directed *traversability graph* $G = (\mathcal{N}, \mathcal{E})$ that defines all the possible movements of the robot that can be considered when computing a solution path.

In order to build the traversability graph we assume that a detailed *map* of the environment is available to the planner. The waypoints are defined on the basis of the locations accessible to the robot, while the transitions between waypoints depend both on the environment, and on the geometry and dynamics of the robot (e.g., if it is holonomic or not). For convenience, and in order to match the time and space discretization, we only consider transitions between points $i, j \in \mathcal{N}$ such that the traveling time from i to j is equal to Δ_t, with the exception of transitions ending at the destination e. Furthermore, and for reasons of reducing the problem complexity, we only consider transitions that describe movements towards the destination.

Taking into account all the previous considerations, we propose the following procedure to construct the traversability graph. First, we define a uniform 2D grid, starting at s, with cells of size $v_a \Delta_t \times v_a \Delta_t$, where v_a is the speed of the robot, and we let $\mathcal{N}$ be the set of corners of the grid cells. Finally, we let $\mathcal{E}$ be the set of arcs (i, j) such that the distance between i and j is equal to $v_a \Delta_t$, or $j = e$. Each arc has an associated travel distance d_{ij}, with $d_{ij} = v_a \Delta_t$ for all $j \neq e$, and $d_{ij} = \|i - j\|$ if $j = e$. From the previous procedure, two important properties result. First, for any path $p = < p_1, p_2, \ldots, p_n >$, the robot passes through waypoint p_i, $1 \leq i \leq n-1$, at precisely time step t_i which, as shown later, simplifies the calculation of $\bar{Q}_{t_i}$. Secondly, any waypoint $i \in \mathcal{N} \setminus \{e\}$ can appear at most once, and always at the same specific position in any feasible path. Figure 3 shows an example of a traversability graph with possible transitions.

The use of paths expressed as sequences of waypoints allows to accommodate for the *uncertainties* intrinsic to real-world scenarios, which can hardly justify the use of paths specified as continuous trajectories. In fact, the actuation of the path would necessarily deviate in practice from what has been precisely

Fig. 3. Traversability graph from a grid consisting of 16 cells. The transitions towards the destination are those between adjacent grid points moving north or east, as indicated by the arrows. In the middle, a feasible path that consists of 9 waypoints is shown. In the right, a feasible path consisting of 5 waypoints.

calculated. Moreover, the discretization allows to reduce the *complexity* of the problem, and enables the use of different levels of granularity to address the computational constraints on the online, repeated path calculations.

We associate to each arc $(i, j) \in \mathcal{E}$ a collision penalty C_{ij}^t that indicates the risk of collisions between a and other robots when a moves from i to j during interval $[t, t+1]$. The value of C_{ij}^t depends on the information about the planned trajectories of robots in the vicinity. Similar to the network rewards, we consider the information that each robot has about the planned trajectories of other, nearby robots. Let (i, j) be an arc in $\mathcal{E}$ and τ_k be the trajectory robot k (among the robots whose trajectories are known). In order to calculate C_{ij}^t, we compute the closest distance d_{ka}^t between k and a during the interval $[t, t+1]$. We define a collision risk function Ω that, for a given distance, assigns a value in $[0, 1]$ representing the risk of collisions, on the basis of a minimum safety distance threshold d_{safe}. The value of $\Omega(d)$ equals to 0 if $d > d_{safe}$. On the contrary, it takes a value equals to $1 - d/d_{safe}$.

Finally, C_{ij}^t is defined as $\max_k \Omega(d_{ka}^t)$ among all robots k for which we have information about their trajectories. By default, the penalty associated to arcs ending at the destination C_{ie}^t is equal to zero, for all t. The rationale behind this lies in the observation that these arcs typically describe movements of long duration, covering larger distances. Let path $p =< p_1, p_2, \ldots, p_n >$, the value of $\bar{Q}_t$ is defined as the network reward of p_t minus the penalty of crossing arc $p_t \to p_{t+1}$, that is, $\bar{Q}_t(p) = R_{p_t}^t - C_{p_t p_{t+1}}^t$.

Given the properties of the traversability graph, each point $i \in \mathcal{N} \setminus \{e\}$ may appear at only one specific time step t_i, and therefore each arc $(i, j) \in \mathcal{E}$ can be crossed only at time t_i. As a consequence, we only need to consider rewards and costs for those specific times. Henceforth, we denote as R_i the network reward at point i and time step t_i. Similarly for C_{ij}.

4.1 Receding Horizon

We need to take into account the fact that predictions for far away points are subject to large errors, due to the mobility of the robots, and the lack of complete information. In order to deal with these challenges, we propose an online *receding horizon* strategy: a procedure in which the planning process is *periodically iterated*. Each *stage* corresponds to a different starting location, from which the robot calculates its path over a limited horizon of T steps. This provides a sequence of waypoints describing a trajectory that passes through at most T waypoints before reaching the destination. The robot then executes the path for a certain time, and recalculates a new path starting from the new current position. The process is iterated over time until the destination is reached.

In order to respect the initial distance limit D_{max}, and at the same time not to consume all the extra distance at the early stages, we adopt the following procedure. First, we derive D_{min} by the shortest path calculation. Based on the defined frequency of replanning and on the known constant velocity of the robot, we compute the maximum number R of replanning stages needed to get to the destination. The extra distance $D_{extra} = D_{max} - D_{min}$ is divided uniformly among the R stages: at each stage only an extra D_{extra}/R is allowed with respect to the shortest path. All extra distance which is not used, becomes available for the subsequent stages. This procedure guarantees the constraints on D_{max} and provides a fair use of the extra distance budget over the stages.

4.2   MIP Formulation

We formulate the network-aware path planning as an *Orienteering Problem* (OP) [18]. The goal is to determine a path in the graph such that the total score collected along the path is maximized. In our case, vertices correspond to waypoints and the score is the networking reward obtained for passing by a waypoint. We also consider costs associated to arcs that we include as penalties in the objective function. The path length, in terms of distance, is limited to D, and in terms of number of waypoints, excluding the destination, limited to T.

The MIP formulation for the path planning problem is as follows:

$$
\max_{\mathbf{x},\mathbf{y}} \sum_{i \in \mathcal{N}} R_i y_i - \beta \sum_{(i,j) \in \mathcal{E}} C_{ij} x_{ij} - \epsilon \sum_{(i,j) \in \mathcal{E}} d_{ij}
$$

$$
\begin{aligned}
s.t. \quad & y_s = y_e = x_{es} = 1 \\
& \sum_{(i,j) \in \mathcal{E}} x_{ij} = \sum_{(j,i) \in \mathcal{E}} x_{ji} = y_i && i \in \mathcal{N} \\
& t_i - t_j + 1 \le (|\mathcal{N}| - 1)(1 - x_{ij}) && (i,j) \in \mathcal{E},\ i,j \notin \{s,e\} \\
& 1 \le t_i \le T && i \in \mathcal{N} \\
& \sum_{(i,j) \in \mathcal{E}} x_{ij} d_{ij} \le D \\
& x_{ij},\ y_i \in \{0,1\} && i,j \in \mathcal{N}
\end{aligned}
$$

The MIP *decision variables* are the following. x_{ij}: binary, equals 1 if arc $(i,j) \in \mathcal{E}$ belongs to the path; y_i: binary, equals 1 if way-point $i \in \mathcal{N}$ belongs to the path.

The objective function consists of three components. The first, maximizes the reward obtained from waypoints in a path. Next, the collision aspects are included as a penalty, weighted by a parameter β. The last component is used to give preference for shorter paths, in case there are many optimal solutions, with ϵ being a small constant. The first set of constraints ensure that paths start and end at the selected initial and ending points. Path continuity is guaranteed by the following set of constraints. The next constraints eliminate subtours [18], set the distance bound, and the binary requirements on the variables, respectively.

Figure 4 shows an example of the resulting trajectories. For illustration purposes, in the example all robots except one remain stationary, and only one robot performs the path planning. The robot must visit three targets (triangles) in sequence, and the path planning is performed between each pair of consecutive targets. In the left, a maximum distance equal to $1.25D_{min}$ is used, while in the right, the robot is allowed to travel at most $1.5D_{min}$. We can appreciate how the resulting trajectories guide the robot through regions with higher reward values, and how the extra distance budget is distributed along the trajectory.

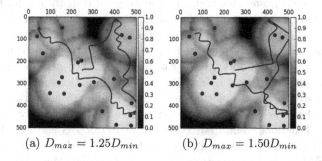

(a) $D_{max} = 1.25D_{min}$ (b) $D_{max} = 1.50D_{min}$

Fig. 4. Example of computed trajectories for one mobile robot inside a $500\,\mathrm{m} \times 500\,\mathrm{m}$ area. In the background, the network reward is displayed as a 2D function. The colorbar in the right indicates the reward values. The robot visits 3 target destinations depicted as triangles. Blue circles depict static robots.

5   Experimental Evaluation

We consider a multi-robot scenario in which each robot needs to visit certain locations on a map in a specific sequence. However, the robots are not required to go directly to the destination: they can make some extra distance in order to improve the local connectivity of the network. On its way, a robot passes other robots that form network topologies and send some data. The goal is to adjust the trajectory of each robot in a way that maximizes the quality of local communication with other robots met on the path. This scenario is designed as representative of many multi-robot tasks where robots need to perform various actions in different locations, and at the same time they need to communicate with others in order to perform the coordination and cooperation. Our aim is to show that, while performing its task, a robot can slightly adjust its trajectory and significantly improve the quality of local wireless connections.

We deploy 30 robots inside an area of 1000×1000 m^2. We assume that robots move at constant speed of $5^m/_s$, in open space, resembling a scenario that involves the use of aerial robots (e.g., quadrotors). To each robot, we assign a list of target destinations, randomly placed inside the area, with the distance between consecutive targets lying between 500 m–1000 m in order to perform a meaningful evaluation of the path planning. To plan the trajectories, we use a time step $\Delta_t = 4$ s, which in turn results into cells of size 20×20 m^2 for the computation of the traversability graph. The receding horizon strategy considers a horizon of length $T = 15$, meaning that robots replan their trajectories every minute. Given the spatial distribution of targets, all robots need to perform several replanning iterations while traveling between targets.

We use the NS-3 network simulator [19] with the following configuration. We simulate 802.11a Wi-Fi networks, with a transmission rate of 6 Mbps. We use a log-distance propagation loss model with default parameters (path loss exponent set to 3.0), corresponding to a transmission range of roughly 120 m. Each robot generates a constant bit rate traffic in a form of broadcast transmissions, at a rate equal to 1.0 Mbps. A robot also broadcasts *HELLO* messages, once every second, including information about its planned trajectory. These messages are re-sent by its direct neighbors, in order to reach the 2-hop neighborhood. Packet size is set to 1000 bytes, and by default, we run simulations for 20 min.

As a first step in the evaluation, we consider different network reward functions and analyze their effect on the communication performance. We consider three reward functions. The first, called *COMPOSED* is the function defined in Sect. 3. The other two, termed *AVG. LQ.* and *SUM LQ.* represent a reward based respectively on the average and the sum of the links' qualities (i.e., PRR). For each simulation run, at each node we calculate the total number of packets received, and the number of packets received from each other single node during the simulation. We consider two evaluation metrics. The first is related to the *fairness* of communication, measured as the variance of the number of packets received from each node. A large value of this metric implies that the data gathered by the robot is unbalanced, in the sense that it received a lot of data from a fewer sources. The second metric is related to the total amount of data gathered, measured as the total number of packets received during the simulation. The metrics are calculated for each robot in the scenario. Figure 5a and b shows the distribution of the values for both metrics among all robots. We can appreciate that both *COMPOSED* and *SUM. LQ.* reward functions induce a better fairness in comparison with the *AVG. LQ.* The function *COMPOSED* provides a larger amount of gathered data, and in overall provides a better balance between both metric, in comparison with the other two reward functions.

Next, we consider different limits on the distance allowed for traveling between destinations. Limits are expressed as a *maximum distance factor*, multiplied by the shortest path distance D_{min}. We considered factors $\{1.0, 1.25, 1.5\}$, and the total amount of packets received by each node as performance metric. Figure 5c shows the distribution of the metric values among all nodes for each scenario. Results show that increasing the distance factor allows the nodes to consistently gather more data during the mission.

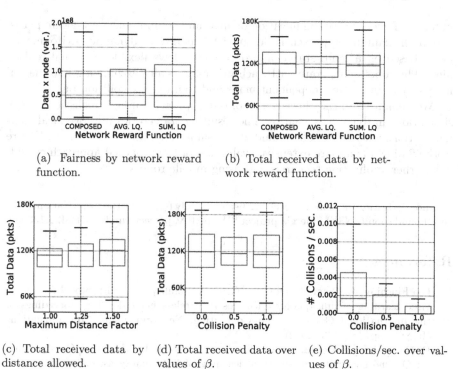

(a) Fairness by network reward function.

(b) Total received data by network reward function.

(c) Total received data by distance allowed.

(d) Total received data over values of β.

(e) Collisions/sec. over values of β.

Fig. 5. Simulation results for different performance metrics.

At last, we evaluate the collision avoidance aspects of the model. To this end, we consider three different values for the β parameter. Specifically, $\beta \in \{0, 0.5, 1.0\}$. We examine two metrics. First, the total amount of data gathered using each value of β. Secondly, we count the number of *potential collisions*, that is, a situation in which the robot is below the safety distance d_{safety} from any other robot. This number is divided by the total simulation time in order to obtain an estimated number of collisions over time. Figure 5d and e shows the distribution of values of both metrics among all nodes for each scenario. For larger values of β, we appreciate a significant reduction in the number of collisions, with a slight decrease of the total amount of data gathered. This shows the implicit trade-off between both, communication and collision prevention objectives, and illustrates the use of β as a means of managing the trade-off between these conflicting goals.

6 Conclusions and Future Work

We illustrate how models based on machine learning techniques can be used for spatial predictions of the quality of wireless links in robotic networks. More specifically, we propose a trajectory optimizer based on mathematical programming that exploits time-dependent spatial predictions to identify good regions in terms of expected communication quality, and compute network-aware paths. Our path planning approach is designed to operate in dynamic environments, being based on a multi-staged approach that allows flexible continual replanning. The robots cooperate with each other by locally exchanging information necessary to build time-dependent maps and to minimize collision risks.

We demonstrate the effectiveness of our approach through realistic network simulations in dynamic scenarios. The resulting trajectories improve local communications with other robots, and at the same time, prevent collisions. Future work will include a more extensive evaluation of the proposed approach considering other application scenarios and using mobile robots.

Acknowledgments. This research has been partially funded by the Swiss National Science Foundation (SNSF) Sinergia project SWARMIX, project number CRSI22.133059.

References

1. Feo Flushing, E., Nagi, J., Di Caro, G.A.: A mobility-assisted protocol for supervised learning of link quality estimates in wireless networks. In: Proceedings of the International Conference on Computing, Networking and Communications (ICNC), pp. 137–143 (2012)
2. Di Caro, G.A., Kudelski, M., Feo Flushing, E., Nagi, J., Ahmed, I., Gambardella, L.: On-line supervised incremental learning of link quality estimates in wireless networks. In: Proceedings of the 12th IEEE/IFIP Annual Mediterranean Ad Hoc Networking Workshop (Med-Hoc-Net), pp. 69–76 (2013)
3. Kudelski, M., Gambardella, L., Di Caro, G.A.: A mobility-controlled link quality learning protocol for multi-robot coordination tasks. In: Proceedings of the IEEE ICRA, pp. 5024–5031 (2014)
4. Kuwata, Y., How, J.P.: Cooperative distributed robust trajectory optimization using receding horizon MILP. IEEE Trans. Control Syst. Technol. **19**(2), 423–431 (2011)
5. Hsieh, M.A., Cowley, A., Kumar, V., Taylor, C.J.: Maintaining network connectivity and performance in robot teams. J. Field Robot. **25**(1–2), 111–131 (2008)
6. Rooker, M.N., Birk, A.: Multi-robot exploration under the constraints of wireless networking. Control Eng. Pract. **15**(4), 435–445 (2007)
7. Hollinger, G.A., Singh, S.: Multirobot coordination with periodic connectivity: theory and experiments. IEEE Trans. Robot. **28**(4), 967–973 (2012)
8. Tardioli, D., Mosteo, A.R., Riazuelo, L., Villarroel, J.L., Montano, L.: Enforcing network connectivity in robot team missions. Int. J. Robot. Res. **29**(4), 460–480 (2010)
9. Ghaffarkhah, A., Mostofi, Y.: Path planning for networked robotic surveillance. IEEE Trans. Signal Process. **60**(7), 3560–3575 (2012)
10. Thunberg, J., Ögren, P.: A mixed integer linear programming approach to pursuit evasion problems with optional connectivity constraints. Auton. Rob. **31**(4), 333–343 (2011)
11. Gil, S., Feldman, D., Rus, D.: Communication coverage for independently moving robots. In: Proceedings of IEEE/RSJ IROS, pp. 4865–4872 (2012)
12. Yan, Y., Mostofi, Y.: Robotic router formation in realistic communication environments. IEEE Trans. Robot. **28**(4), 810–827 (2012)

13. Grøtli, E.I., Johansen, T.A.: Path planning for UAVs under communication constraints using SPLAT! and MILP. J. Intel. Robot. Syst. **65**(1–4), 265–282 (2011)
14. Fink, J., Ribeiro, A., Kumar, V.: Robust control of mobility and communications in autonomous robot teams. IEEE Access **1**, 290–309 (2013)
15. Malmirchegini, M., Mostofi, Y.: On the spatial predictability of communication channels. IEEE Trans. Wireless Commun. **11**(3), 964–978 (2012)
16. Lindhe, M., Johansson, K.H.: Adaptive exploitation of multipath fading for mobile sensors. In: Proceedings of the IEEE ICRA, pp. 1934–1939 (2010)
17. Feo Flushing, E., Kudelski, M., Nagi, J., Gambardella, L., Di Caro, G.A.: Poster abstract: link quality estimation - a case study for on-line supervised learning in wireless sensor networks. In: Langendoen, K., Hu, W., Ferrari, F., Zimmerling, M., Mottola, L. (eds.) Real-World Wireless Sensor Networks. LNEE, vol. 281, pp. 97–101. Springer, Heidelberg (2014)
18. Vansteenwegen, P., Souffriau, W., Van Oudheusden, D.: The orienteering problem: a survey. Eur. J. Oper. Res. **209**(1), 1–10 (2011)
19. NS-3. Discrete-event network simulator for Internet systems (2013). http://www.nsnam.org

Responsibility Area Based Task Allocation Method for Homogeneous Multi Robot Systems

Egons Lavendelis[(✉)]

Department of System Theory and Design, Riga Technical University,
Latvia, Riga
egons.lavendelis@rtu.lv

Abstract. The paper presents task decomposition and allocation method for multi-robot systems for area coverage tasks. The method is based on the notion of responsibility area which is the part of the environment that is considered to be atomic task which is allocated to a single robot. The responsibility areas are defined based on the equality of the needed amount of work for their processing. The amount of work is calculated based on the particular area and obstacles in it. The task allocation is done in the way that the most suitable responsibility areas are sequentially added to each robot. The main criterion for the task allocation is the distance from the responsibility area to the particular robot. Still the indexing mechanism is introduced to make the robots to process the environment region by region without leaving unprocessed responsibility areas. The method is implemented and tested in the multi-robot system for vacuum cleaning of large areas that cannot be cleaned by a single vacuum cleaning robot.

Keywords: Task allocation · Multi-robot systems · Area coverage tasks · Responsibility area

1 Introduction

Currently various autonomous robots for different purposes exist, for example, autonomous vacuum cleaning robots (Palacin et al. 2004), (Young et al. 2009), agricultural robots for different tasks in precise agriculture (Satish Kumar and Sudeep 2007), (Yamaguchi et al. 2010), various painting robots (Ashlock et al. 2003) etc. At the moment these robots implement sufficient algorithms for autonomous execution of their missions. For example, a vacuum cleaning robot is intelligent enough to autonomously and decently clean particular area. Still, all current solutions have significant physical limitations causing them to be unable to effectively do more complex and larger tasks. Example of such tasks is a large area that cannot be cleaned by a single vacuum cleaning robot because of time and resource limitations.

One of the possible solutions is to use multiple robots for the particular task. During the last years intensive research in the area of such systems has been done (Doroftei et al. 2012), (Lavendelis et al. 2012), (Liekna et al. 2013), (Mancini et al. 2011). Still, one can conclude that the current state of the art in deployed solutions is more focused on single robot systems while the multi-robot applications are still in close-to-market

© Springer-Verlag Berlin Heidelberg 2015
M. Garcia Pineda et al. (Eds.): ADHOC-NOW Workshops 2014, LNCS 8629, pp. 232–245, 2015.
DOI: 10.1007/978-3-662-46338-3_19

research or prototyping stages. One of the reasons of this fact is lack of reliable frameworks of systems development and maintenance. This issue is more related to robot software part because the current achievements in mechanics and electronics provide a wide variety of possible technical approaches and solutions for particular problems of the service robotics domain. One part of the multi-robot system development framework is a set of methods to implement various mechanisms that complement approaches from single robot systems to enable distributed problem solving. For example, some work has been done in the area of simultaneous location and mapping in multi-robot systems (Niktenko et al. 2013), (Andersone 2012). Another part is task decomposition and allocation to robots. To maintain the autonomy of the robotic systems it is desired that the task allocation mechanism is distributed in the sense that there is no centralized planning and task allocation element. Unfortunately, the task decomposition and allocation mechanisms depend on the problem domain, type of the task and capabilities of different robots. The aim of the paper is analyse the state of the art in task allocation mechanism research and propose new method for area coverage task decomposition and allocation in the homogeneous multi-robot system.

The remainder of the paper is organized as follows. Section 2 outlines the developed multi robot system and defines the problem of task allocation. Section 3 gives an overview of related work in the area of task allocation mechanisms in multi-robot systems. Section 4 proposes the method of task allocation based on responsibility areas. Section 5 concludes the paper.

2 Multi-robot System

The current research done at Riga Technical University aims to develop a hardware and software platform that can be added to the existing and commercially produced single robotic systems enabling them to work together for larger tasks accomplishment. The first prototype is built for vacuum cleaning robots iRobot Roomba (IRobot 2014). Depending on the particular modification of the robot each Roomba is capable to clean 60–100 square meters. At the moment there are no solutions for much larger areas like warehouses, supermarkets, etc., because existing robots lack abilities to communicate, share knowledge and as a consequence also work together. So the objective of the development is to create a multi-robot management software that would enable the existing vacuum cleaning robots to work together to clean larger areas than each of the robots is capable to do alone.

From the mathematical viewpoint the domain of vacuum cleaning belongs to the task of area coverage (Choset 2001) and all the vacuum cleaning robots have the same capabilities, i.e., the multi-robot system is homogeneous, thus the remainder of the paper focuses on the task allocation methods in homogeneous multi-robot systems for area coverage tasks.

The management software is deployed on the server and on the hardware platform that is put on top of the Roomba robots. The aim is that the platform is not too expensive and also energy efficient. Thus the software run at the robot side must not be computational demanding and the most complex tasks should be processed at the server side.

The management software implements a set of methods enabling the robots to work together. The communication method is implemented to provide communications between server and robots. Task decomposition and allocation method is created to divide the whole task into subtasks and allocate the latter to particular robots. Planning method is developed for planning the execution of tasks and driving trajectories of robots in the environment to avoid collisions among robots and between robots and obstacles. Finally mapping and localisation methods based on data from multiple robots have been implemented (Nikitenko et al. 2013), (Andersone et al. 2013). Due to the scope of the paper only the task decomposition and allocation method is analysed.

The task decomposition and allocation mechanism has the following information available:

- The maximum area s that can be processed (cleaned in case of vacuum cleaning system) by a single robot;
- The map of the environment:
  - The size of the environment x*y where x is horizontal distance from the farthest left point to the farthest right point and y is the distance between top and bottom points of the map (see Fig. 1).
  - The occupancy grid of the environment showing which cells contain obstacles and which ones are free. In some implementations the occupancy grid contains uncertain knowledge in the form of probabilities for the cells to contain obstacles. In this case the method will use a threshold of the probability to determine if the cell is considered as occupied. Localization and mapping methods are used to build the occupancy grid (Andersone et al. 2013), (Nikitenko et al. 2013).

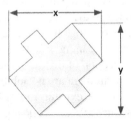

Fig. 1. The measurements of the environment used in the task decomposition and allocation

The contribution of the paper is to provide methods that take the whole task as an input and give the tasks allocated to robots as an output. To solve the task two methods have been developed. The first method is used to divide the area into subareas (from now on called responsibility areas) and allocate them to the robots available in the system. The environment should be divided into areas that need as equal as possible amount of work to process them. In case of environment without any obstacles it means that the areas will be of equal size, but in case of obstacles occupying some part of the environment the division must be done taking into consideration the obstacles, because massive obstacles significantly change the amount of work that is needed to process the corresponding area. The second method assigns the responsibility areas defined by the first method to the robots. The assignment is done to minimize the total time and resources needed to do all the subtasks.

3 Related Work

The task decomposition usually is domain specific. Thus this section will concentrate on the task allocation mechanisms and analyse the existing mechanisms that can be used to allocate tasks in the multi-robot systems.

Different task allocation mechanisms for distributed systems exist. Some of these mechanisms have origins in multi-agent system research while others have been developed especially for multi-robot systems. Historically the first task allocation mechanism is Contract-Net protocol (FIPA 2002) whose initial idea was developed in 1980ies. The protocol is well known and widely used to allocate single task in the multi-agent domain. Other options coming from the multi-agent research are different auction based protocols, like English, Dutch and Vickrey auction protocols (Wooldridge 2009). These protocols are good option if there is a single task or tasks can be allocated sequentially. Unfortunately, as proved in (Liekna et al. 2012) sequential allocation allows choosing the most suitable performer only for one particular task, while such a method is not optimal if all the set of tasks is considered, because the first task may be allocated to some robot that should do other tasks in optimal allocation. To summarize, the task allocation mechanisms that are allocating tasks sequentially based only on the information about the current task in the area coverage tasks may lead to extra distances travelled.

The multi-robot system research has resulted in different specific task allocation mechanisms. The Murdoch mechanism proposed by B. P. Gerkey and M.J. Mataric (Gerkey 2003), (Gerkey and Mataric 2003) is based on the idea that messages are addressed by contents not by recipients. Messages are sent not to the list of robots, but to the robots that are interested in particular topic. Despite of completely novel addressing mechanism in task allocation protocol domain, the core mechanism of the protocol is the same as for the sequential execution of the Contract Net protocol. The executor of each task is found only on the basis of the information about the task and about the available executors (robots), but not on the basis of all other tasks. Thus Murdoch can provide more efficient message addressing and passing, but does not improve the resulting allocation.

Traderbots developed by A. Stenz and B.M. Dias (Dias 2004), (Zlot and Stenz 2005) is economics and market based approach. The mechanism is grounded in the concepts of cost, income and profit as well as in revenue and cost functions that are calculated for each individual robot. The idea is that robots may sell tasks to other robots if the other ones can do the tasks more efficiently (with higher profit). So the external income is not the only one that provides rewards to robots – they can use other robots to get rewarded for execution of their tasks. As a result of the local profit maximization by each robot the profit of the system is maximized because if there will be any robot that can do the task more efficiently, then the particular robot would give the task to the most suitable one. In the calculation of the profit the resources used and opportunities missed are taken into consideration as well. The main conclusion about the Traderbots approach, is the fact that it allows robots to reallocate their tasks to other robots, but it cannot be directly applied to the initial problem when the set of subtasks must be assigned to the group of robots.

The literature analysis resulted in the conclusion that the existing mechanisms have significant drawbacks in the allocation of multiple tasks at the same time. They may result in a situation when the tasks are not performed by the most appropriate robots. It is usually caused by the fact that existing mechanisms are sequentially processing tasks and trying to find the most appropriate robot for each task. Thus they are taking decisions that disable the possibility to use particular robot for other tasks without knowing about these tasks. Other possibility is to find the most suitable task for each robot that enables usage of the information about all the responsibility areas during each allocation process. As the number of free robots usually is smaller than number of responsibility areas and robots are located closer to each other than all the available responsibility areas, this approach will lead to the task allocations with smaller extra distances travelled by the robots. Thus paper presents a specific task allocation method for homogeneous multi-robot systems for area coverage tasks.

4 The Method of Responsibility Area Based Task Allocation

The proposed method is based on a static decomposition of the whole area into sub-areas and dynamic allocation of them to robots. After having collected the initial data about the environment and constructing initial map (after the initialization phase of the system when the task of the robots is to collectively create a map of the environment (Lavendelis et al. 2012)) the environment is divided into responsibility areas. This task is done by using the algorithm for responsibility area definition described in the next subsection. After this task the responsibility areas normally stay unchanged. Still, if there are significant changes in the environment, the division may be redone (it must be manually initiated by the user of the system). Each responsibility area is assigned to definite robot when it is needed to clean the particular area. It is done using the algorithm given in the Sect. 4.2.

4.1 Definition of Responsibility Areas

The responsibility areas are defined based on empirical and heuristics based method. The method contains the following steps:

1. Determine the number of cells in the occupancy grid containing obstacles. Unfortunately, many map building methods, for example (Andersone et al. 2013), will not be capable to identify all the occupied cells in the map, for example, if mapping is done only based on distance and bumper sensors, then cells that are in the middle of the obstacle will never be identified as occupied (see Fig. 2) and as a consequence the number of occupied cells will be underestimated. Therefore the calculation of total number of obstacles uses robotic platform and domain specific empirical coefficient that expresses the part of the obstacles that are mapped. Based on the number of obstacles, the workable area s_t is calculated using Eq. 1.

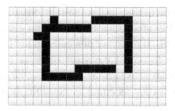

Fig. 2. A typical example of massive obstacle representation in the occupancy grid

$$s_t = x * y - o * k * s_o, \text{ where} \tag{1}$$

s_t – the workable area;

o – the number of occupied cells in the map;

k – coefficient expressing the part of occupied cells that has been successfully identified. The experiments with vacuum cleaning robots Roomba showed that the value of the coefficient is around 2 meaning that approximately 50 % of occupied cells are marked as obstacles in the map;

s_o – area of a single occupancy grid cell;

x and y are the same as in the Sect. 2.

2. Calculate the minimal number of responsibility areas a based on the maximum size that a single robot can clean:

$$a = \frac{s_t}{s}, \text{ where} \tag{2}$$

a – minimum number of responsibility areas. It is rounded upwards.

s – the maximum area that single robot can efficiently clean

s_t is the same as in Eq. 1.

3. If the calculated value of a is less than the number of robots n in the system, then a = n. This step can be done only if the number of robots is known and constant, otherwise it is omitted.

4. Find the number of tiles into which each axis must be divided. To ensure that the responsibility areas are as close to squares as possible the proportion of x and y is used. The X axis is divided into b that is the closest integer to the value $\sqrt{ax/y}$, that a can be divided by. The Y axe will be divided into c = a/b tiles. There is one problem in this step with prime numbers and numbers that can be divided only into unequal multipliers. For example, if the number of elements is 38, then it can be divided only into multipliers 1 and 38 or 2 and 19 which will result in long and narrow responsibility areas that are not suitable for existing robots. The problem is solved by checking if the proportions of the responsibility area sides $\frac{x/b}{y/c}$ and $\frac{y/c}{x/b}$ are greater than some constant, which is determined empirically and in current implementation of the multi-robot system's management software set to the value of 3. If the proportion is greater than 3, then a is increased by one and this step is restarted. This check guarantees that the responsibility areas will have sides with proportion less than 3.

If the map of the environment does not contain information about obstacles, then the algorithm for responsibility area generation is trivial – each axis can be just divided into b and c equal tiles. Still, if there are obstacles in the map, then one of the goals of the division algorithm is to adjust the sizes of areas so that the amount of work needed for cleaning all areas is as similar as possible, i.e., the cleanable areas that are not occupied with obstacles are as equal as possible. It is ensured by the following steps of the algorithm.

5. The map of the environment is divided into the previously calculated number of columns b. It is done based on the number of cells containing obstacles in different parts of the map. The columns are defined sequentially starting from the left side of the map. The definition of each (i-th) column is done in the following steps:

   (a) Calculate the width of the i-th column w_c^{init} if the remaining part of the map would be divided into columns of equal width (see Fig. 3):

   $$w_c^{init} = \frac{w_r}{b - i}, \text{ where} \tag{3}$$

   w_c^{init} – proportionally calculated width of the column;
   w_r – the width of the remaining part of the map;
   b – the number of columns calculated in the step 4;
   i – the index of the column (the first value is 0).

   (b) Find the number of cells containing obstacles o_c in the column if it had the width w_c^{init}. Here and in all subsequent steps the number of obstacles in some area is calculated by finding the cells of the occupancy grid that overlap with the particular area and counting ones that have been marked as obstacles.

   (c) Find the number of cells o_r containing obstacles in the remaining part of the map.

   (d) Calculate the width of the column based on the relative number of the obstacles in the column with the initial width in accordance with the Eq. 4.

   $$w_c = w_c^{init} * \frac{1 - o_r * k*s_0/s_r}{1 - o_c * k*s_0/s_c}, \text{ where} \tag{4}$$

   s_r – area of the remaining part of the map;
   s_c – area of initially calculated column;
   o_r – the number of cells in the remaining part containing obstacles;
   o_c – the number of cells in the column containing obstacles;
   k and s_0 are the same as in Eq. 1, while w_c^{init} is the same as in Eq. 3.

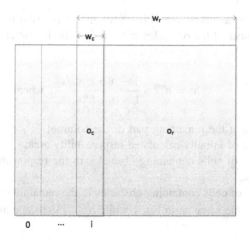

Fig. 3. Calculation of the column width

6. Create the column and calculate its main characteristics:
   (a) Calculate the actual number of cells o_c containing obstacles in the column based on the final width of the column.
   (b) Calculate the number of responsibility areas c_i to divide the column into by using Eq. 5. The result is rounded upwards. Additionally, if c_i is calculated to be less than 1, then it is 1.

$$c_i = \frac{w_c * y - o_c * k * s_o}{s}, \text{ where} \tag{5}$$

   s_o and k and y are the same as in Eq. 1 and o_c and w_c are the same as in Eq. 4.
7. Divide the column into the responsibility areas by sequentially adding them from the bottom of the column. Each area is defined by doing the following steps:
   (a) Calculate the height h_a^{init} of the j-th responsibility area if the remaining part of the column would be divided into areas of equal height:

$$h_a^{init} = \frac{h_r}{c_i - j}, \text{ where} \tag{6}$$

   h_a^{init} – initially (proportionally) calculated height of the responsibility area;
   h_r – the height of the remaining part of the column;
   j – the index of the responsibility area in the column (the first value is 0);
   c_i – the number of responsibility areas in the column (calculated in the step 6b).

   (b) Calculate the actual number of cells o_a containing obstacles in the responsibility area with the initial height and number of cells o_r containing obstacles in the remaining part of the column.

c) Calculate the height h_a (see Fig. 4) of the responsibility area based on the relative number of the obstacles in the area with the initial width by using the Eq. 7.

$$h_a = h_a^{init} * \frac{1 - o_r * k*s_o/s_r}{1 - o_a * k*s_o/s_a}, \text{ where} \tag{7}$$

s_r – the area of the remaining part of the column;
s_a – the area of initially calculated responsibility area;
o_a – number of cells containing obstacles in the responsibility area with the initial height.
o_r – number of cells containing obstacles in the remaining part of the column.
k and s_o are the same as in Eq. 1 while h_a^{init} is the same as in Eq. 6.

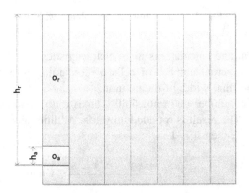

Fig. 4. Calculation of the height of the responsibility area

(d) Check the height of the area according to the initially calculated height. Too big difference of the responsibility area's height from its initially calculated value, may be resulted from over adjusting because of one large obstacle in the particular area. The check is done using the Eq. 8:

$$h_a > k_h * h_a^{init}, \text{ where} \tag{8}$$

k_h – coefficient expressing how big the difference from the proportional size may be without recalculating more exact size. The value of the coefficient is empirical and in current implementation of the method its value is 1.8.
If the Eq. 8 gives a positive result, then the final height h_a is recalculated based on the concentration of obstacles in the area with the height of h_a by using Eq. 9.

$$h_a = h_a^{init} * \frac{1 - o_r' * k^* \, s_o/s_r'}{1 - o_a' * k^* \, s_o/s_a'}, \text{where} \tag{9}$$

$s_{r'}$– the area of the remaining part of the column;

s_a' – the area of previously calculated responsibility area;

$o_{a'}$ – the number of cells containing obstacles in the responsibility area with the previously calculated height h_a;

$o_{r'}$ – the number of cells containing obstacles in the remaining part of the column;

s_o, k, and h_a^{init} are the same as in Eq. 7.

(e) Create the responsibility area with coordinates $(x - w_r, \, y - h_r + h_a, \, x - w_r + w_c, \, y - h_r)$, where w_r and h_r are the width and height of the remaining parts and add it to the list of areas. Calculate the precise number of cells o_a marked as obstacles in the defined responsibility area. Decrease the number of obstacles in the remaining part of the column and decrease the remaining height of the column.

$$h_r = h_r - h_a \tag{10}$$

$$o_r = o_r - o_a \tag{11}$$

8. Decrease the remaining width of the map by the width of the column:

$$w_r = w_r - w_c \tag{12}$$

After executing the described algorithm the environment is divided into responsibility areas that are atomic units in task allocation and can be allocated to any of the vacuum cleaning robots. The next subsection describes the task allocation mechanism.

4.2   Responsibility Area Allocation

At the moment, when there is a robot that is free and ready to do some task (its battery is charged, it has no technical problems, etc.) and there is at least one responsibility area that needs to be cleaned, the task allocation must be carried out. The main principle is to allocate the closest responsibility area to the robot to minimize the extra distances travelled by robots without doing particular tasks. Still the Fig. 5 illustrates a simple single robot example showing that the distance cannot be the only one principle of task allocation. Depending on the exact finishing position inside the responsibility area the choice of the closest responsibility area can lead to some responsibility areas left unprocessed (white rectangles in Fig. 5) while all the neighbors are processed (grey in Fig. 5), resulting in the situation when all robots will be in other regions of the environment and one of them will have to come back to clean some isolated area, that will lead to extra work done for travelling long distances to clean isolated areas (dashed lines in Fig. 5).

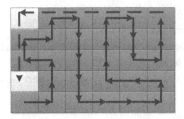

3	3	4	4	9	9
2	2	4	4	8	8
2	2	5	5	8	8
1	1	5	5	7	7
1	1	6	6	7	7

Fig. 5. Task allocation based only on distance **Fig. 6.** Index allocation to responsibility areas

To ensure that the robots first fully clean one region of the environment and only then move to other regions, indexes are assigned to the responsibility areas. Indexes join the responsibility areas into groups that are processed one after another by the same robot. All responsibility areas of the same group are neighbors (they either have common sides or corners). Thus 2 by 2 responsibility areas are included into the same group and as a consequence up to four responsibility areas have the same index. As shown in Fig. 6 some groups may be smaller due to the actual number of elements in columns and rows.

After grouping the responsibility areas and assigning indexes, the following algorithm is used to allocate task to one particular robot:

1. Find all the responsibility areas waiting for processing that are not more than 1.5 diagonals of the particular responsibility area far from the robot. The constant 1.5 diagonals gives task allocation where robots do not travel too big distances to the next responsibility area without a need to do so.
2. If at least one responsibility area is found during the step 1, then the index difference is calculated between each of the areas found and the responsibility area that was previously processed by the robot. The area with the lowest index difference is assigned to the robot. The index of previously processed responsibility area for the robots that just have become active (just put into the system, just ended charging, etc.) must be set to the value corresponding to the index of the responsibility area where the base is located.
3. If there are more than one responsibility area with the same minimum index difference, then the one with the smallest distance from the current position of the robot, is chosen.
4. If no responsibility areas are found during the first step, then the closest responsibility area waiting for processing is assigned to the robot without any reference to indexes.

4.3 Results

The proposed task allocation algorithm has been implemented in the management software of the multi-robot system for vacuum cleaning of large areas. The current software is suitable for any number of robots working together to clean the same area. Currently the system has been tested for up to 5 robots in the close to real environment, e.g. a hall with a possibility to create different configurations of obstacles. For example in one of the tests several obstacles were put in the middle of the room to test how the tasks will be divided and allocated to the robots. The Fig. 7 shows the screenshot from the implemented software with the division of the tasks. The black cells denote obstacles, the white ones are known to be free while the grey ones are unexplored. The responsibility areas are shown as rectangles with the index in the middle. Two robots are also shown in this Figure (denoted by numbers 3 and 5). As it can be seen the division algorithm created larger areas if they contain more obstacles. The example of such areas is in the middle of the environment where bottom left area from the group with the index 10 contains the most obstacles and thus is the largest one. The task allocation depends on the initial positions of robots. Several experiments with the software showed that robots do not make long journeys from one responsibility area to another and thus do not waste time and resources.

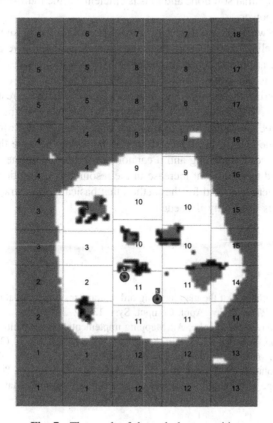

Fig. 7. The result of the task decomposition

5 Conclusions

The paper presents area coverage task decomposition and allocation method for the homogeneous multi-robot system based on obstacle concentration. The main novelty of the method is the fact that it sequentially finds the most suitable task for each particular robot instead of finding the most appropriate robot for each task. Thus the task allocation method is applicable for domains where the number of subtasks is larger than the number of robots and all robots have the same capabilities, i.e. the system is homogeneous. Additionally, the task decomposition method is suitable for area coverage tasks in domains where the areas should be defined to require the same amount of work and as a consequence the sizes of the responsibility areas should vary based on the obstacles in the particular area.

The experiments in close to real environment show that the heuristic approach used in the method provides the division of the environment into the responsibility areas that are of similar size in terms of workload needed to clean them. Together with the proposed task allocation mechanism the proposed methods lead to task decomposition and allocation that do not require long and unnecessary journeys from one responsibility area to another in majority of the close to real environments. So the proposed method is not computation and communication consuming, but at the same time provides close to optimal solutions and thus is efficient for the multi-robot systems with low computation power.

The only case when the method gave the solution that led to significant waste of resources into traveling extra distances was the situation with massive obstacles that the robot has to drive around to reach the neighbour responsibility area. The example of such an obstacles is the internal walls of the office premises. The current implementation was considering the straight line distances to the neighbour responsibility areas instead of real ones. Thus one of the future works is to use the path planning algorithm inside the task allocation algorithm to find real distances between robots and responsibility areas. In the current implementation it is not done, because the path planning algorithm is resources consuming and it cannot be used to determine all the distances, because it will lead to significant increase in the resource consumption during the task allocation. Thus there is a need for the method that basically uses straight line distance and the real path is used only if needed.

References

Andersone, I.: The influence of the map merging order on the resulting global map in multi-robot mapping. Sci. J. RTU. 5. Ser. Appl. Comput. Syst. **13**, 22–28 (2012)

Andersone, I., Liekna, A., Nikitenko, A.: Mapping implementation for multi-robot system with glyph localisation. Sci. J. RTU. 5. Ser. Appl. Comput. Syst. **14**, 67–72 (2013)

Ashlock, D., et al.: A note on general adaptation in populations of painting robots. In: The 2003 Congress on Evolutionary Computation, CEC 2003, vol. 1, pp. 46–53, 8–12 Dec 2003

Choset, H.: Coverage for robotics - a survey of recent results. Annals Math. Artif. Intell. **31**, 113–126 (2001)

Dias, M.B.: TraderBots: a new paradigm for robust and efficient multirobot coordination in dynamic environments. Technical report. Doctoral dissertation, Robotics Institute, Carnegie Mellon University, January 2004

Doroftei, D., De Cubber, G., Chintamani, K.: Towards collaborative human and robotic rescue workers. In: 5th International Workshop on Human-Friendly Robotics (HFR2012), 18–19 October 2012

FIPA 2002, FIPA Contract Net Interaction Protocol Specification. Foundation for Intelligent Physical Agents. http://www.fipa.org/specs/fipa00029/ (2002). Accessed 21 March 2014

Gerkey, B.P.: On Multi-robot task allocation. Ph.D. Dissertation. University of Southern California Computer Science Department, Aug 2003

Gerkey, B.P., Mataric, M.J.: A framework for studying multi-robot task allocation. In: Schultz, A.C., et al. (eds.) Multi-robot systems: from swarms to intelligent automata, vol. II, pp. 15–26. Kluwer Academic Publishers, The Netherlands (2003)

IRobot Roomba specification. http://www.irobot.com/us/learn/home/roomba.aspx (2014). Accessed 21 March 2014

Lavendelis E., et al.: Multi-agent robotic system architecture for effective task allocation and management. In: Recent Researches in Communications, Electronics, Signal Processing and Automatic: Proceedings of the 11th WSEAS International Conference on Signal Processing, Robotics and Automation (ISPRA '12), pp 167–174, United Kingdom, Cambridge, 22–24 February 2012

Liekna, A., Lavendelis, E., Grabovskis, A.: Experimental analysis of contract net protocol in multi-robot task allocation. Sci. J. RTU. 5. Ser. Appl. Comput. Syst. 213, 6–14 (2012)

Liekna, A., Lavendelis, E., Ņikitenko, A.: Challenges in development of real time multi-robot system using behaviour based agents. In: Omatu, S., Neves, J., Corchado Rodriguez, J.M. (eds.) Distributed Computing and Artificial Intelligence. Advances in Intelligent Systems and Computing, pp. 587–598. Springer, Heidelberg (2013)

Mancini A., et al.: Coalition formation for unmanned quadrotors. In: Proceedings of the 7th International ASME/IEEE Conference on Mechatronics and Embedded Systems and Applications, pp. 315–320, September 2011

Ņikitenko, A., et al.: Single robot localisation approach for indoor robotic systems through integration of odometry and artificial landmarks. Appl. Comput. Syst. 14(2013), 50–58 (2013)

Palacín, J., et al.: Building a mobile robot for a floor-cleaning operation in domestic environments. IEEE Trans. Instrum. Measur. 53(5), 1418–1424 (2004)

Satish Kumar, K.N., Sudeep, C.S.: Robots for precision agriculture. In: Electronic Proceedings of 13th National Conference on Mechanisms and Machines (NaCoMM07), Bangalore, India, 12–13 December 2007

Wooldridge, M.: An Introduction to Multi Agent Systems, 2nd edn, p. 484. Wiley, Chichester (2009)

Yamaguchi, Y., et al.: Development of an intelligent robot for an agricultural production ecosystem (viii) – improvement of predator-prey model and analysis of the activity of snail in paddy by image processing. J. Fac. Agric. Kyushu Univ. 55(1), 101–105 (2010)

Young, J.E., et al.: Toward acceptable domestic robots: applying insights from social psychology. Int. J. Soc. Robot. 1(1), 95–108 (2009). Springer, The Netherlands

Zlot, R.M., Stentz, A.: Complex task allocation for multiple robots. In: Proceedings of the International Conference on Robotics and Automation, pp. 1515–1522, April 2005

Key Factors for a Proper
Available-Bandwidth-Based Flow Admission
Control in Ad-Hoc Wireless Sensor Networks

Muhammad Omer Farooq[1] and Thomas Kunz[2]([✉])

[1] Institute of Telematics, University of Luebeck, Luebeck, Germany
farooq@itm.uni-luebeck.de
[2] Department of Systems and Computer Engineering,
Carleton University, Ottawa, Canada
tkunz@sce.carleton.ca

Abstract. In this paper, first, we present our simulation studies that help to outline key factors for a proper available-bandwidth-based flow admission control in ad-hoc Wireless Sensor Networks (WSNs). In most cases, WSNs use the IEEE 802.15.4 standard, therefore our simulation studies are based on the same standard. The identified key factors are: (i) the overheads (back-off, retransmission, contention window, ACK packet, and ACK waiting time) associated with the unslotted IEEE 802.15.4 Carrier Sense Multiple Access Collision Avoidance (CSMA-CA) MAC layer protocol reduce the amount of available bandwidth, (ii) the impact of the MAC layer overheads on a node's available bandwidth is a function of the number of active transmitters and data traffic load within the interference range of the node, (iii) contention count on a node that is not on a flow's data forwarding path is a function of the number of active transmitters (along the flow's data forwarding path) within the interference range of the node, and (iv) a flow's intra-flow contention count on a node (along the flow's data forwarding path) depends on the hop-count distance of the node from the source and the destination nodes, and the node's interference range. Second, we present a survey of state-of-the-art flow admission control algorithms for ad-hoc wireless networks. The survey demonstrates that the state-of-the-art flow admission control algorithms do not completely consider the key identified factors or make incorrect assumptions about them. Third, we propose techniques that an available-bandwidth-based flow admission control algorithm can use to incorporate the key identified factors. Hence, the work presented in this paper can serve as a basis of a more effective available-bandwidth-based flow admission control algorithm for ad-hoc wireless networks.

Keywords: Flow admission control · Ad-hoc Wireless Sensor Networks · Measurement-based bandwidth estimation · Quality of Service (QoS)

1 Introduction

Real-time multimedia applications generate inelastic data requiring soft bandwidth guarantee and bounded delay. The soft bandwidth and bounded delay

© Springer-Verlag Berlin Heidelberg 2015
M. Garcia Pineda et al. (Eds.): ADHOC-NOW Workshops 2014, LNCS 8629, pp. 246–260, 2015.
DOI: 10.1007/978-3-662-46338-3_20

requirements of such applications are invariably called their Quality of Service (QoS) requirements. Excessive data (w.r.t. the available bandwidth) inside a network can cause congestion, and congestion increases packet drop rate and end-to-end packet delivery delay. Therefore, to restrict applications' data inside a network within a network's manageable limits (so that the QoS requirements of the real-time applications can be satisfied), flow admission control algorithms are used [6,8].

In wireless networks bandwidth is a shared resource. The most common assumption is that the bandwidth available to a node is shared within the interference range of the node, and nodes within a two hops distance can cause interference [7]. We also hold this assumption throughout the paper. The shared nature of the bandwidth in wireless networks results in the following phenomena: (i) the data generation rate of nodes within the interference range of a node inside a network affects the available bandwidth at the node [7] and (ii) intra-flow and inter-flow contention [7]. Our study demonstrates that the IEEE 802.15.4 unslotted CSMA-CA MAC layer overheads (back-off, retransmission, contention window, ACK packet, and ACK waiting time) reduce the amount of available bandwidth, therefore a good bandwidth estimator must consider the complete impact of the MAC layer on the available bandwidth. Moreover, the MAC layer overhead at a node depends on the total number of transmitters and the data traffic load within the interference range of the node. Therefore, an effective flow admission control algorithm must incorporate a mechanism to proactively estimate the impact of the MAC layer overhead on a node's available bandwidth, before deciding about a new flow's (flow requesting admission) admission request. Furthermore, we demonstrate that before admitting a new flow, a flow admission control algorithm must determine the correct intra-flow and inter-flow contention counts. Our study demonstrates that the contention count on a node that is not along a flow's data forwarding path, but is within the interference range of transmitters along the flow's data forwarding path is a function of the number of transmitters within the interference range of the node.

Apart from the MAC layer, the transport layer can also impact the amount of available bandwidth at a node, e.g., the congestion control and flow control algorithms of the Transmission Control Protocol (TCP). In this research paper, we assume that there are no congestion control and flow control algorithms working at the transport layer.

The remainder of this paper is organized as follows. In Sect. 2, we present our simulation studies that help to outline key factors for a proper available-bandwidth-based flow admission control in ad-hoc wireless networks. Section 3 surveys state-of-the-art flow admission control algorithms for ad-hoc wireless networks. Section 4 presents our techniques that an available-bandwidth-based flow admission control algorithm can use to take into account the key identified factors. Finally, we conclude this research paper in Sect. 4.

2   Experimental Studies

We performed our experimental studies using the Cooja WSN simulator [4]. The general simulation parameters are shown in Table 1.

Table 1. General simulation parameters

Parameter	Value
MAC layer	Unslotted CSMA-CA
MAC layer reliability	Enabled
Radio duty cycling algorithm	No duty cycling
Radio model	Unit disk graph model
MAC layer queue size	10 frames
Channel rate	250 kbps
Node transmission range	50 m
Node carrier sensing range	100 m
Total frame size	127 bytes
Simulated node type	Tmote sky

2.1   Impact of the MAC Layer Overhead on the Available Bandwidth

To measure the IEEE 802.15.4's unslotted CSMA-CA MAC layer overhead (back-off and retransmission), we conducted multiple simulations. We created different simulation scenarios by varying the number of active transmitters and data traffic load on the IEEE 802.15.4 communication channel. In our experiments, we use an ad-hoc network topology, furthermore we assume that the maximum number of transmitters within the interference range of a node is not more than eight (the MAC layer overhead results presented in this paper can be extended easily to consider more than 8 transmitters). To estimate the MAC layer overhead (back-off and retransmission), we consider the aggregate data rate and the number of transmitters, but there are other parameters that may affect the MAC layer overhead such as the packet size and the nature of data traffic (burst, constant bit rate). We expect that, beyond the aggregate data rate and the number of transmitters, other parameters will only have a modest impact. We created 7 different simulation scenarios. In the simulation scenarios, we increase the number of transmitters from 2 to 8, and each scenario has sub-scenarios. In each sub-scenario, we vary the offered data load inside a network from 8 kbps to 64 kbps (results can be extended to consider higher data traffic load). Each transmitter is within the transmission range of the other transmitters, and all the transmitters transmit to the same destination node. The destination node is also within the transmission range of the transmitters. To determine the mean value of the IEEE 802.15.4's unslotted CSMA-CA MAC layer overhead along with the 95 % confidence interval, each sub-scenario is repeated 25 times. The back-off overhead is measured in time, but Fig. 1 reports the MAC layer overhead in bps. We converted the mentioned overhead to bps by multiplying the accumulated time duration (during each second) a node spends in the back-off mode with the channel rate. The following conclusions can be drawn from Fig. 1.

(a) The MAC layer protocol overheads consume bandwidth, therefore an available bandwidth estimator should consider the amount of bandwidth consumed during the MAC layer protocol's operation.
(b) The mean IEEE 802.15.4's unslotted CSMA-CA MAC layer overhead increases with an increase in the aggregate data load on the IEEE 802.15.4 communication channel. There is only one exception, i.e., in case of two transmitters, the mean overhead decreases with an increase in the aggregate data load from 32 kbps to 48 kbps. This is quite counter-intuitive and further work is required to explore this in more detail.
(c) If the aggregate data load on the IEEE 802.15.4 communication channel is less than or equal to 32 kbps, the increase in the number of transmitters does not affect the mean overhead.

Cooja emulates the Contiki operating system's [1] unslotted CSMA-CA MAC layer implementation, and Contiki unslotted CSMA-CA uses a constant contention window size, therefore we can derive the contention window overhead by knowing the number of additional packets a node intends to transmit. Moreover, as the MAC layer waits for a constant period of time to receive an ACK for the transmitted data frame, an estimate of the overhead associated with MAC layer ACKs can also be derived from the number of additional packets a node intends to transmit. Consequently, the total MAC overhead can be obtained by adding these constant factors to the results plotted in Fig. 1.

2.2 The MAC Layer Overhead Impact on Non-relaying Nodes

To demonstrate a new flow's impact on the MAC layer overhead at nodes that do not relay the new flow's data, but are within the interference range of transmitters along the new flow's data forwarding path, we create two simulation scenarios, using the network topology shown in Fig. 2. In Scenario 1, node C transmits 10 kbps to node D (10 data packets per second). In Scenario 2, in addition to the flow from node C to node D, node A transmits 10 kbps to node B (10 data packets per second) and node E transmits 10 kbps to node F (10 data packets per second). In both scenarios, we keep track of the mean MAC layer overhead at node C. We repeat each simulation scenario 10 times. One thing to notice is that node C is neither on the data forwarding path of node A's flow nor it is on the data path of node E's flow, but it is within the interference range of nodes A and E. Moreover, node C is also transmitting data.

The results shown in Table 2 are obtained after considering the complete MAC layer overhead (i.e., including the constant components mentioned above). For Tmote Sky motes, Contiki uses a constant contention window duration, and

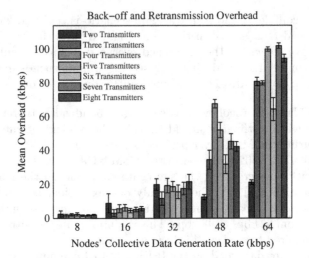

Fig. 1. Data load vs. mean back-off and retransmission overhead

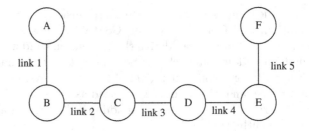

Fig. 2. Simulated network topology

the duration is equivalent to 7812 bits (considering the channel rate of 250 kbps). Therefore, before the transmission of a data frame, the MAC layer waits for a time duration that is equivalent to 7812 bits. The size of the IEEE 802.15.4 ACK frame is 40 bits. After a data frame is transmitted, a node waits for a time duration that is equivalent to 260 bits to receive an ACK frame. Considering the overheads in the both scenarios, the contention window overhead is ($7812 \times 10 = 78.12$ kbps), minimum ACK frame overhead is ($40 \times 10 = 0.40$ kbps), and minimum ACK waiting time overhead is ($260 \times 10 = 2.60$ kbps). Therefore, the minimum MAC layer overhead excluding the back-off and retransmission overheads at node C is (78.12 kbps $+ 0.40$ kbps $+ 2.60$ kbps $= 81.12$ kbps). The results shown in Table 2 in addition include the back-off and retransmissions overheads.

The results shown in Table 2 demonstrate that, with the admission of the new flows, the total mean MAC layer overhead has increased at node C, and the difference is statistically significant. Therefore, a good flow admission control

Table 2. The total mean MAC layer overhead at node C

Scenario	Mean overhead	95 % confidence interval
1	91.67 kbps	90.20–93.14 kbps
2	103.55 kbps	100–107.16 kbps

Table 3. Mean data activity as measured by nodes

Node ID	Mean data activity	95 % confidence interval
A	31.41 kbps	31.21–31.61 kbps
B	41.35 kbps	41.16–41.53 kbps
C	51.85 kbps	51.53–52.17 kbps
D	43.48 kbps	43.33–43.63 kbps
E	33.18 kbps	32.87–33.49 kbps
F	22.85 kbps	22.64–23.06 kbps

algorithm must consider the additional MAC layer overhead at nodes which are not on a new flow's data forwarding path, but are within the interference range of transmitters along the data forwarding path, before deciding about the new flow's admission request. This factor must only be taken into account for nodes which are transmitting/relying data, if a node is not transmitting data there is no inter-flow contention.

2.3 Determining Maximum Intra-flow Contention Count

It has been claimed in [3,5,8] that the maximum possible intra-flow contention count on a node along the data forwarding path is 4. To verify the claim, we carried out a simulation-based experiment. In our experiment, using the topology shown in Fig. 2, node A is the source node and node F is the sink node. Node A transmits at the rate of 10 kbps (10 data packets per second) to the sink node. Node A starts the transmission at a simulation time of 5 s and terminates the data packets transmission at a simulation time of 105 s. In our experiment, we measured the data activity at nodes inside the network via wireless channel-sensing throughout the duration of node A's flow. We repeated the experiment 10 times, and the mean data load observed by the nodes while node A is transmitting data packets is shown in Table 3 along with the 95 % confidence interval. The end-to-end flow throughput was perfect, i.e., 10 kbps.

Table 3 demonstrates that the mean data load observed by node C is approximately 50 kbps, which is 5 times the transmission rate of node A. The data loads observed by nodes A, B, D, E, and F are approximately 30 kbps, 40 kbps, 40 kbps, 30 kbps, and 20 kbps respectively. Therefore, the contention counts at nodes A, B, D, E, and F are 3, 4, 4, 3, and 2 respectively. The maximum contention count is 5: node C is two hops away from the source node A, therefore

Table 4. Data activity as measured by node G

Scenario	Mean data activity	95 % confidence interval
1	10.20 kbps	10.05–10.35 kbps
2	20.36 kbps	20.06–20.66 kbps
3	31.90 kbps	31.25–32.55 kbps
4	42.95 kbps	42.25–43.65 kbps

the flow's contention due to the upstream nodes transmission is two times the bandwidth required by the flow. Moreover node C is more than two hops away from the destination node F, therefore the flow's contention due to the transmission of downstream nodes is two times the required bandwidth, and node C also relays the flow's data, hence the maximum intra-flow contention count is 5. This also shows that the intra-flow contention count on a node depends on the node's hop-count distance from the source and the destination nodes and the interference range of the node. The mean data load observed at nodes is larger than 10× the contention count, this is due to retransmitted data/control frames.

2.4   Determining Correct Contention Count on Nodes Not on a New Flow's Data Forwarding Path

There may be nodes inside a network that are not on a new flow's data forwarding path, but other flows' data is being generated/relayed by those nodes, and some of the transmitters on the new flow's data forwarding path are within the interference range of the nodes. Therefore, it is necessary to determine the new flow's correct contention count on those nodes, otherwise end-to-end QoS requirements of admitted flows may be compromised.

We performed a number of simulations to show that the contention count on a node that is not on a flow's data forwarding path, but is within the interference range of transmitter(s) along the data forwarding path, is a function of the number of transmitters within the interference range of the node. We added one more node in the network shown in Fig. 2, and created 4 different simulation scenarios by changing the location of the additional node inside the network. In these simulations, node G measures the data activity using the wireless channel-sensing technique. Each simulation scenario is repeated 10 times. Figure 3 shows the modified network topologies for different simulation scenarios. Table 4 shows the mean data activity measured by node G during the duration of node A's flow.

In Scenario 1 node G is within the interference range of one transmitter, i.e., node E, and Table 4 shows that the contention count at node G in this case is approximately 1. Similarly, in Scenarios 2, 3, and 4, node G is within the interference range of 2, 3, and 4 transmitters, and Table 4 demonstrates that the contention count at node G in these cases is 2, 3, and 4 respectively. Node A was transmitting at the rate of 10 kbps, and the end-to-end throughput was 10 kbps. The data activity measured by node G in different simulation scenarios

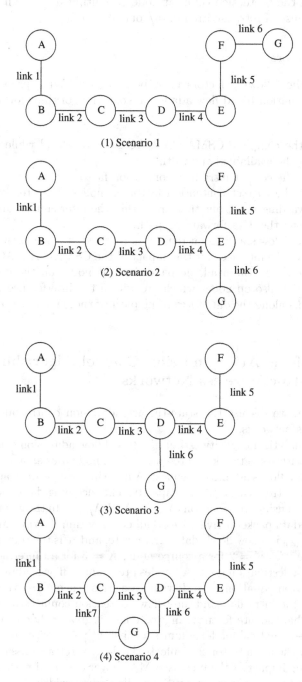

Fig. 3. Simulated network topologies

is greater than the contention count at node G multiplied by the node A's flow data rate, because of retransmitted data/control frames.

2.5   Key Factors

In summary, the following factors must be considered for a proper available-bandwidth-estimation-based flow admission control algorithms for ad-hoc wireless networks.

(a) Consider the complete CSMA-CA MAC layer overhead while periodically estimating the available bandwidth.
(b) Determine the correct intra-flow contention factor.
(c) Determine the correct contention factor on nodes that are not on a data flow's forwarding path, but that are within the interference range of transmitters along the data forwarding path.
(d) When a new flow's admission request is received at a node, the node must proactively take into account the impact of the CSMA-CA MAC layer overhead not only on the available bandwidth at nodes on the data forwarding path, but also on nodes which are within the interference range of the transmitters along the data forwarding path (if those nodes are transmitting data).

3   State-of-the-Art Admission Control Algorithms for Ad-Hoc Wireless Networks

In this section, we discuss the state-the-art admission control algorithms for ad-hoc wireless networks.

In [6], an analytical capacity estimation based flow admission control scheme for multi-hop wireless networks is presented. Each node uses an analytical model to decide about a flow's admission request. A node inside a network accepts a flow if λ_{new} is smaller than the available capacity. The incoming data packets arrival rate is calculated using the equation ($\lambda_{new} = \lambda + K\lambda_{flow}$). In the given equation λ represents the data packet arrival rates of all nodes within the transmission range of the node, λ_{flow} is a new flow's data arrival rate, and K is the contention count. This technique uses $K = 2$ for a source node, $K = 3$ for an intermediate node, and $K = 1$ for a destination node. All nodes processing a flow's admission request evaluate the given equation. The downsides of this scheme are: there are cases in which both the intra-flow and inter-flow contention count estimation will be wrong (given that the interference range of a node is greater than its transmission range), the mathematical model assumes a constant packet size, assumes that at the MAC layer the main factor that affects the delay is retransmissions, and does not consider the impact of the increased MAC layer overhead with an increased data traffic load inside a network on the available bandwidth.

In [8], a distributed flow admission control for assuring QoS in ad-hoc wireless networks is presented. It is claimed in [8] that before deciding about a flow's admission request both intra-flow and inter-flow contention have been taken into account. The authors of this work claim that the maximum intra-flow contention count on an intermediate node along the data forwarding path is 4. Inter-flow contention is considered by matching a flow's required bandwidth with the minimum available bandwidth within the interference range of a node deciding about the flow's admission request. An estimate of the available bandwidth is provided through a wireless channel-sensing mechanism that uses both the virtual and physical carrier sensing mechanisms of the IEEE 802.11 MAC layer protocol. The proposed flow admission control algorithm uses a two-pass signaling mechanism to reserve resources along the data forwarding path. The drawbacks associated with this technique are: incomplete inter-flow contention, i.e., each node only checks the minimum available bandwidth within its interference range. In some scenarios, the calculated intra-flow contention count will be wrong. Also, the impact of an increased MAC layer overhead with an increased data traffic load inside a network is not considered.

CapEst [2] is a measurement-based link capacity estimator for wireless networks. It monitors the service time of data packets at each link, and based on this measurement an estimate of a link's capacity is made, hence it is not a MAC layer specific method. CapEst does not consider the increased MAC layer overhead with an increased data load inside a network.

In [7], a contention-aware flow admission control for ad-hoc wireless networks (CACP) is presented. Flow admission control is performed based on the available bandwidth estimate. An estimate of the available bandwidth is provided through the wireless channel-sensing mechanism, and it considers back-off periods as idle periods. Here the assumption is that the back-off periods are negligible even if the channel is saturated. The algorithm considers both intra-flow and inter-flow contention counts in a distributed manner. The drawbacks associated with this scheme are: the impact of the MAC layer on the available bandwidth is not considered and the impact of the MAC layer overhead on the available bandwidth with an increased data traffic load inside a network is not considered.

In [5], an available bandwidth-based flow admission control algorithm (ABE) for ad-hoc wireless networks is presented. An estimate of the available bandwidth is provided through the wireless channel-sensing mechanism considering both virtual and physical carrier sensing, and different types of IEEE 802.11 CSMA-CA MAC layer inter-frame spacings. It is argued in [5] that measuring the channel activity considering the time spent in virtual and physical carrier sensing, and different inter-frame spacing results in an overestimate of the available bandwidth. This happens due to the non-synchronization of sender and receiver nodes in an ad-hoc wireless network. Therefore, a mathematical model is presented that takes into account the collision probability to estimate the actual available bandwidth. Hence, it probabilistically takes into account the future back-off overhead through a mathematical model. In [5], nodes periodically broadcast control packets, called HELLO packets. The collision probability is derived from the number of HELLO messages a node has received over the number of HELLO packets the node expected to receive during the last measurement interval. The flow admission control algorithm uses one hop and two hop neighbor information to calculate the intra-flow contention, and the authors claim that the maximum intra-flow contention count on a node is 4. Inter-flow contention is taken into account by determining the minimum available bandwidth within the interference range of a node when deciding about a flow's admission request. The downsides of this technique are: with an increased data traffic load inside a network only additional back-off overhead is considered, additional retransmission and contention window overheads are ignored. The intra-flow contention count estimator does not always provide the right contention count, and the inter-flow contention count estimator is too simple as it only considers minimum available bandwidth within the interference range of a node. Finally, the collision probability is derived without considering the future data traffic load and the number of transmitters.

RABE is a probabilistic mathematical model used to consider the complete impact of the IEEE 802.11 CSMA-CA MAC layer on the available bandwidth. The drawback of this scheme are: it assumes a fixed packet size, the impact of the number of transmitters on the additional MAC layer overhead is not considered.

Table 5 summarizes our evaluation of state-of-the-art available bandwidth-based flow admission control algorithms for ad-hoc wireless networks. Table 5 demonstrates that none of these algorithms take into account all the identified

Table 5. Evaluation of state-of-the-art available-bandwidth-based flow admission control algorithms for ad-hoc Wireless Networks

Algorithm	MAC layer effects on the available bandwidth	Intra-flow contention	Contention non-relaying nodes	Add. MAC layer overhead	Add. MAC layer overhead on non-relaying nodes
CapEst [2]	Yes	No	Partially correct	No	No
RABE [3]	Yes	No	No	Yes	No
ABE [5]	Yes	Partially correct	Partially correct	Partially correct	No
Analytical-Capacity-based [6]	Yes	Partially correct	Partially correct	No	No
CACP [7]	No	Yes	Yes	No	No
Distributed Admission Control [8]	Yes	Partially correct	Partially correct	No	No

key factors. Hence, there is a need for an available bandwidth-based flow admission control algorithm for ad-hoc wireless networks that considers all these factors.

4 Techniques for Taking into Account the Identified Factors

In this section, we propose techniques that an available-bandwidth-based flow admission control algorithm can use to take into account the identified factors.

4.1 A Technique for Measuring the IEEE 802.15.4 Unslotted CSMA-CA MAC Layer Impact on the Available Bandwidth

To consider the MAC layer impact on the available bandwidth, we propose a technique that requires hooks into the MAC layer implementation code. A node must keep track of the following per time unit: (i) total back-off duration, (ii) total number of retransmitted bits, (iii) total ACK frames waiting duration, and (iv) total size of transmitted ACK frames. The total back-off and ACK waiting durations are measured in time, and these overhead should be converted to bps by multiplying them with the channel rate. Afterwards, all the listed MAC layer overheads should be added, and the result must be subtracted from the total channel rate.

4.2 Deciding the Intra-flow Contention Count

We suppose that nodes within the two hops distance can cause interference. Therefore, we propose that a node can decide about the correct intra-flow contention count by knowing nodes within its two-hop neighborhood. We demonstrate this with the help of an example. Let us suppose $A \leftrightarrow B \leftrightarrow C \leftrightarrow D \leftrightarrow E$ represents an ad-hoc network topology. We further suppose that A is the source node and E is the destination node, and C has to decide about the correct intra-flow contention count. As C knows all the nodes within its two-hop neighborhood, therefore C checks the existence of the source node A within its two-hop neighborhood, and in this case the source node is two hops away from C, therefore the intra-flow contention count due to the upstream nodes is 2. Afterwards, C checks the existence of the destination node E within its two-hop neighborhood, in this case the destination node is two hops away from C, therefore the intra-flow contention due to the downstream nodes is 1 (as the destination node does not relay data). Furthermore, C is also acting as a relaying node, hence the total contention count at node C will be 4, i.e., intra-flow contention due to the downstream nodes + intra-flow contention due to the upstream nodes + contention due to node C. It is important to note that, if either the source node or the destination node is not within the two-hop neighborhood, we assume the maximum intra-flow contention count of 2 in that direction.

4.3   Deciding Contention Count on Non-relaying Nodes

Typically, when a flow's admission request arrives at a node, the contention on such nodes is typically only considered by determining the minimum available bandwidth within the interference range of the node (hereafter, we refer to this technique as locally estimating the contention count). Hence, the algorithm is assuming a contention count of 1 for all such nodes. This technique suffers from the following problem. If a common node (node that is not on a new flow's data forwarding path) is within the interference range of more than one transmitter (nodes on the data forwarding path of a new flow), the contention count on the node is equal to the number of transmitters within the interference range of the node. Hence, a flow admission algorithm may wrongly admit a flow. Therefore, we propose the following to decide the contention count on non-relaying nodes. Whenever a node receives an admission request message, the node stores the information in the admission request message along with the bandwidth required by a flow, in an internal data structure. Afterwards, the node broadcasts a control message called bandwidth increment message. In the control message, the node informs its direct neighbors about its increased bandwidth usage due to the new flow. After broadcasting the control message, the node waits for a small period of time before forwarding the admission request message to the next hop along the data forwarding path. Upon reception of the bandwidth increment message, each direct neighbor of the node calculates its available bandwidth by considering the increased bandwidth usage information, and performs other important checks for a proper flow admission control. If a node decides that it has enough bandwidth to bear the interference caused by the new flow, it updates bandwidth usage information in its internal data structure. Afterwards, direct neighbors rebroadcast the control message so that the increased bandwidth usage information of the node (ideally) reaches all nodes within its interference range. Each two hop neighbor processes the control message in the same way as a one hop neighbor did, but it does not rebroadcast the control message. If any of the nodes within the interference range of the node determines that it does not have enough bandwidth to accommodate the interference caused by the new flow, it unicasts an admission reject message to the node. It is possible that a node receives multiple copies of the same bandwidth increment message, hence a duplicate detection mechanism is required to detect these scenarios. Note that with this contention estimation algorithm the node only needs to consider the downstream nodes' intra-flow contention as due to the bandwidth increment message the nodes on the forwarding path have already considered the upstream nodes contention.

4.4   Estimating the Additional MAC Layer Overhead

To estimate the impact of a new flow on the MAC layer overhead, we propose a method that is based on empirically collected data, e.g., the data shown in Fig. 1. The two important parameters that affect the MAC layer overhead are: (i) the total data traffic load within the interference range of a node, (ii) and the number of transmitters within the interference range of a node. A node can determine the total data traffic load and the number of transmitters within its interference range through a control message, i.e., each node inside a network must broadcast a control message containing information about the node's data generation/relaying rate (except control messages data rate). This information should be propagated within the two-hop neighborhood of the node. Using this method, each node inside a network can determine the total data traffic load and the number of transmitters within the interference range of the node. This information along with the linear interpolation technique can be used to estimate the impact of the new flow on the MAC layer overhead using the empirically collected data. Note that such an approach, when based on carefully collected experimental data with sufficient fine-grained resolution, will correctly capture the random access nature of the MAC. As the data load in the network increases, the additional MAC overhead will increase linearly, but will eventually drop off as the link reaches saturation levels.

4.5 Estimating the Additional MAC Layer Overhead on Non-relaying Nodes

Whenever a node receives the bandwidth increment message, it can estimate the additional MAC layer overhead in a similar manner, as discussed in Sect. 4.4.

5 Conclusions and Future Work

The state-of-the-art flow admission control algorithms for ad-hoc wireless networks do not correctly take into account all essential factors for a proper flow admission control, hence they may incorrectly admit flows. Such incorrect flow admission decisions may result in compromising the QoS requirements of real-time multimedia applications. To satisfy real-time multimedia flows' QoS requirements, this paper highlighted essential factors through simulation-based studies that must be taken into account by a proper available-bandwidth-based flow admission control algorithm for ad-hoc wireless sensor networks. Moreover, in this paper, we presented techniques that can be used by an available-bandwidth-based flow admission control to take into account the identified factors. In future, we plan to incorporate the identified factors using the presented techniques in an available-bandwidth-based flow admission control algorithm.

References

1. Dunkles, A., Gronvall, B., Voigt, T.: Contiki: a lightweight and flexible operating systems for tiny networked sensors. In: 9th Annual IEEE International Conference on Local Computer Networks (2004)
2. Jindal, A., Psounis, K., Liu, M.: Capest: a measurement-based approach to estimating link capacity in wireless networks. IEEE Trans. Mob. Comput. **11**(12), 2098–2108 (2012)

3. Nam, N.V., Guerin-Lassous, I., Victor, M., Cheikh, S.: Retransmission-based available bandwidth estimation in IEEE 802.11-based multihop wireless networks. In: 14th ACM Iternational Conference on Modeling, Analysis and Simulation of Wireless and Mobile Systems (2011)
4. Osterlind, F., Dunkels, A., Eriksson, J., Finne, N., Voigt, T.: Cross-level sensor network simulation with Cooja. In: 31st IEEE Conference on Local Computer Networks (2006)
5. Sarr, C., Chaudet, C., Chelius, G., Lassous, I.G.: Bandwidth estimation for IEEE 802.11-based ad hoc networks. IEEE Trans. Mob. Comput. 7(10), 1228–1241 (2008)
6. Xu, Y., Deng, J., Nowostawski, M.: Quality of service for video streaming over multi-hop wireless networks: admission control approach based on analytical capacity estimation. In: IEEE Conference on Intelligent Sensors, Sensor Networks and Information Processing (2013)
7. Yang, Y., Kravets, R.: Contention-aware admission control for ad hoc networks. IEEE Trans. Mob. Comput. 4(4), 363–377 (2005)
8. Youn, J.S., Packl, S., Hong, Y.G.: Distributed admission control protocol for end-to-end QoS assurance in ad hoc wireless networks. EURASIP J. Wirel. Commun. Netw. 1(163), 18 (2011)

OpenCV WebCam Applications in an Arduino-Based Rover

Valeria Loscrí[1], Nathalie Mitton[1(✉)], and Emilio Compagnone[2]

[1] Inria, Villeneuve-d'Ascq, France
nathalie.mitton@inria.fr
[2] University of Calabria, Rende, Italy

Abstract. In this work we design and implement Arduino-based Rovers with characteristics of re-programmability, modularity in terms of type and number of components, communication capability, equipped with motion support and capability to exploit information both from the surrounding and from other wireless devices. These latter can be homogeneous devices (i.e. others similar rovers) and heterogeneous devices (i.e. laptops, smartphones, etc.). We propose a Behavioral Algorithm that is implemented on our devices in order to supply a proof-of-concept of the effectiveness of Detection task. Specifically, we implement "Object Detection" and "Face Recognition" techniques based on OpenCV and we detail the modifications necessary to work on embedded devices. We show the effectiveness of controlled mobility concept in order to accomplish a task, both in a centralized way (i.e. driven by a central computer that assign the task) and in a totally distributed fashion, in cooperation with other Rovers. We also highlight the limitations of similar devices required to accomplish specific tasks and their potentiality.

Keywords: Rover · OpenCV · WebCam applications

1 Introduction

In the last few years, we have witnessed a technological development in robots, computing and communications, that have allowed the design of Network Robotic Systems (NRS) [14]. A NRS is composed by a robotic unit, able to communicate and cooperate both, with other similar units and with other interconnected devices. A very interesting feature of this type of systems is the possibility to "use" the same robots to accomplish different and complex tasks and the same set of robots may perform the same task in different conditions [10]. There have been increasing interests in deploying team of robots able to cooperate and to self-coordinate, to fulfill also complex tasks such as target tracking [3], disaster recovery [8], etc. NRS concept will marry in a very natural fashion with the concept of Swarm Robotics, where many robotic units are considered to cooperatively accomplish very complex tasks [11,13]. Beni [2] provides a very effective

This work has been partially supported by the FP7 VITAL.

M. Garcia Pineda et al. (Eds.): ADHOC-NOW Workshops 2014, LNCS 8629, pp. 261–274, 2015.
DOI: 10.1007/978-3-662-46338-3_21

and precise definition of what a swarm of robots: "The group of robots is not just a group. It has some special characteristics, which are found in swarms of insects, that is, decentralized control, lack of synchronization, simple and (quasi) identical members". In this work we overcome the concept of members that have to be similar by considering devices that are able to communicate with any interconnected node. In any case, we outline how the devices are able to cooperate and self-coordinate also without a central unit that drives them.

In this paper, we build an Arduino-based platform, that presents some important characteristics such as re-programmability, modularity in terms of type and number of components and communication capability. Our devices are also equipped with motion capability and are able to exploit the concept of controlled mobility, by reaching specific target locations. Our robots can leverage information both from the surrounding and from other interconnected devices (i.e. others similar robots, smartphones, laptops, etc.). Based on the acquired information, our devices will behave in a specific fashion, in order to accomplish a specific task assigned either by a central unit (e.g. a central computer where a human may ask the accomplishment of specific tasks) or by another robot. These actions are realized through a very simple Behavioral Algorithm that will be implemented onto the devices and will be tested in different scenarios, in order to verify the effectiveness and to highlight the critical aspect of the system. The robot will react based on external stimulus, acquired through a WebCam or communication equipment. In order to make the analysis of the acquired images feasible, we referred to a well-known vision tool, *Open Source Computer Vision Library*, OpenCV.

The main contributions of this work can be summarized as follows:

- Modifications of specific tools such as OpenCV libraries originally developed for more powerful devices (e.g. computer) to adapt them to distributed and constrained environments;
- Development of a testbed on real Arduino-based platforms;
- Proposal of a very simple Behavioral Algorithm to be implemented as a proof-of-concept of the developed tools and platform.

The rest of the paper is organized as follows. Section 2 describes the simple Behavioral Algorithm and the basic rules the robot accomplishes for searching target. Section 3 presents the tools we use to acquire data and how to exploit this information. Section 4 details how we modified existing tools in order to allow effective actions in a distributed and constrained environment. Section 5 presents the platform setup and implementation of the Behavioral Algorithm integrated with the modified libraries. Finally, we conclude this work in Sect. 6.

2   The Behavioral Algorithm

This section details our Behavior Algorithm for mobile robots. The main goal is to search for possible targets, identify them and thus, cover them by having robots reach them, in a distributed way. Each robot runs the same algorithm independently of the others and cooperate to fulfill a list of tasks, *a list of targets to reach*.

First of all it is worth defining what a target is. In this context, a target represents an object with specific characteristics in terms of shape and/or color, that has to been detected by the robot. The specific characteristics are defined by a human controller through a central computer, but our project can be easily modified, by including a dynamic definition of the target also defined by other inter-connected devices. Initially, each robot is assigned with a target. A same target can be assigned to several robots.

The algorithm is mainly composed of two phases: the Searching Phase and the Approaching Phase. During both phases, an underlying obstacle avoidance process is running. It i run on every robot when moving. More details about the obstacle avoidance implementation are given in Sect. 4.3. The details of the

---

Algorithm 1. Behavioral Algorithm

• **Local variables:** TargetFound = FALSE; TaskCompleted = FALSE;

```
 1: while the task list is not empty do
 2:     Move to next task on the list
 3:     while (TargetFound = FALSE) AND (TaskCompleted = FALSE) do
 4:         {Searching Phase}
 5:         while (TargetFound = FALSE) AND (TaskCompleted = FALSE) do
 6:             Δt ← Random().
 7:             Move forward for Δt at speed s.
 8:             Listen to other robots.
 9:             if Target reached by another robot then
10:                 TaskCompleted← TRUE.
11:             else
12:                 Scan environment.
13:                 if Target identified then
14:                     TargetFound←TRUE.
15:                 else
16:                     Change direction.
17:                 end if
18:             end if
19:         end while
            {Approaching Phase}
20:         while (TargetFound = TRUE) AND (TaskCompleted = FALSE) do
21:             Δt ← Random()
22:             Move toward target for Δt at speed s
23:             if Target is reached then
24:                 TaskCompleted← TRUE.
25:                 Advertise other robots.
26:             else
27:                 Listen to other robots.
28:                 if Target reached by another robot then
29:                     TaskCompleted← TRUE.
30:                 else
31:                     Check false positive.
32:                     if False positive identified then
33:                         TargetFound← FALSE.
34:                     end if
35:                 end if
36:             end if
37:         end while
38:     end while
39: end while
```

Behavioral Algorithm are given in the following pseudo-code, on Algorithm 1. Our algorithm will terminate when all tasks assigned to the robot are completed.

The robot starts with the task on top of its list. It first enters the Searching Phase (Lines 19–37 in Algorithm 1) which consists in locating the target. If the robot does not identify the pre-assigned target in its environment (in its vision field), it moves forward for an arbitrary time Δt (e.g. 1 or 2 s.) by following a Random Way Path (RWP)[6]. More precisely, it travels a prefixed distance then it stops and checks for the presence of the target within its 'new' environment. It then repeats this process while the target is not identified. Once the target is identified, the robot switches to the Approaching Phase (Lines 4–19 in Algorithm 1). The robot simply heads to the direction of the target till either reaching it or realizing that the target is a "false positive", due to for example to lights. In order to detect these "false positives", a specific mechanism is set up to allow the robot to make a new search without being influenced by the previous results as described in Sect. 4.2. It then switches back to the Searching Phase. Once the target is reached, the robot advertises other robots though its communication channel. Upon reception of this message, other robots abort their task to move to the next one. This latter phase can be revisited as a dynamic change of the task, based on the input that one of them broadcasts.

3   Background: Software and Hardware Used in This Work

In this section, we give a brief description of the hardware and software tools later used in this paper for our different implementations.

3.1   Hardware

In order to evaluate our contributions, we use the following hardware platform. An Arduino module [1] is set on an aluminum robotic platform with 4 wheels. Arduino allows us to re-program our node and to make it able to interact with the external environment. Since our devices are equipped with motion capabilities, we apply on them an ultrasound sensor in order to avoid obstacles (See Sect. 4.3). Moreover, we consider a WebCam to exploit images for task accomplishment. We used Open Source Computer Vision Library, OpenCV [12] libraries together with Computer Vision in order to "extract" meaningful data from the images. A very important component of our devices is the Inertial Movement Unit (IMU), a platform with an accelerometer, magnetometer and gyroscope. The IMU makes the Rover navigable. In effect, there exist other "easier" and well-performing solutions to realize the "navigation" function, such as stepper [7], but the ratio cost/efficiency of our solution is better than the other available. Another fundamental component of our nodes is the miniPC, that we modified in order to use a Linux platform instead of the Android available. The choice of this OS (Operating System) is related to the possibility to better control the platform with high level program languages, to increase the computational capability and to control external devices (i.e. those already mentioned), through an USB interface.

Fig. 1. HSV model.

3.2 Software

In order to make our devices able to perform some specific tasks, we focused on Object Detection techniques. More specifically, we consider a Computer Vision library, named OpenCV [12]. OpenCV has been developed by Intel and supplies a set of high level functions to acquire images and computer vision in real time. An important feature of OpenCV is that it allows the execution on different platforms (Linux, windows, etc.). We mainly focus on two specific techniques:

- *Face Recognition* based on the Cascade Classifiers method;
- *Object Detection* based on the HSV Model [5].

In the following, we detail these two techniques.

Object Detection via HSV. Hue, Saturation and Value (HSV) consists in a re-deployment of points of the RGB Cartesian Space into a cylindric space. The main goal of this operation is a more intuitive analysis of the colors in its components, since in the RGB, the analysis is very complex. In Fig. 1 we show how the HSV model works. In the figure, in each cylinder, the angle around the central axis represents the *tonality*, the distance from the central axis is the *saturation*, and the distance from the basis of the cylinder is the *value*.

Figure 2 shows the difference from an optical point of view obtained when we apply the HSV model on an RGB model. The evaluation of this difference has been very important for the usage of the HSV model as Object Detection. In order to make possible the detection operation, it is necessary to consider a color filtering phase based on the function inRange of OpenCV.

This filtering phase is paramount in the Object Detection, since we consider the form-color combination. It is worth noticing that the exact values for color filtering are not known and for that it is necessary the use of a specific tool that allow the extraction of HSV values from an image (see Fig. 3).

It is also important to notice that H channels range from 0 to 360° and in OpenCV this range is comprised from 0 to 180°. Through the function findContour in OpenCV, it is possible to detect elementary geometrical forms as squares, rectangles, circle, etc. and it is also possible to define proper own geometrical

Fig. 2. RGB vs HSV.

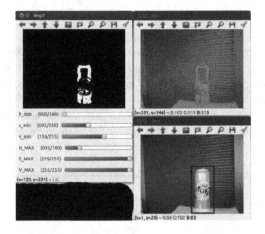

Fig. 3. Identification process of a can.

forms. In Fig. 3 we show the result we obtained by considering as objective the form-color detection of a can.

Face Recognition Based on Cascade Classificators. Face Recognition is an example of high-level recognition and is one of the most complicated examples of pattern individuation. The mathematical representation of a face is really complicated when compared to elementary geometric as triangles, circles, etc. In order to be able to realize the goal of Face Recognition, we need to formulate it as a *Machine Learning* problem, where an initial training phase is mandatory. In OpenCV, it is also possible to find some configuration to detect specific parts of a face, stored in XML files. If something different is necessary, it is possible to train the Classifier by using a proper own set of images with the specific object that need to be detected. The training phase needs to be realized in a very accurate way, since many different factors need to be taken into account, such as the brightness, quality, color, etc. More specifically, the training phase we considered is called *HaarTraining*, [4]. This procedure requires a lot of

Fig. 4. Identification of an human face. @Lena.

computational resources, since the training phase requires the analysis of many images to extract the information necessary to create a classifier. The operation is based on some intermediate phases, based on the analysis of "negative" and "positive" images. Specifically, the "positive" images are those that contain inside the target that has to be identified. The "negative" images are those without the target to be identified. The term "cascade" is referred to the fact that the classifier is the result of various simple classifiers (phases), that are applied in a sequential way either to a frame or to an image. In practice, the output of a classifier is the input of the next classifier and so on. In Fig. 4, we can observe the result of searching a human face inside an image, by applying the Cascade Classifiers method. It is thus very efficient.

4   Adaptation of Tools

This section presents how the tools described in Sect. 3 are modified to make the robots able to detect in an effective way a target object by considering information coming both from the surrounding (i.e. WebCam) and the other Rovers (by enabling communication among them). In this section, we will detail how the tools described above have been modified in order to be suitable with our embedded devices. It is worth noting, that the libraries of OpenCV have been conceived for powerful systems like laptops and computer, whereas we are considering embedded devices with limited computational resources. We focus on two specific techniques:

- Object Detection based on HSV model;
- Face Recognition based on Cascade Classifiers.

   The first fundamental step, for both techniques, consists on the choice of the right filter to analyze images/frames. In OpenCV various filters are available such as Canny, Gaussian, Kalman, ConDensation, etc. In order to test the different filters on our Rovers, we modeled them as Python models and we implemented them on the devices.

4.1 HSV Object Detection

The Object Detection task consists in a first phase named Detection Phase. The frame coming from the WebCam passes through a filter and has to be opportunely resized, due to the scarce computational resources of our device. After the resizing, the frame will be converted from RGB to HSV. It is also necessary to determine the characteristic values of a color. This problem has been faced by realizing a suitable interface as shown in Fig. 7.

Resizing an image makes the detection more difficult, but by iteratively applying a specific OpenCV filter that enlarges the image, we can overcome this issue. Unfortunately, the impurities of the color spectrum will be enlarged too. In order to solve this additional problem, we apply the *GaussianBlur* filter that makes the image more homogeneous. After these basic operations, the target definition algorithm can be integrated in the Behavioral Algorithm and is ready to be tested. The test phase shows an additional problem related to the "retrieving" of the right HSV values from the WebCam. This data generates a stream video that has to be sent to the GUI. By taking the limited resources of our device into account, we had to apply a MPEG encoding to the stream, whereas originally data was YUV. This encoding change alters the correct HSV values. It is also worth recalling that the HSV technique is based on the brightness and contrast concepts. In order to obtain a mechanism able to work in every environmental condition, we need powerful hardware, but we will show in Sect. 5, that in certain conditions (places with low light and with constant brightness), the algorithm implemented on our Rover works in a very effective way.

4.2 Cascade Classifiers Face Recognition

In order to realize the Face Recognition task, we formulate it as a *Machine Learning* problem and we consider a training phase named *HaarTraining* [9]. In the training phase, many factors need to be taken into account in a very accurate way, such as brightness, quality of the image, color, etc. The *Haar-Training* procedure requires a lot of computational resources, since the training phase is based on the analysis of many images in order to have the necessary information to realize a classifier. Concerning this training phase, we can distinguish the analysis of "positive" images (that contains the target that has to be identified) and "negative" images (where the target is not detected). In order to perform the Face Recognition task, we consider a suitable OpenCV Classifier. Also in this case, as in the previous application of HSV, we need to manage the image by resizing it and we convert the RGB model to the GRAY model. This latter step is necessary in order to decrease the amount of information to process and to improve the speed of the analysis. The next step consists in the integration of the data into the Behavioral Algorithm. After this integration, we performed some tests and we noticed that the movements of the Rovers and data of the Task seemed misaligned. Once again, the scarce processing resources have made a correct processing of data frames, impossible. In order to fix this problem, we reduced the number of frames processed in the unit of time. Instead of considering a continuous data flow, we limited the number of frames to 15, for each operation. In practice, we analyzed 15 frames at the center, 15 frames at left and 15 frames at right. Moreover, we included a "release" phase for the WebCam after each acquiring, that implies the presence of "settling" frames, where the WebCam exploits its focus functions to define the image. In Sect. 5, we experimentally determine the number of frames useful for this purpose and that have to be discarded from the analysis of the image. Another kind of issue is related to the "false positive" detection, namely some object that is not a face is wrongly identified as a target. In order to detect false positives, we modified the Recognition Algorithm, by considering "trustable" a detection where at least 3 over 10 frames recognize an object as target.

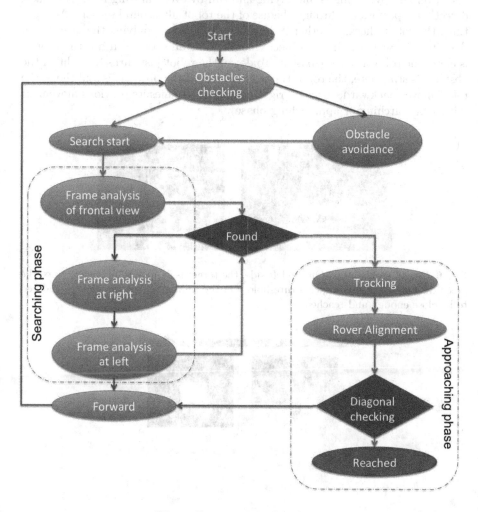

Fig. 5. Rover actions flowchart.

In summary, we can claim that the tools we considered for managing multimedia data, opportunely modified in order to work in an effective way with the Rovers, are really effective for detection tasks.

4.3 Implementation

In this section we describe how the various phases of the Behavioral Algorithm have been implemented into our Rover by using our adapted tools. The main actions performed by our devices are detailed in Fig. 5.

Obstacle Detection and Bypassing. As mentioned in Sect. 2, an obstacle detection and bypassing is underlying and run by every moving robot. Obstacle detection is performed through the use of the robot ultrasound sensors. At every time, the robot checks whether it detects an object by applying the mechanism described in Sect. 4.1. If this object (identified as an obstacle if not the target) is close, the robot has to bypass it. To do so, the robot just turns $\alpha°$ right. If the obstacle is still there, the robot turns $2 \times \alpha°$ left. If the obstacle is still there, the robot moves backward, turns $\alpha°$ right again and resumes its previous movement (either in searching or approaching phase).

Fig. 6. Target reachability. On the left side, the target is supposed unreached since the green diagonal length is under a threshold. On the right side, the target is considered to be close enough and reached.

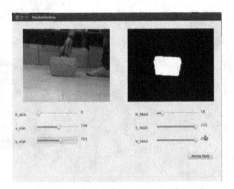

Fig. 7. HSV Gui interface. The orange object is the target assigned.

Searching Phase. After a preliminary check of the obstacle presence, the Rover is able to perform the Searching Phase, by acquiring the data frames in a suitable way. Data acquisition is performed as Whether the robot detects the target, it will start to reach it. The searching phase is performed through the processing of the WebCam frames. From the analysis of the frames, the device makes a comparison with the target, and will be able to establish whether the target has been found or not. Objets are detected by using adapted tool described in Sect. 4.1 and identified by using the face recognition tool as explained in Sect. 4.2.

Approaching Phase. In this phase, the robot simply heads in direction of the target till either reaching it or realizing that the target is a "false positive". To detect "false positive", at every step, the robot simply runs the face recognition tools detailed in Sect. 4.2 again to check whether it confirms the detected object in the right target. The target is considered as reached based on the size of objects on the WebCam pictures. The robot stops when the diagonal length of the identified target is larger than a threshold as shown on Fig. 6.

Once the target is reached, the robot needs to advertise its peers to allow them to abort their current task and move to the next one in the list. To do so, the robot simply light on its red leds.

5   Testbed Description and Results

This section describes the performance of the Behavioral Algorithm implemented on our Rovers in order to accomplish Detection and Covering Tasks.

We consider two scenarios: the first one is based on the use of the HSV technique and the last is a Face Recognition Task in which the Rovers are also able to communicate to each other in order to exchange information about the tasks.

5.1   HSV Scenario

In this scenario the task, robots are assigned the detection of a specific object, namely the Rover has to move to towards the target. In Fig. 7 one can observe the generation of the HSV Values, by using the tools of the HSV Gui.

When the HSV Task starts, the Behavioral Algorithm allows the process of the Searching Phase and Approaching Phase and the Rover tries to accomplish the assigned task. In this work, we were able to test the effectiveness of the algorithm implemented, by considering an environment with low light and constant brightness conditions. In Fig. 8 we can observe a snapshot of the experiments conducted. In order to test the effectiveness of the algorithm, we considered many different scenarios, by keeping similar brightness conditions, and the Rover was able to accomplish its assigned task in all the tests.

Fig. 8. The target is reached.

5.2   Face Recognition Scenario

In this scenario the target to be detected is a human face.

The goal of the test is the localization of the face and the first Rover that identifies and reaches the Target, sends an "alert" message to the others, in order to "attract" them. The other devices are required to dynamically change their task in this case. In Fig. 9(a), we show the Laboratory room where the test has been realized.

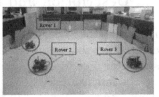

(a) The room where the tests have been realized.

(b) Searching Phase for the three Rovers.

(c) Identification of the Target.

Fig. 9. The phases to search and identify a target

We realized some tests to verify the exact number of frames used by the WebCam as "settling" frames, and we verified that 5 frames are necessary and have to be discarded. In practice, the first 5 frames cannot be analyzed.

In the first phase, the Rovers start the Target Searching Phase (see Fig. 9(b)).

In the specific test we are considering, the Rover 1 is the first one that detects the target, and then, it starts the Approaching Phase as shown in the Fig. 10(a), whereas Rovers 2 and 3 continue in their searching phase. When Rover 1 reaches the target, it broadcasts a message to Rovers 2 and 3 by switching on its red led, in order to be visually identified as a target (see Fig. 10).

(a) The led is switched on.   (b) The others two Rovers cooperate with the first one and reach it.

Fig. 10. Cooperation among the Rovers.

When Rovers 2 and 3 receive the message, they are able to dynamically move to the next task and, their objective will be modified (see Fig. 10(b)).

We repeated this kind of tests many times, by changing the initial position of the Rovers and by modifying the position of the target. We noticed, that the only problem related with this kind of experiments was an excessive luminosity. In fact, in this case the limited number of frames elaborated by the devices is not enough accurate.

6 Conclusions and Future Works

In this work we faced the real implementation of mobile devices based on Arduino platforms and equipped with many sensors and a WebCam. The devices are able to accomplish Detection Tasks of objects and faces. The target objects can be identified by exploiting the combination of shape/color. On the Rovers we implemented a Behavioral Algorithm, where the commands related with the detection have been integrated. In order to make effective the tools considered for the detection purpose, we opportunely modified them, considering the scarce resources available and the movement of the Rovers. We realized a proof-of-concept of our Behavioral Algorithm and we have shown its effectiveness in different scenarios. The main issue is related with the brightness conditions of the environment and with rapid changing of the luminosity, but the use of more powerful computational devices allows to overcome easily this kind of matter. As future work, we will consider quad-core mini-pc, and we will test our Behavioral Algorithm in more variable conditions of luminosity.

References

1. http://www.arduino.cc/. Accessed 23 August 2013
2. Beni, G.: From swarm intelligence to swarm robotics. In: Şahin, E., Spears, W.M. (eds.) Swarm Robotics 2004. LNCS, vol. 3342, pp. 1–9. Springer, Heidelberg (2005)

3. Blzovics, L., Csorba, K., Forstner, B., Hassan, C.: Target tracking and surrounding with swarm robots. In: 2012 IEEE 19th International Conference and Workshops on Engineering of Computer-Based Systems, pp. 135–141 (2012)
4. http://note.sonots.com/SciSoftware/haartraining.html
5. http://en.wikipedia.org/wiki/HSL_and_HSV
6. Hyytiä, E., Virtamo, J.: Random waypoint model in n-dimensional space. Oper. Res. Lett. **33**(6), 567–571 (2005)
7. Jinasena, K.K., Meegama, R.G.N.: Design of a low-cost autonomous mobile robot. Int. J. Robot. Autom. (IJRA) **2**(1) (2011)
8. Kuntze, H.-B., Frey, C.W., Tchouchenkov, I., Staehle, B., Rome, E., Pfeiffer, K., Wenzel, A., Wollenstein, J.: SENEKA - sensor network with mobile robots for disaster management. In: Proceedings of IEEE Conference on Technologies for Homeland Security (HST), pp. 13–15, November 2012
9. Lienhart, R., Maydt, J.: An extended set of haar-like features for rapid object detection. In: IEEE ICIP 2002, vol. 1, pp. 900–903, September 2002. http://www.lienhart.de/ICIP2002.pdf
10. Lundh, R., Karlsson, L., Saffiotti, A.: Autonomous functional configuration of a network robot system. Robot. Auton. Syst. **56**, 819–830 (2008)
11. Nagi, J., Di Caro, G.A., Giusti, A., Gambardella, L.: Convolutional support vector machines for quantifyingl the visual learning and recognition progress in swarm robotic systems. In: Proceedings of the 11th International Conference on Machine Learning and Applications (ICMLA 2012), Boca Raton, Florida, 12–15 December 2012
12. opencv.org
13. Pinciroli, C., Trianni, V., O'Grady, R., Pini, G., Brutschy, A., Brambilla, M., Mathews, N., Ferrante, E., Di Caro, G., Ducatelle, F., Stirling, T., Gutierrez, A., Gambardella, L.M., Dorigo, M.: ARGoS: a modular, multi-engine simulator for heterogeneous swarm robotics. In: Proceedings of the IEEE/RSJ International Conference on Intelligent Robots and Systems (IROS), San Francisco, USA, 25–30 September 2011
14. Sanfeliu, A., Hagita, N., Saffiotti, A.: Robotics and autonomous systems. Elsevier Robot. Auton. Syst. **56**, 793–797 (2008)

A Generalized Data Preservation Problem in Sensor Networks – A Network Flow Perspective

Bin Tang[1]([⊠]), Rajiv Bagai[2], FNU Nilofar[2],
and Mehmet Bayram Yildirim[3]

[1] Department of Computer Science, California State University,
Dominguez Hills, USA
btang@csudh.edu
[2] Department of Electrical Engineering and Computer Science,
Wichita State University, Wichita, USA
rajiv.bagai@wichita.edu, nilu23@gmail.com
[3] Department of Industrial and Manufacturing Engineering,
Wichita State University, Wichita, USA
bayram.yildirim@wichita.edu

Abstract. Many emerging sensor network applications require sensor node deployment in challenging environments that are remote and inaccessible. In such applications, it is not always possible to deploy base stations in or near the sensor field to collect sensory data. Therefore, the overflow data generated by some nodes is first offloaded to other nodes inside the network to be preserved, then gets collected when uploading opportunities become available. In this paper, we study a *generalized data preservation problem* in sensor networks, whose goal is to minimize the total energy consumption of preserving data inside sensor networks, given that each node has limited battery power. With an intricate transformation of the sensor network graph, we demonstrate that this problem can be modeled and solved as a minimum cost flow problem. Also, using data preservation in sensor networks as an example, we show that seemingly equivalent maximum flow techniques can result in dramatically different network performance. Much caution thus needs to be exercised while adopting classic network flow techniques into sensor network applications, despite successful application of network flow theory to many existing sensor network problems. Finally, we present a load-balancing data preservation algorithm, which not only minimizes the total energy consumption, but also maximizes the minimum remaining energy of nodes that receive distributed data, thereby preserving data for longer time. Simulation results show that compared to the existing techniques, this results in much evenly distributed remaining energy among sensor nodes.

Keywords: Data preservation · Network flow · Sensor networks

© Springer-Verlag Berlin Heidelberg 2015
M. Garcia Pineda et al. (Eds.): ADHOC-NOW Workshops 2014, LNCS 8629, pp. 275–289, 2015.
DOI: 10.1007/978-3-662-46338-3_22

1   Introduction

Data preservation is critical in sensor networks that are deployed in challenging environments, such as underwater or ocean sensor networks [1,2], acoustic sensor networks [3], and sensor networks monitoring volcano eruption and glacial melting [4,5]. Due to limited accessibility in these environments, it is not possible to deploy a base station with power outlet near or inside the sensor network to collect the data. Meanwhile, each sensor node has limited storage capacity and a finite battery supply. Sensor nodes close to the event of interest constantly generate large amounts of sensory data, which can quickly exhaust their limited storage capacity. We refer to these sensor nodes with exhausted data storage as *source nodes*. Other sensor nodes that still have available storage are referred to as *destination nodes*. In order to prevent data loss, the overflow data generated at source nodes needs to be offloaded to some destination nodes before uploading opportunities (such as data mules [6,7] or low rate satellite link [8]) become available. We refer to this process as *data preservation in sensor networks*.

Previous research on data preservation [9] has two major limitations. First, it assumes that each node has "enough" battery power, so data items can always be offloaded from source to destination nodes using shortest paths (in terms of number of hops) between them. In this paper, we study a more challenging and general problem, wherein each node has limited battery power, therefore shortest paths may not always be viable. By defining and solving a *generalized data preservation problem*, we demonstrate that even with low energy levels of sensor nodes, optimal data preservation is still achievable. In particular, by fine-tuning the costs and capacities of edges of the flow network transformed from the sensor network graph, we show that the generalized data preservation problem is equivalent to the minimum cost flow problem, which can be solved optimally and efficiently [10]. Second, works such as [9] only focus on total energy consumption in data preservation and do not pay attention to load-balancing of energy consumption of individual sensor nodes. Once a node storing data items depletes its energy, the data preservation fails. Therefore, maintaining load-balancing among sensor nodes is critical for data preservation. We achieve load-balancing by maximizing the minimum remaining energy among all destination nodes. In contrast, Hou et al. [11] do not minimize the total energy consumption of data preservation, while Patel et al. [12] provide separate solutions for minimizing the routing cost and maximizing the minimum remaining energy.

Network flow theory [10] has been adopted to solve many fundamental problems in sensor networks, including data gathering [12–14], data aggregation [15], and clustering [16]. Classic network flow problems including maximum flow, minimum cost flow and multi-commodity flow have all been employed in sensor network research (Sect. 2 contains a review of such work). Using data preservation as an example, however, we demonstrate that much caution needs to be exercised while adopting classic maximum flow techniques into sensor network applications, as seemingly equivalent techniques can result in dramatically different network performance. In particular, we show that different maximum flow algorithms, viz. the Ford-Fulkerson Algorithm and the Edmonds-Karp

Algorithm [10], which differ only in time complexity, yield dramatically different energy consumption in sensor networks.[1]

We also empirically compare and analyze the performance of our chosen maximum flow algorithm (i.e., Edmonds-Karp Algorithm) and minimum cost flow algorithm, for various network scenarios. Based on simulation results, we draw some conclusions as to what extent and how well the network flow algorithms can be applied to solve sensor network problems.

Paper Organization. The rest of the paper is organized as follows. In Sect. 2 we give an overview of sensor network research that adopts network flow algorithms. In Sect. 3 we formulate and solve the generalized data preservation problem. We also solve a related problem that finds the maximum number of data items that can be offloaded, and show that two classic maximum flow algorithms yield very different performance. Section 4 proposes load-balancing data preservation algorithm. Section 5 gives some analysis of the simulation results. We conclude the paper in Sect. 6 and discuss possible future work.

2   Related Work

The network flow algorithms adopted in sensor network research include maximum flow [11,13,14,17], minimum cost flow [9,12,18], and multi-commodity flow [15]. Below we give a brief review.

Maximum Flow Problem: Hong et al. [14] study store-and-gather problems in sensor networks, and show that these are essentially flow maximization under vertex capacity constraint, which reduces to a standard maximum flow problem. Bodlaender et al. [13] study integer maximum flow in wireless sensor networks with energy constraint. They show that despite the efficiency of traditional maximum flow methods, integer maximum flow in sensor networks is indeed strongly NP-complete and in fact APX-hard [19], which means it is unlikely to have a polynomial time approximation scheme. Xue et al. [17] let sensory data from different sensor nodes have different priorities, and study how to preserve data inside the network with highest priorities. They model the problem as a maximum weighted flow problem, wherein different flows have different weights. Hou et al. [11] study the data preservation problem in intermittently connected sensor networks and design a maximum flow based algorithm to maximize data preservation time in the network. Besides, they observe that due to energy constraints at sensor nodes, it is possible that not all overflow data items can be preserved. They propose a Modified Edmonds-Karp Algorithm to find out if this is the case. We show that with more intricate transformation of the flow network, Edmonds-Karp Algorithm can be applied directly without being modified.

Minimum Cost Flow Problem: Patel et al. [12] minimize the energy cost of sending data packets from sensor nodes to base stations while satisfying the

---

[1] We focus here only on the data preservation problem in sensor networks [9,11,17], but our findings are applicable to many other sensor network applications as well.

capacity limits of wireless links, and propose a routing protocol based on the minimum cost flow algorithm. Ghiasi et al. [16] study a so called balanced k-clustering problem in sensor networks, wherein each of the k clusters is balanced (in terms of number of sensor nodes) and the total distance between sensor nodes and master nodes is minimized. They show that the k-clustering problem can be modeled as a minimum cost flow problem. Tang et al. [9] formulate the energy-efficient data redistribution problem in data-intensive sensor networks as a minimum cost flow problem and present a distributed algorithm.

Multi-commodity Flow Problem: Xue et al. [15] study energy efficient routing for data aggregation in wireless sensor networks, with the goal of maximizing the lifetime of the network. The resulting model is a multi-commodity flow problem, where each commodity represents the data generated from a sensor node. Since multi-commodity flow problem is NP-hard, they propose a fast ϵ-approximation algorithm, and extend their algorithm for multiple base stations.

In contrast to existing research, our work takes a new perspective by studying how different network flow algorithms could have different effect on sensor network performance such as energy consumption. We believe this is an important effort – as shown in Sect. 3.2, network flow modeling does not necessarily take into account resource consumption/allocation in a network-specific context.

3   Generalized Data Preservation Problem

3.1   System Model and Problem Formulation

System Model. We model the sensor network as an undirected graph $G(V, E)$, where $V = \{1, 2, ..., |V|\}$ is the set of $|V|$ nodes, and E is the set of $|E|$ edges. There are p source nodes, denoted as V_s. Without loss of generality, let $V_s = \{1, 2, ..., p\}$. Source node i is referred to as SN i. Let d_i denote the number of overflow data items SN i needs to offload. Let $q = \sum_{i=1}^{p} d_i$ be the total number of data items to be offloaded in the network. Let c_i be the available free storage space (in terms of number of data items) at sensor node $i \in V$. Note that a source node does not have available storage space.

Sensor node i has a finite and unreplenishable initial energy E_i. We adopt first order radio model [20] as the energy model for wireless communication. In this model, for R-bit data over distance l, the transmission energy $E_t(R, l) = E_{elec} \times R + \epsilon_{amp} \times R \times l^2$, and the receiving energy $E_r(R) = E_{elec} \times R$. E_{elec} is the energy consumption per bit on the transmitter circuit and receiver circuit, and ϵ_{amp} calculates the energy consumption per bit on the transmit amplifier. For densely and uniformly deployed sensor nodes, we can approximate that $E_t(R, l) = E_r(R)$. To be consistent with the assumption in [9] that energy consumption of sending a data item from a source node to a destination node equals the number of hops it traverses, we assume that for each node, sending or receiving a data item each costs 0.5 unit of its energy.

Problem Formulation. Let $D = \{D_1, D_2, ..., D_q\}$ denote the set of q data items to be offloaded in the entire network. Let $S(i) \in V_s$, where $1 \le i \le q$, denote the *source node* of data item D_i. An *offloading function* is defined as $r : D \to V - V_s$, indicating $D_i \in D$ is offloaded from $S(i)$ to its *destination node* $r(i) \in V - V_s$. Let $P_i : S(i), ..., r(i)$, referred to as the *offloading path* of D_i, be the sequence of distinct sensor nodes along which D_i is offloaded from $S(i)$ to $r(i)$ (note that the offloading path is not necessarily the shortest path between source node and destination node). Let $\mathcal{E}_i$ be the energy consumption of offloading D_i from $S(i)$ to $r(i)$ following P_i ($\mathcal{E}_i$ equals the number of nodes on P_i minus one). Let E_i' denote i's energy level after all the q data items are offloaded. The objective of generalized data preservation problem is to find an offloading scheme r and a set of paths $\mathcal{P} = \{P_1, P_2, ..., P_q\}$, to send each of the q data items to its destination node, such that the total energy consumption in this process is minimized, i.e. $\min_{r,\mathcal{P}} \sum_{1 \leq i \leq q} \mathcal{E}_i$, under the energy constraint: $E_j' \geq 0, \forall j \in V$, and the storage capacity constraint: $|\{i \mid r(i) = j, 1 \leq i \leq q\}| \leq c_j, \forall j \in V$.

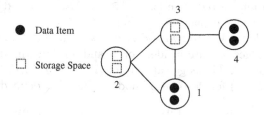

Fig. 1. A small sensor network of 4 nodes.

Example 1. Figure 1 gives an example of the generalized data preservation problem in a small sensor network of 4 nodes. Nodes 1 and 4 are source nodes, with 2 and 2 data items to offload, respectively. Nodes 2 and 3 are destination nodes, with 2 and 2 available storage spaces, respectively. The initial energy of each node is 10. The optimal solution is that node 1 offloads its two data items to node 2, while node 4 offloads its two data items to node 3, resulting in minimum total energy consumption of 4. Other solutions are non-optimal.

3.2 Maximum Flow Algorithms to Determine the Maximum Number of Offloaded Data Items

If the energy levels get low, not all the overflow data items at source nodes can be offloaded. Therefore, a related question is: *What is the maximum number of data items that can be offloaded given that the energy levels of sensor nodes are low?* For example, in the sensor network of Fig. 1, if the initial energy level of each node is 0.5 (instead of 10), source node 1 and 4 can each only offload 1 data item, and destination node 2 and 3 can each receive and store 1 data item.

Lemma 1. *In an optimal solution that maximizes number of offloaded data items, a SN does not relay data unless it finishes offloading all its own data items, a destination node does not relay data unless its own storage is full.*

Proof: By way of contradiction, assume that in the optimal solution, there is a SN B that serves as a relaying node before finishing offloading all its data items. That is, there exists in the optimal solution an offloading path $P : A, ..., B, ..., C$, which offloads the overflow data items of SN A to destination node C, while SN B still has its own overflow data items to offload. Assume that a data items are offloaded from A, and B still has b amounts of data items to offload. We therefore can select the path $P' : B, ..., C$, along which B offloads $\min(a, b)$ data items to C (A offloads $\max(0, a - b)$ data items to C along P). This strategy achieves the same maximum amount of offloaded data, while costing less energy than the optimal solution. The argument for destination nodes is similar.   □

To find the maximum amount of data items offloaded, we first transform the undirected graph $G(V, E)$ into a new directed graph $G'(V', E')$ as follows:

I. $V' = V \cup \{s, t\}$, where s is the new source node and t is the new sink node.
II. Replace each undirected edge $(i, j) \in E$ with two directed edges (i, j) and (j, i). Set their edge capacities as infinity.
III. Split each node $i \in V$ into two nodes: *in-node* i' and *out-node* i''. All incoming directed edges of i are incident on i' and all outgoing directed edges of i emanate from i''. The edge capacity of (i', i'') is $f(E_i, d_i)$ for source node SN i, and $f(E_i, c_i)$ for destination node i. $f(x, y)$ is defined as:

$$f(x, y) = \begin{cases} 2x & \text{if } (x < y/2), \\ x + y/2 & \text{otherwise.} \end{cases} \tag{1}$$

IV. Connect s to in-node of SN $i \in V_s$ with an edge of capacity d_i. Connect out-node of destination node $j \in V - V_s$ to t with an edge of capacity c_j.

Therefore $|V'| = 2|V| + 2$ and $|E'| = 2|E| + 2|V|$. Figure 2(a) shows the transformed graph G' of the sensor network in Fig. 1.

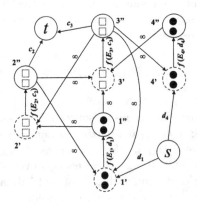
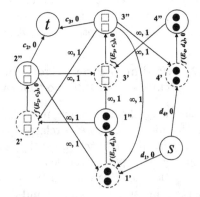

(a) Maximum flow problem.   (b) Minimum cost flow problem.

Fig. 2. The transformed graphs of the sensor network in Fig. 1.

Theorem 1. *Finding the maximum number of offloaded data items in $G(V, E)$ is equivalent to finding the maximum flow in $G'(V', E')$.*

Proof: First we explain the rationale for $f(x, y)$ in Eq. 1. We focus on source nodes, but the same analysis works also for destination nodes. According to Lemma 1, each source node offloads its own data items before relaying data items for others. Specifically, for SN i,

- when $E_i < d_i/2$, SN i does not have enough energy to offload all its d_i data items, since it needs 0.5 unit of energy for each data item. But in Fig. 2(a), each unit of flow reduces one unit of edge capacity, signifying one unit of energy cost for SN i. Therefore the edge capacity of (i', i'') in Fig. 2(a) is set as $2 \times E_i$, which is the largest amount of offloaded data items allowed by E_i.
- when $E_i \geq d_i/2$, SN i has enough energy to offload all its d_i data items. We set the edge capacity of (i', i'') in Fig. 2(a) as $E_i + d_i/2$. If it does offload all d_i data items, the edge capacity of (i', i'') becomes $E_i + d_i/2 - d_i = E_i - d_i/2$, which is exactly SN i's energy after it offloads all its d_i data items. Otherwise,[2] since it will not serve as relaying nodes either according to Lemma 1, adding $d_i/2$ upon E_i does not increase SN i's ability to offload more data.

Now if the value of the maximum flow from s to t in Fig. 2(a) is f, with f_i amount of flow on edge (s, i') and $\sum_{i=1}^{p} f_i = f$, there must be f_i amount of net flow out of SN i, meaning SN i offloads f_i amount of its own data items. On the other hand, if SN i can offload its f_i number of data items ($f_i \leq d_i$) following an offloading path P_i from SN i to a destination node, then in Fig. 2(a), it can offload f_i units of flow from s to t, without violating the capacity conditions of edges, giving maximum $\sum_{i=1}^{p} f_i$ amount of flow.   □

However, above maximum flow modelling does not fully address data preservation from the efficient resource allocation perspective. As we will see next, different maximum flow algorithms could result in very different energy consumption in sensor networks.

Comparing Ford-Fulkerson and Edmonds-Karp. Both Ford-Fulkerson and Edmonds-Karp are classic maximum flow algorithms. In each iteration of Ford-Fulkerson, an arbitrary augmenting path is selected in the residual graph to push flow from source to sink, whereas in Edmonds-Karp, a shortest augmenting path is selected. The time complexity of Ford-Fulkerson is $O(|E'|C)$, where C is the value of a maximum flow in G'. The time complexity of Edmonds-Karp is $O(|V'||E'|^2)$. Note that Hou et al. [11] designed a modified Edmonds-Karp maximum flow algorithm, called MEA, to determine the maximum number of data items offloaded. Our findings are an improvement upon theirs. First, Theorem 1 shows that with intricate specification of the edge capacity of the transformed graph (i.e., Eq. 1), any maximum flow algorithm can be directly applied to the transformed graph without modification. More fundamentally, we observe

---

[2] Note that this is possible when destination nodes around SN i do not have enough energy to store all d_i data items.

that when being applied to solve sensor network problems, different classic maximum flow algorithms, namely Ford-Fulkerson algorithm and Edmonds-Karp algorithm, could result in very different energy consumption, even though both achieve maximum amount of flow and only differ in time complexity. This is not explored in [11].

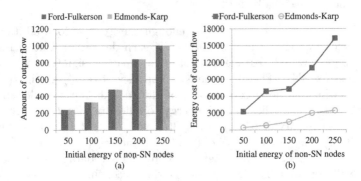

Fig. 3. Comparison between the Ford-Fulkerson and the Edmonds-Karp algorithms of the flow amounts and costs of their output flows for 1000 data items at SN nodes, over different initial energy levels of non-SN nodes.

To illustrate, we create a 10×10 grid network (with 100 nodes). One node is randomly selected as the source node with 1000 data items to offload. For each destination node, its storage capacity is 20, and its initial energy levels is varied as 50, 100, 150, 200, 250.[3] Fig. 3(a) shows that both algorithms offload the same number of data items because both achieve maximum flow. However, Fig. 3(b) shows that Edmonds-Karp costs much less energy than Ford-Fulkerson does, and the difference gets larger with the increase of energy levels.

3.3   Minimum Cost Flow Algorithm

The minimum cost flow problem [10] is the following: Given a graph in which each edge has a capacity and a cost, some nodes are supply nodes and some are demand nodes, and total supply equals total demand; the goal is to find flows from supply nodes to demand nodes with minimum total cost while the capacity constraint at each edge is satisfied. To find the optimal algorithm for generalized data preservation, we first transform undirected graph $G(V, E)$ into another new directed graph $G''(V'', E'')$. Much of the transformation is the same as the one in Sect. 3.2. Additionally, for any edge (i, j) in G, we set the costs of corresponding edges (i'', j') and (j'', i') in G'' to be 1. We set the costs of all other edges in G'' as 0. Finally, we set both the supply at s and the demand at t as $\sum_{i=1}^{p} d_i$, the total number of data

---

[3] We assign the source node large enough energy so that all 1000 data items can be offloaded. Otherwise, the amount of data offloaded in the network (i.e., the effect of maximum flow algorithms) is mainly limited by the energy level of source node, making the comparison less interesting.

items to be offloaded in the entire network. Figure 2(b) shows the transformed network graph G'' corresponding to the sensor network in Fig. 1.

Theorem 2. *The generalized data preservation problem in $G(V, E)$ is equivalent to the minimum cost flow problem in $G''(V'', E'')$.*

Proof: We show that a minimum cost flow from s to t in G'' solves the generalized data preservation problem in G optimally. Specifically, we show that, (a) it offloads all the data items from their source nodes to some destination nodes, and (b) it incurs minimum energy cost in this process.

A minimum cost flow from s to t must include d_i amount of flow on edge (s, i') in G'' ($1 \leq i \leq p$), since the capacity of (s, i') is d_i and the total amount of flow from s to t is $\sum_{i=1}^{p} d_i$. This signifies that in G, d_i amount of data items are offloaded from SN i. For any data item D_i in G, its corresponding flow in G'' goes from s to $S(i)'$, $S(i)''$, ..., $r(i)'$, $r(i)''$, and ends at t, indicating that D_i is finally stored at destination node $r(i)$. Besides, the capacity of edge (i'', t) being c_i, the storage capacity of destination node i, guarantees that in G, a destination node never stores more than its storage capacity allows.

For an edge (i, j) in G, sending a data item between i and j costs one amount of energy, which is accurately captured in G'', wherein the costs of corresponding edges (i'', j') and (j'', i') are 1 while costs of others are all 0. The minimum cost of sending $\sum_{i=1}^{p} d_i$ amount of flow from s to t in G'' is therefore incurring minimum amount of energy cost offloading d_i amount of data from SN i ($1 \leq i \leq p$). □

Time Complexity. There are various polynomial algorithms to solve minimum cost flow problem. In this paper, we use the algorithm and implementation by Goldberg [21, 22] due to its practical nature. This algorithm has the time complexity of $O(|V''|^2|E''|log(|V''|\mathcal{C}))$, where $|V''|$, $|E''|$, and $\mathcal{C}$ are, respectively, the number of nodes, edges, and the maximum capacity of an edge in graph G''. Since $|V''| = 2|V| + 2$ and $|E''| = 2|E| + 2|V|$, the time complexity of the minimum cost flow algorithm is therefore $O(|V|^2|E|log(|V|\mathcal{C}))$.

4   Data Preservation Problem With Load Balancing

Problem Formulation. The goal of data preservation problem with load-balancing is first to minimize the total energy consumption in data preservation; then among the minimum total energy consumption solutions, to find the one that maximizes the minimum remaining energy among all the destination nodes. Specifically, we find an offloading scheme r and a set of paths $\mathcal{P} = \{P_1, P_2, ..., P_q\}$, to offload each of the q data items to its destination node, such that the total energy consumption in this process is minimized, i.e. $\min_{r, \mathcal{P}} \sum_{1 \leq i \leq q} \mathcal{E}_i$, and the minimum remaining energy among all the destination nodes is maximized, i.e., $\max_{r, \mathcal{P}} \min_{1 \leq i \leq q} E'_{r(i)}$.

Finding All Shortest Paths Between Nodes In Grid Networks. Before presenting the load-balancing algorithm (Algorithm 2), we first find all the shortest paths between any given two nodes in a grid network (Algorithm 1). There is extensive research on finding the k shortest simple paths in a directed weighted graph [23,24]. In this paper we use a grid network for clarity of presentation, and design a much simpler recursive algorithm.[4] The algorithm works as follows. In the base case, when the source and destination nodes are the same, it returns only one path with just that node. Otherwise, this algorithm returns paths obtained by appending the source node to all shortest paths from the nodes that are one step closer to the destination node, on each of the x and y axes.

Algorithm 1. Finding All Shortest Paths Between Two Nodes In Grids.
Input: The coordinates of two nodes: (x_1, y_1) and (x_2, y_2)
Output: Set of all shortest paths between them

$AllShortestPaths(x_1, y_1, x_2, y_2)$

1. $xStep = \begin{cases} 1 & \text{if } x_1 < x_2 \\ -1 & \text{if } x_1 > x_2 \\ 0 & \text{otherwise} \end{cases}$

2. $yStep = \begin{cases} 1 & \text{if } y_1 < y_2 \\ -1 & \text{if } y_1 > y_2 \\ 0 & \text{otherwise} \end{cases}$

3. **if** $(xStep == 0$ and $yStep == 0)$ **RETURN** $\{\langle(x_1, y_1)\rangle\}$;
4. $S = \phi$;
5. **if** $(xStep \neq 0)$
   $S = S \cup \{(x_1, y_1) :: P \mid P \in AllShortestPaths(x_1 + xStep, y_1, x_2, y_2)\}$;
6. **if** $(yStep \neq 0)$
   $S = S \cup \{(x_1, y_1) :: P \mid P \in AllShortestPaths(x_1, y_1 + yStep, x_2, y_2)\}$;
7. **RETURN** S.

Time Complexity of Algorithm 1. Let $X = |x_1 - x_2|$, and $Y = |y_1 - y_2|$. There are $C_{(X+Y)}^X$ shortest paths between (x_1, y_1) and (x_2, y_2), where $C_{(X+Y)}^X$ is the number of ways of selecting X items from a set of $(X + Y)$ items. Finding each shortest path requires $O(X + Y)$ append operations. Therefore, the time complexity of Algorithm 1 is $O((X + Y)C_{(X+Y)}^X)$.

Data Preservation With Load-balancing. Load-balancing data preservation algorithm (Algorithm 2) works as follows. First, we use minimum cost flow algorithm (Sect. 3.3) to achieve minimum energy consumption as well as to find the destination nodes of all data items (line 1). Then, for each data item, we find all the shortest paths between its SN and the destination node using Algorithm 1 (line 5). Finally, among all the shortest paths for this data item, we choose the one whose minimum energy-node has the maximum energy as its offloading path (line 7–11). This way it ensures that destination nodes with less energy do not relay data items, hence saving energy and preserving their stored data for longer

---

[4] However, we are aware of that [23,24] present much more efficient algorithms using complex data structures, which are difficult to implement.

time. Although we can not prove the optimality of this algorithm, we show in Sect. 5 that it performs better than the one without load-balancing.

Algorithm 2. Load-balancing Data Preservation Algorithm.
Input: The sensor network graph and set of data items D
Output: Set of offloading paths: $\{P_j | D_j \in D\}$
1.   Run minimum cost flow algorithm;
2.   **for** each source node with coordinate (x_1, y_1)
3.   **for** each of its data item
4.   Let the coordinate of its destination node be (x_2, y_2);
5.   Get all the shortest paths from $AllShortestPaths(x_1, y_1, x_2, y_2)$;
6. $MaxMinEnergy = 0$;
7.   **for** each such shortest path P_j
8. $MinEnergy[j] = $ minimum energy of nodes in P_j;
9.   **if** $(MinEnergy[j] > MaxMinEnergy)$
                        $MaxMinEnergy = MinEnergy[j]$;
10.   **end for;**
11.   Find path, say P_k, with $MaxMinEnergy$, as its offloading path;
12.   **end for;**
13.   **end for;**
14.   **RETURN** all the offloading paths.

Time Complexity of Algorithm 2. The length of any shortest path in a grid of $|V|$ nodes is at most $2\sqrt{|V|}$, since $\sqrt{|V|}$ is the number of nodes along either x or y axis. The largest number of shortest paths between any two nodes is therefore $C_{2\sqrt{|V|}}^{\sqrt{|V|}}$. The time complexity of the minimum cost flow algorithm is $O(|V|^2|E|log(|V|\mathcal{C}))$, which is less than $C_{2\sqrt{|V|}}^{\sqrt{|V|}}$. Therefore, the time complexity of Algorithm 2 is $O(q \times 2\sqrt{|V|} \times C_{2\sqrt{|V|}}^{\sqrt{|V|}})$.

5   Performance Evaluation

We first compare the network performance of the load-balancing data preservation algorithm with the minimum cost flow data preservation algorithm in [9], which does not employ load-balancing technique. We then compare energy consumption of data preservation using maximum flow algorithm and minimum cost flow algorithm. Since it is shown in Sect. 3.2 that Edmonds-Karp algorithm costs less energy than Ford-Fulkerson does, we adopt Edmonds-Karp as the maximum flow algorithm in the comparison.

Comparing load-balancing data preservation and minimum cost flow-based data preservation. The sensor network is a 10×10 grid network. We randomly choose two nodes in the network as SNs. The storage capacity of each destination node is 10, and the initial energy of each node (including the source

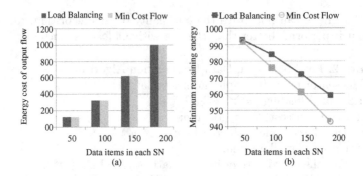

Fig. 4. Comparison between the load-balancing data preservation and the minimum cost flow-based data preservation.

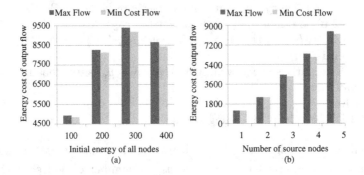

Fig. 5. Comparison between the Edmonds-Karp maximum flow algorithm and the minimum cost flow algorithm.

nodes) is 1000. Figure 4(a) shows that both algorithms cost the same amount of energy. However, Fig. 4(b) shows that with the increase of the number of data items at source nodes, the minimum remaining energy of the destination nodes given by load-balancing algorithm gets larger than that given by minimum cost flow algorithm. This indicates that data preservation would fail much later with load-balancing algorithm than without load-balancing.

Comparing maximum flow and minimum cost flow. Since minimum cost flow algorithm gives minimum energy consumption for data preservation, and Edmonds-Karp costs much less energy than Ford-Fulkerson does, we ask the following question: *How close does Edmonds-Karp perform w.r.t. minimum cost flow?* Below we compare their performances. We set the network size as 15×15 and randomly choose 5 nodes as SNs, each with 400 data items. The storage capacity of each destination node is 10. Figure 5(a) shows the energy consumption comparison by varying the initial energy level of each node. When initial energy equals 100, not all the data items can be offloaded. We thereby use Edmonds-Karp to first find the maximum amount of data items that can be offloaded by each SN, and use that information as input for minimum cost flow algorithm. It shows that the total energy consumption by Edmonds-Karp is larger than that of minimum cost flow. When initial energy is 200 and 300, all the data items can be offloaded from their SNs. When initial energy gets to 400, since each destination node has enough energy to either store or relay the data items, there are more shorter offloading paths available, therefore the total energy consumption decreases for both algorithms. In all cases, the total energy consumption by Edmonds-Karp is larger than that of minimum cost flow. Figure 5(b) uses the same parameters as in Fig. 5(a), except that now we fix the energy level of all the nodes as 200, and change the number of SNs. It shows that when there are only one or two SNs, Edmonds-Karp and minimum cost flow have similar performance. However, when the number of SNs increases, minimum cost flow costs less energy than Edmonds-Karp does. In all, minimum cost flow performs better than Edmonds-Karp in more stressed scenarios (i.e., more data items to offload).

6 Conclusion and Future Work

We study a generalized data preservation problem to preserve data inside sensor networks, considering that each node has limited battery power. We show that this problem can be modeled and solved as a minimum cost flow problem, which is solvable in polynomial time. By examining how different network flow algorithms can affect sensor network performance, we take a step further to view network flow problems from the perspective of efficient resource allocation, and study their applicability to sensor network scenarios. Our ongoing and future works are as follows. First, instead of a grid network, we will adopt a randomly generated sensor network for further study. Second, it would be interesting to prove if Edmonds-Karp costs the *minimum* amount of energy for data preservation, among all maximum flow algorithms. That is, when each edge has the same unit cost, is Edmonds-Karp a minimum cost maximum flow?[5] Third, we hope to study the problem using a more general energy model, wherein the energy consumption of sending data from one node to another depends on not only the size of the data but also the distance between nodes.

Acknowledgment. This work was supported in part by the NSF Grant CNS-1116849.

References

1. Vasilescu, I., Kotay, K., Rus, D., Dunbabin, M., Corke, P.: Data collection, storage, and retrieval with an underwater sensor network. In: Proceedings of SenSys 2005, pp. 154–165 (2005)
2. Li, S., Liu, Y., Li, X.: Capacity of large scale wireless networks under gaussian channel model. In: Proceedings of MOBICOM 2008, pp. 140–151 (2008)

---

[5] Note that minimum cost maximum flow problem is to find a maximum flow that has the minimum cost among all the maximum flows, considering that each edge has both a capacity and a cost.

3. Luo, L., Cao, Q., Huang, C., Wang, L., Abdelzaher, T., Stankovic, J.: Design, implementation, and evaluation of enviromic: a storage-centric audio sensor network. ACM Trans. Sens. Netw. 5(3), 1–35 (2009)
4. Werner-Allen, G., Lorincz, K., Johnson, J., Lees, J., Welsh, M.: Fidelity and yield in a volcano monitoring sensor network. In: Proceedings of OSDI 2006, pp. 381–396 (2006)
5. Martinez, K., Ong, R., Hart, J.: Glacsweb: a sensor network for hostile environments. In: Proceedings of SECON 2004, pp. 81–87 (2004)
6. Jain, S., Shah, R., Brunette, W., Borriello, G., Roy, S.: Exploiting mobility for energy efficient data collection in wireless sensor networks. MONET 11(3), 327–339 (2006)
7. Jea, D., Somasundara, A., Srivastava, M.B.: Multiple controlled mobile elements (data mules) for data collection in sensor networks. In: Prasanna, V.K., Iyengar, S.S., Spirakis, P.G., Welsh, M. (eds.) DCOSS 2005. LNCS, vol. 3560, pp. 244–257. Springer, Heidelberg (2005)
8. Mathioudakis, I., White, N.M., Harris, N.R.: Wireless sensor networks: applications utilizing satellite links. In: Proceedings of the IEEE 18th International Symposium on Personal, Indoor and Mobile Radio Communications (PIMRC 2007), pp. 1–5 (2007)
9. Tang, B., Jaggi, N., Wu, H., Kurkal, R.: Energy efficient data redistribution in sensor networks. ACM Trans. Sens. Netw. 9(2), 1–28 (2013)
10. Ahuja, R.K., Magnanti, T.L., Orlin, J.B.: Network Flows: Theory, Algorithms, and Applications. Prentice Hall, Englewood Cliffs (1993)
11. Hou, X., Sumpter, Z., Burson, L., Xue, X., Tang, B.: Maximizing data preservation in intermittently connected sensor networks. In: Proceedings of IEEE MASS 2012, pp. 448–452 (2012)
12. Patel, M.S., Venkatesan, R.C.: Energy efficient capacity constrained routing in wireless sensor networks. Int. J. Pervasive Comput. Commun. 2, 69–80 (2006)
13. Bodlaender, H.L., Tan, R.B., Dijk, T.C., Leeuwen, J.: Integer maximum flow in wireless sensor networks with energy constraint. In: Proceedings of the 11th Scandinavian Workshop on Algorithm Theory, SWAT 08, pp. 102–113 (2008)
14. Hong, B., Prasanna, V.K.: Maximum data gathering in networked sensor systems. Int. J. Distrib. Sens. Netw. 1, 57–80 (2005)
15. Xue, Y., Cui, Y., Nahrstedt, K.: Maximizing lifetime for data aggregation in wireless sensor networks. Mob. Netw. Appl. 10(6), 853–864 (2005)
16. Ghiasi, S., Srivastava, A., Yang, X., Sarrafzadeh, M.: Optimal energy aware clustering in sensor networks. Sensors 2(7), 258–269 (2002)
17. Xue, X., Hou, X., Tang, B., Bagai, R.: Data preservation in intermittently connected sensor networks with data priorities. In: Proceedings of IEEE SECON 2013, pp. 65–73 (2013)
18. Ha, R.W., Ho, P.H., Shen, X.S., Zhang, J.: Sleep scheduling for wireless sensor networks via network flow model. Comput. Commun. 29, 2469–2481
19. Papadimitriou, C., Yannakakis, M.: Optimization, approximation and complexity classes. J. Comput. Syst. Sci. 43, 425–440 (1991)
20. Heinzelman, W., Chandrakasan, A., Balakrishnan, H.: Energy-efficient communication protocol for wireless microsensor networks. In: Proceedings of HICSS (2000)
21. Goldberg, A.V.: An efficient implementation of a scaling minimum-cost flow algorithm. J. Algorithms 22(1), 1–29 (1997)
22. Goldberg, A.V.: Andrew Goldberg's network optimization library. http://www.avglab.com/andrew/soft.html

23. Hershberger, J., Maxel, M., Suri, S.: Finding the k shortest simple paths: a new algorithm and its implementation. ACM Trans. Algorithms **3**(4), 1–19 (2007)
24. Eppstein, D.: Finding the k shortest paths. SIAM J. Comput. **28**(2), 652–673 (1998)

SHERPA: An Air-Ground Wireless Network for Communicating Human and Robots to Improve the Rescuing Activities in Alpine Environments

Md Arafatur Rahman[1,2(✉)]

[1] Faculty of Computer Systems and Software Engineering,
University Malaysia Pahang, Gambang, Pahang, Malaysia
[2] Laboratorio Nazionale di Comunicazioni Multimediali (CNIT), Naples, Italy
arafatur.rahman@unina.it

Abstract. Robot-based rescue systems are envisioned now-a-days as a promising solution for saving human lives after the avalanche accidents in alpine environments. To this aim, a European project named "Smart collaboration between Humans and ground-aErial Robots for imProving rescuing activities in Alpine environments (SHERPA)" has been launched. Robots with smart sensors and mobility feature are needed for achieving the goal of this project, therefore, the SHERPA networks need to consider two degrees of freedom: one is throughput for transmitting realtime images and videos and another is range for mobility. In this paper, we design a wireless network infrastructure with the objective to communicate human and robots during the rescue mission in alpine environments. Firstly, we study about the network components, scenario and topology according to this environment. Then we design the network infrastructure for communicating among network components by taking account of the two degrees of freedom. Finally, the performance of the network is analyzed by means of numerical simulations. The simulation results reveal the effectiveness of the proposal.

Keywords: Air-ground robotic networks · Alpine scenarios · WiMAX

1 Introduction

Introducing robotic platforms in a rescue system is envisioned now-a-days as a promising solution for saving human lives after the avalanche accidents in alpine environments. With the popularity of winter tourism, the winter recreation activities has been increased rapidly. As a consequence, the number of avalanche accidents is significantly raised. According to the statistics provided by the Club Alpino Italiano, in 2010 about 6,000 persons were rescued in alpine accidents in Italy with more than 450 fatalities and about thirty thousand rescuers involved, and with a worrying increasing trend of those numbers [1]. In 2010

© Springer-Verlag Berlin Heidelberg 2015
M. Garcia Pineda et al. (Eds.): ADHOC-NOW Workshops 2014, LNCS 8629, pp. 290–302, 2015.
DOI: 10.1007/978-3-662-46338-3_23

the Swiss Air Rescue alone conducted more than ten thousand missions by helicopters in Switzerland with more than 2,200 people that were recovered in the mountains [2]. Conveying those numbers to a global scale immediately gives the significance of the problem and the relevance of the real-world scenario.

Many features of the real-world scenario in which robotic platforms could provide an added value to potentially save lives during the rescue mission. To this aim, a European project named "Smart collaboration between Humans and ground-aErial Robots for imProving rescuing activities in Alpine environments (SHERPA)" has been launched [3]. Robots with smart sensors and mobility feature are needed for achieving the goal of this project, therefore, the SHERPA networks need to consider two degrees of freedom: throughput and range. Since the robots need to transmit realtime pictures and videos about the targeted area, high throughput need to be assured. On the other hand, since the air-robots need to move their territory in terms of kilometers, the high range of the network also need to be considered.

In this paper, we design a wireless network infrastructure with the objective of enabling wireless communications among human and robots as envisioned by the SHERPA project. More in details, firstly, we study about the network components (i.e., air and ground robots, and human) network scenario (i.e., the place where the network components are resiting) and two-tier topology according to this environment. Then we design the network infrastructure for communicating among network components by taking account of the two degrees of freedom i.e., throughput and range. Finally, the performance of the network is analyzed by means of numerical simulations. The simulation results reveal the effectiveness of the proposal.

The rest of the paper is organized as follows. In Sect. 2, we provide the related work, while in Sect. 3, we describe the overview of the network. In Sect. 4, we discuss about the network requirements and in Sect. 5, we present the considered technology. In Sect. 6, we provide the performance evaluation and finally, in Sect. 7, we conclude the paper.

2   Related Work

There are several works that address the infrastructure design for wireless mesh, sensor and ad-hoc networks [4–12]. However, there is not enough literature for designing infrastructure in alpine environments specially for communicating between human and robots during the rescue mission, only [13] addresses this issue. In [4,5], the authors design infrastructure for wireless mesh networks. In [6], the design of a wireless communications network for advanced measuring infrastructure is proposed. In [7], the authors study the application of wireless sensor networks and infrastructure design for security and privacy purpose. In [10], an infrastructure is designed for remote environmental monitoring systems.

There are some works that consider RFID for designing infrastructure in these types of network. For example, in [14], an infrastructure to work with RFID embedded in a service oriented architecture is proposed; in [15], the authors present a RFID based secure infrastructure for intelligent building management service and In [16], the authors propose an integrated architecture featured by the optimized coexistence and cooperation between Wireless Sensor Network (WSN) and RFID infrastructures.

There are some other works that consider ZigBee for designing infrastructure in these networks. In [19,20], the authors design a monitoring and control system based on ZigBee wireless sensor network. In [21,22], the author study ZigBee based wireless infrastructure for reliability intra-vehicular communications. In [23], the authors design and analysis a robust broad-cast scheme for the safety related services of the vehicular networks. In [24], the authors evaluate the ZigBee standard specially for cyber-physical systems, which is a class of engineered systems that features the integration of computation, communications, and control.

These traditional technologies can not fulfil the requirements of the SHERPA networks because of the necessity of the two degrees of freedom in terms of high throughput and extended range. Therefore, in this paper we assess the feasibility of WiMAX technology for enabling wireless communications among human and robots in alpine environments, which is different than all the aforementioned works.

3   Overview of the Network

In this section, first, we discuss about the network components, then demonstrate the scenario where these components are utilized. Thirdly, we provide the topology of the network and finally present the network architecture.

3.1   Network Components

Several actors/robots, such as Small-scale Rotary-Wing Unmanned Aerial Vehicles (RW-UAVs), Ground rover (GR), Fixed-Wing UAVs (FW-UAVs) and RMAX rotary-wing UAVs are involving to cooperate a human rescuer for accomplish a common goal. These actors are called the network components, as shown in Fig. 1. They are working like a team and all the members of the team are briefly discussed in the following:

Human Rescuer: A human rescuer obtains information form the robotic platform and utilizes them to accomplish the team goals. Human rescuer is called "Busy Genius" in the team. He can also communicate to the other rescuers who are involving in the same mission. He may carry the SHERPA Box that contains: (i) main computational and communication hardwares to communicate with other SHERPA actors using WiFi, WiMAX, Xbee and GSM/UMTS networks; (ii) docking/rechargeable station for the small scale UAVs; (iii) storage for the rescuer.

Ground Rover: It can carry the SERPA Box. It has ability to reach wild areas and to overtake the big natural obstacles. It can also be used as a communication relay among the SHERPA actors. It plays the role of an "Intelligent Donkey" of the team.

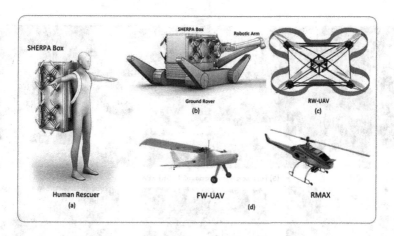

Fig. 1. The Components of the SHERPA Network.

RW-UAVs: It is characterized by limited autonomy and on-board intelligence but with incomparable capabilities in terms of capturing data (like visual information) from privileged or highly manoeuvrable positions, and following the rescuer in inaccessible areas. Its radius of action is necessarily confined in the neighborhood of the hosting ground rover due to the limited duration of the batteries. Employing multiple RW-UAVs, any tasks can be parallelized, boosting the efficiency of their mission. It plays the role of a "Trained Wasps" of the team.

FW-UAVs and RMAX: The FW-UAVs is characterized by matchless eagle-eyed capabilities that allow it to patrol large areas with a limited amount of energy. The RMAX has ability to carry remarkable payload. It can deliver the SHERPA Box and fly in the critical weather conditions. The high-altitude information captured by these vehicles enables optimization and coordination of the local activities of the team. Their radius of action is necessarily confined in the neighborhood of the rescuer. They play the role of a "Patrolling Hawks" of the team.

3.2 Scenarios

The scenario of the alpine environment is very hostile, as shown in Fig. 2. It is very difficult to move the actors during the rescue mission due to the obstacles, slopes and bad weather (wind, fog, rain). Initially the human reaches to the targeted palace along with the Ground Rover (GR), as depicted in Fig. 2(b). Then he starts the rescue mission. The RW-UAV is utilized for getting information of the small territory where the human can not reach, as demonstrated in Fig. 2(c). For the long distance area, FW-UAVs and RMAX are used, as shown in Fig. 2(d). GR is acting like a base station. The rescuer will get all the necessary information about the alpine environment through GR.

(a) Possible usage scenarios of the SHERPA team

(b) Rescuer and Ground Rover in the scenario

(c) RW-UAVs in the scenario

(d) FW-UAVs And RMAX in the scenario

Fig. 2. The Scenario of the SHERPA Network.

3.3   Topology

Network topology is a schematic description of the arrangement of a network, including its nodes and connecting lines. In this network, we proposed two-tier topology, which is the combination of mesh and star topology, as shown in Fig. 3. We briefly discuss about these tiers in the following:

The first tier topology focuses only the intra-team communications. There is a central entity (e.g., GR) to which all the actors are directly connected. Each actor is indirectly connected to others through the GR.

The second tier topology focuses only the inter-team communications. All the GRs are connected each others through mesh connectivity. An actor can communicate with other actors from differen team through GRs.

3.4   Network Architecture

Figure 4 depicts the architecture of the SHERPA network. It is consisting of two tiers, such as information acquisition and information distribution tier. In the first tier, all the actors are collecting information from the alpine environments and in the second tier, the information is distributed among the components through the network connectivity. A rescuer can utilize these aggregate information for achieving his final goal.

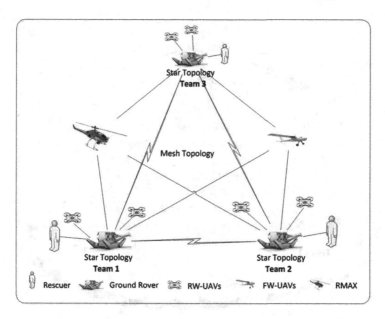

Fig. 3. The Two-tier Topology of the SHERPA Network.

4   Requirements

The most important issue to be addressed is the choice of the wireless network, through which the various actors interact with each others. Such a choice is challenging because the SHERPA network has to provide the two degrees of freedom, i.e., high throughput and range for high data rate and node mobility, respectively.

The key requirement for the proper functioning of the entire system is to be able to transmit real time pictures and video in high definition with the other members of the rescue team. It also requires that the network is established for their connection supports with a data rate that is the order of 1 Mbps, and the latency of the data is relatively low. Therefore, high throughput needs to be assured by the considered wireless technology.

Regarding the mobility, it is necessary to take into account that each actor is characterized by a highly varying speed of movement [17,18], in fact it goes from 30 m/s for FW-UAVs and RMAX, 10 m/s for RW-UAVs, 1.5 m/s the GR, and pedestrian movement for rescuer. The trajectories are also very different, for example the RMAX must be able to communicate with the other teams even it is placed at distances up to 5 km. Thus, it is an important requirement for the considered technology to provide high mobility features for the network components, and to allow for the maintenance of the connections even if the distances involved among the actors are in the order of 1 km. Furthermore, it is simple to note that, the different trajectories can be hardly predicted due to

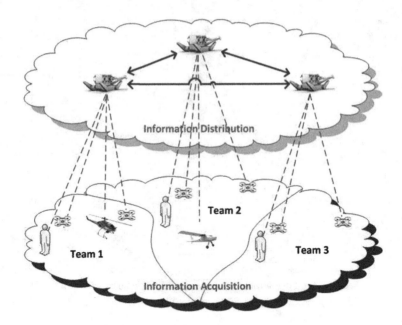

Fig. 4. The Two-tier Architecture of the SHERPA Network.

the movements of the actors. It is not also possible to increase the range of the various standards through the use of directive antennas.

Another interesting feature is the power consumption of the actors. Since the various actors being battery powered, the power available for the operations of data transmission/reception is limited. Therefore, the energy expenditure in terms of power should not be high.

Other requirements need to be taken into account when selecting the standard: *the operation band of the standard*, as the operation in the licensed band is a limit due to the verification of the availability of use and the associated cost; *the degree of diffusion of standard*, in fact, the recent standards are expensive and also lower reliability compared to the established standards.

5   Considered Technology

According to the network requirements, WiMAX standard is suitable for SHERPA network compared to the other wireless standards, such as WiFi, ZigBee, XBee and LTE, because of the following attractive features:

- *Flexibility*: it is able to support both Point-to-MultiPoint and mesh systems;
- *Safety*: it implements several techniques of encryption, authentication and security against intrusion;
- *Preparation for the management of quality of service (QoS)*: it uses different management methods depending on the types of traffic, and then characterized by specific needs;

- *High Throughput*: it ensures a high throughput using the modulation scheme defined by the IEEE 802.16 features, thanks to the good spectral efficiency of the signals;
- *Easy Installation*: it does not require special equipments for establishing the network;
- *Mobility*: it allows connections in mobile environments up to 120 km/h;
- *Cost*: low cost causes the rapid spread of this standard;
- *Coverage*: it has a capacity of very wide coverage, over 10 km. Since it is not possible that the line of sight is present in wide coverage area, the IEEE has developed and released version 802.16e that works for non line of sight communications. However, the performance is significantly reduced compared to line of sight communications.

6 Performance Evaluation

In order to evaluate the performance of the network, we have carried out couple of experiments through a discrete event simulation software, OPNET, with the relative packages for the WiMAX module. We have also adopted several metrics for measuring the performance, such as:

- Throughput: it is the total data traffic (packets/s) successfully delivered to the WiMAX MAC layer of the receiver and sent to the higher levels;
- Delay: the time spent by a packet to reach its destination (this metric accounts only for the data packets successfully received).

6.1 Experiment 1

In this sub-section, we evaluate the performance of the network by varying the number of actors/robots in the scenario. We also investigate how performance varies in presence of different service classes, such as Gold and Silver class. The adopted simulation set is defined as follows: the number of actors set is {10, 20, 30, 40, 50, 60, 70, 80}, the max and min traffic rate in Gold and Silver classes are 5 Mbps and 1 Mbps, and 1 Mbps and 0.5 Mbps, respectively, the latency is 30 ms, the speeds of the Human, RW-UAVs, FW-UAVs and RMAX are 1.5 m/s, 10 m/s, 30 m/s, respectively, the radius of the GR is 1 Km. Since GR is the base station of the network, we consider it as static for reducing the design complexity.

It is plotted the throughput and delay with the varying number of actors, as shown in Figs. 5 and 6. In terms of throughput, we note that in the both Gold and Silver classes the performance initially increases and then it reaches to the saturated point, and again performance declines. This is reasonable because more number of actors transmits more packets. However, after the saturation point the network becomes congested that causes declination of the performance. We also observe that the Silver class outperforms the Gold class, since the data load of the Gold class is higher than Silver class.

In terms of delay, the performance decreases with the increasing number of actors. This is because, the more number of actors needs to wait more time for

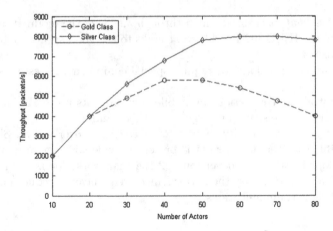

Fig. 5. Throughput vs varying number of Actors in Experiment 1.

utilizing the channel that causes delay. The Silver class again outperforms the Gold class because of the same reasoning in Fig. 5.

6.2   Experiment 2

In this sub-section, we evaluate the performance of the network by varying the radius of the Ground Rover. The adopted simulation set is defined as follows: the number of actors set is {40, 70}, the MAC service class is Gold, the max and min traffic rates in Gold class, are 5 Mbps and 1 Mbps, the latency is 30 ms, the speeds of the GR, Human, RW-UAVs, FW-UAVs and RMAX are 0 m/s,

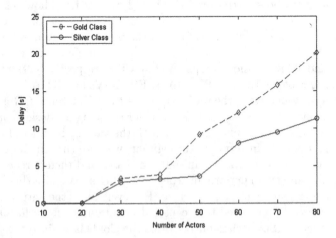

Fig. 6. Delay vs varying number of Actors in Experiment 1.

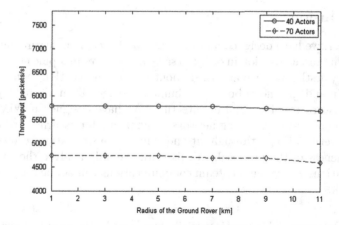

Fig. 7. Throughput vs varying Radius of the Ground Rover in Experiment 2.

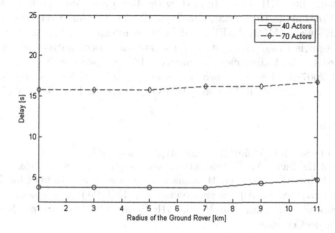

Fig. 8. Delay vs varying Radius of the Ground Rover in Experiment 2.

1.5 m/s, 10 m/s, 30 m/s, respectively, the radius of the GR set is {1, 3, 5, 7, 9, 11} km.

It is plotted the throughput and delay with the varying radius of Ground Rover, as shown in Figs. 7 and 8. In terms of throughput and delay, we note that the performance is slightly decreased when the radius is near about 10 km. This is because of the WiMAX standard, which is suitable for more than 10 km. We also observe that the less number of the actors performs better because of the same reasoning in experiment 1.

From the above discussions, we can conclude that the requirements of the network are satisfied by utilizing WiMAX standard. We will also investigate the performance by introducing the concept of cognitive radio in this scenario as a future work [25–30].

7 Conclusion

In this paper, we have designed an infrastructure for communicating human and robots during rescue mission in order to save human lives in alpine environments. To this end, firstly, we have analyzed about the communication requirements of different air and ground robots, and human rescuer. Then we have proposed two-tier network topology and architecture. We have designed a WiMAX network for communicating among network components for assuring two degree of freedom in terms of high throughput and range. The simulation results confirm the effectiveness of the proposal. In this work, we only analyze the single team communications whereas multi-team communications will be the future direction of this work.

Acknowledgments. This work is partially supported by the project "Smart collaboration between Humans and ground-aErial Robots for imProving rescuing activities in Alpine environments (SHERPA)" funded by the European Community under the 7th Framework Programme (01/02/2013 to 31/01/2017), "Mobile Continuos Connected Comprehensive Care (MC3CARE)", "DRIVEr monitoring: technologies, methodologies, and IN-vehicle INnovative systems for a safe and ecocompatible driving (DRIVE IN²)" founded by the Italian national program Piano Operativo Nazionale Ricerca e Competitivit 2007–2013 and the project,"Sviluppo di Tecniche di Comunicazione di Sistemi Embedded Distribuiti" founded by POR Campania FSE 2007/2013.

References

1. Website of the Club Alpino Italiano: http://www.cai.it/
2. Website of the Swiss Air Rescue: http://www.rega.ch/en/home.aspx
3. Smart collaboration between Humans and ground-aErial Robots for imProving rescuing activities in Alpine environments (SHERPA) is a funded project by the European Community under the 7th Framework Programme. http://www.sherpa-project.eu/sherpa/
4. Saleem, S., Johnson, T., Ramasubramanian, S.: Design of a self-forming, self-healing small-medium infrastructure wireless mesh network. In: Proceedings of 10th Annual IEEE Communications Society Conference on Sensor, Mesh and Ad Hoc Communications and Networks (SECON), pp. 252–254 (2013)
5. Weiyi, Z., Jiang, X.: ReLoAD: resilient location area design for internet-based infrastructure wireless mesh networks. In: Proceedings of IEEE Global Telecommunications Conference (GLOBECOM 2011), pp. 1–5 (2011)
6. Castellanos, R.E., Millan, P.: Design of a wireless communications network for advanced metering infrastructure in a utility in Colombia. In: Proceedings of IEEE Colombian Communications Conference (COLCOM), pp. 1–6 (2012)
7. Buttyan, L., Gessner, D., Hessler, A., Langendoerfer, P.: Application of wireless sensor networks in critical infrastructure protection: challenges and design options (Security and Privacy in Emerging Wireless Networks). In: Proceedings of IEEE Wireless Communications, pp. 44–49 (2010)
8. Spiegel, C., Viessmann, A., Burnic, A., Hessamian-Alinejad, A., Waadt, A., Bruck, G.H., Jung, P.: Platform based design of terminals and infrastructure components for cognitive wireless networks. In: Proceedings of IEEE 66th Vehicular Technology Conference, pp. 2065–2069 (2007)

9. Abu-Sharkh, O.M.F.: Cross-layer design for supporting infrastructure and ad-hoc modes integration in MIMO wireless networks. In: Proceedings of Wireless Advanced (WiAd), pp. 110–115 (2011)
10. Fan Yang, Y., Gondi, V., Hallstrom, J.O., Kuang-Ching, W., Eidson, G., Post, C.J.: Wireless infrastructure for remote environmental monitoring: deployment and evaluation. In: Proceedings of International Conference on Selected Topics in Mobile and Wireless Networking (MoWNeT), pp. 68–73 (2013)
11. Cacciapuoti, A.S., Caleffi, M., Paura, L.: A theoretical model for opportunistic routing in ad hoc networks. In: Proceedings of International Conference on Ultra Modern Telecommunications Workshops (ICUMT 2009), pp. 1–7 (2009)
12. Cacciapuoti, A.S., Caleffi, M., Paura, L.: Optimal constrained candidate selection for opportunistic routing. In: Proceedings of IEEE Global Telecommunications Conference (GLOBECOM 2010), pp. 1–5 (2010)
13. Marconi, L., Melchiorri, C., Beetz, M., Pangercic, D., Siegwart, R., Leutenegger, S., Carloni, R., Stramigioli, S., Bruyninckx, H., Doherty, P., Kleiner, A., Lippiello, V., Finzi, A., Siciliano, B., Sala, A., Tomatis, N.: The SHERPA project: Smart collaboration between humans and ground-aerial robots for improving rescuing activities in alpine environments. In: Proceedings of IEEE International Symposium on Safety, Security, and Rescue Robotics (SSRR), pp. 5–8 (2012)
14. Lopez, B., Melendez, J., Contreras, O., Bueth, D., Wissel, H., Haertle, M., Friederike, F.L., Grosser, O.S.: Location of medical equipment based on a maintenance service oriented infrastructure and RFID technology. In: Proceedings of European Workshop on Smart Objects: System, Technologies and Applications, pp. 1–8 (2010)
15. Byunggil, L., Howon, K.: Design and implementation of a secure IBS platform using RFID and sensor network. In: IEEE Tenth International Symposium on Consumer Electronics, pp. 1–4 (2006)
16. Pileggi, S.F., Palau, C.E., Esteve, M.: On the convergence between wireless Ssnsor network and RFID: industrial environment. In: Proceedings of 8th International Symposium on Modeling and Optimization in Mobile, Ad Hoc and Wireless Networks, pp. 430–436, (2010)
17. Cacciapuoti, A.S., Calabrese, F., Caleffi, M., Di Lorenzo, G., Paura, L.: Human-mobility enabled wireless networks for emergency communications during special events. Elsevier Pervasive Mob. Comput. **9**, 472–483 (2013)
18. Cacciapuoti, A.S., Calabrese, F., Caleffi, M., Di Lorenzo, G., Paura, L.: Human-mobility enabled networks in urban environments: is there any (mobile wireless) small world out there? Elsevier Ad Hoc Netw. **10**, 1520–1531 (2012)
19. Yiming, Z., Xianglong, Y., Xishan, G., Mingang, Z., Liren, W.: A design of green-house monitoring and control system based on ZigBee wireless sensor network. In: Proceedings International Conference on Wireless Communications, Networking and Mobile Computing, pp. 2563–2567 (2007)
20. Chen, L., Yang, S., Xi, Y.: Based on ZigBee wireless sensor network the monitoring system design for chemical production process toxic and harmful gas. In: Proceedings of International Conference on Computer, Mechatronics, Control and Electronic Engineering, pp. 425–428 (2010)
21. Rahman, M.A.: Reliability analysis of zigbee based intra-vehicle wireless sensor networks. In: Sikora, A., Berbineau, M., Vinel, A., Jonsson, M., Pirovano, A., Aguado, M. (eds.) Nets4Cars/Nets4Trains 2014. LNCS, vol. 8435, pp. 103–112. Springer, Heidelberg (2014)

22. Rahman, M.A.: Design of wireless sensor network for intra-vehicular communications. In: Proceedings of 12th International Conference on Wired and Wireless Internet Communications, pp. 1–13 (2014)
23. Ma, X., Zhang, J., Yin, X., Trivedi, K.S.: Design and analysis of a robust broadcast scheme for VANET safety-related services. IEEE Trans. Veh. Technol. **61**, 46–61 (2012)
24. Xia, F., Vinel, A., Gao, R., Wang, L., Qiu, T.: Evaluating IEEE 802.15.4 for cyber-physical systems. EURASIP J. Wirel. Commun. Netw. **2011**, 1–15 (2011)
25. Cacciapuoti, A.S., Caleffi, M., Paura, L., Savoia, R.: Decision maker approaches for cooperative spectrum sensing: participate or not participate in sensing? IEEE Trans. Wirel. Commun. **12**, 2445–2457 (2013)
26. Cacciapuoti, A.S., Caleffi, M., Paura, L.: Reactive routing for mobile cognitive radio ad hoc networks. Elsevier Ad Hoc Netw. **10**, 803–805 (2012)
27. Rahman, M.A., Caleffi, M., Paura, L.: Joint path and spectrum diversity in cognitive radio ad-hoc networks. EURASIP J. Wirel. Commun. Netw. **2012**(1), 1–9 (2012)
28. Cacciapuoti, A.S., Calcagno, C., Caleffi, M., Paura, L.: CAODV: routing in mobile Ad-hoc cognitive radio networks. In: Proceedings of IEEE IFIP Wireless Days, pp. 1–5 (2010)
29. Cacciapuoti, A.S., Caleffi, M., Paura, L.: Widely linear cooperative spectrum sensing for cognitive radio networks. In: Proceedings of IEEE Global Telecommunications Conference (GLOBECOM 2010), pp. 1–5 (2010)
30. Cacciapuoti, A.S., Caleffi, M., Izzo, D., Paura, L.: Cooperative spectrum sensing techniques with temporal dispersive reporting channels. IEEE Trans. Wirel. Commun. **10**, 3392–3402 (2011)

Design and Implementation of the Vehicular Network Testbed Using Wireless Sensors

Jovan Radak[✉], Bertrand Ducourthial, Véronique Cherfaoui,
and Stéphane Bonnet

UMR CNRS 7253 Heudiasyc, Université de Technologie de Compiègne, BP 20529,
60205 Compiègne Cedex, France
{jovan.radak,bertrand.ducourthial,veronique.cherfaoui,
stephane.bonnet}@hds.utc.fr

Abstract. Testbeds are indispensable tools in research and development process in wireless networks technologies. They show us how our solution is going to work in a real environment. In the recent years there is a growing trend in the development of testbeds aimed to be used as tools for both research and verification of the results obtained theoretically and using simulators. We are presenting an experimental vehicular network testbed based on cheap, off-the-shelf wireless sensors that are gathering environmental data, temperature, humidity and luminosity. These sensors are connected to road-side units (RSUs) running the Linux operating system and dedicated software distribution, Airplug. This complete system (wireless sensors, RSUs and Airplug software distribution) can be used for simulation, emulation and experiments in vehicular networks but also for any other type of wireless network. We use this system to gather environmental data and then reuse collected data in different emulation and experimental scenarios. We are showing the usefulness of our wireless sensors testbed and possible scenarios of its usage in emulation and real experiments.

Keywords: Wireless sensors · Roadside units · Testbed · Emulation · Experiments · Airplug · Airbox

1 Introduction

Practical implementation is one of the final steps in research and development for the novel solutions in wireless ad-hoc networks (and in general any research and development process). This step is performed on a real hardware platform and it is usually the most critical one. The problem lies in the complexity of the hardware platforms that cannot be completely taken into account in the process of modeling and simulating the wireless ad-hoc networks. Thus, a dedicated hardware platform should be used for implementation and testing of the solution previously developed for some specific problem.

Different kinds of high-quality network simulators (like ns2, ns3, Omnet++ – just to name a few) are currently in use as primary tools for the development and

© Springer-Verlag Berlin Heidelberg 2015
M. Garcia Pineda et al. (Eds.): ADHOC-NOW Workshops 2014, LNCS 8629, pp. 303–316, 2015.
DOI: 10.1007/978-3-662-46338-3_24

testing of solutions for wireless ad-hoc networks. Network simulators have a large user base who develop different kinds of libraries suiting specific problems they are addressing. It is relatively easy to find a library for the specific problem that we want to tackle (for example tens of different energy efficient MAC layers for wireless sensor networks) and to run the simulation for our targeted instance (for example 50 fixed sensor nodes with 20 mobile nodes having moving patterns that correspond to the movement of the buyer in the supermarket). But, the main problem remains in the limitations of the models used in different simulators (their accuracy and completeness) and the lack of compatibility between imposed models, hardware platforms and real scenarios that are planned to be used in implementation.

Emulation using network simulators [12] or using a dedicated platform [13] takes the middle ground between simulation and experiments. It usually includes a mix of the real hardware or data retrieved from the real hardware and simulation. In this way emulation can be viewed either as an enhanced simulation or as some kind of experiment with a limited number of hardware elements. Indeed this is an approximation of the real experiment that guarantees the flexibility of the simulation with more realistic data but at the cost of limited hardware usage. This approach is advocated because of its practical usefulness and flexibility in rapid application development and testing. However, the main problem, previously mentioned for the network simulators, remains – limitations of the used models and gap between it and the dedicated platform planned for implementation.

We are presenting our testbed for vehicular ad-hoc networks. This testbed is developed to be used in conjunction with the previously developed tools for simulation and testing of dynamic wireless networks. We are using cheap, off-the-shelf, wireless sensors that are gathering environmental data – temperature, humidity and light. Our solution is based on the wireless Xbee sensors[1] (produced by Digi international), Airbox units, dedicated hardware based on the IGEP[2] developments boards and the Airplug software distribution, developed in our laboratory. Airplug is a modular and flexible software platform developed for the simulation, emulation and testing of dynamic wireless networks (and more generally distributed systems) that can be used on any Linux-compatible platform. Thus, our testbed is not limited to this specific embedded architecture (like Airbox), it can be used on any Linux-compatible embedded platform or desktop computer that fulfills a small subset of requirements. Airplug also allows us to reuse previously gathered data in such a way that we can either emulate an entire experiment on a single computer or recreate experiments using dedicated hardware.

The solution that we are proposing in this article is specific in four aspects:

- **Fully developed solution** – it consists of hardware and software elements that can be used as-is without any modifications. Currently we are using environmental wireless sensors that measure temperature, humidity and luminosity.

---

[1] http://www.digi.com/products/wireless-modems-peripherals/wireless-range-extenders-peripherals/xbee-sensors.

[2] https://www.isee.biz/products/igep-processor-boards.

- **Mobile platform** – our current testbed is fixed and developed close to our laboratory but using our guidelines, hardware and software distribution it can be developed on any site – indoors or outdoors, using different hardware platforms.
- **Modular platform** – our platform is modular in both physical implementation and program support. Additional hardware elements can be easily added and they do not depend on the manufacturer as long as they comply with general communication standards used in our platform. The Airplug platform is compatible to any Linux system, so any hardware platform (Linux compatible) can be used to extend current physical implementation. Also, additional program support can be developed using development guidelines for our software platform.
- **Flexible platform** – this platform is developed to be used in experiments on vehicular networks, but it is flexible and can be used in other types of experiments with devices like UAVs (unmanned air vehicles) or robots as long as the communication modules of these devices are compatible to those used in our platform.

In this article we are going to present our solution for vehicular network testbed emphasizing the usage of wireless sensors in it. We will also present program support for these wireless sensors and possible applications of the testbed equipped with environmental sensors. The rest of this article is organized as follows: Sect. 2 presents the work related to our solution, overview of testbeds used for wireless ad-hoc and sensor networks and the approaches in network emulation, Sect. 3 presents the general hardware architecture of our testbed, while Sect. 4 presents the Airplug software distribution and specific application support for the use of wireless Xbee sensors. Section 5 provides example of the data gathered using our testbed and possible scenarios of the use of our solution. Finally, Sect. 6 presents our conclusion of this article and ideas for the future work.

2   Related Work

In recent years we have seen great effort to develop experimental infrastructures for wireless ad-hoc networks and the Internet of things paradigm that will allow researchers to test their solutions in a real world environment. Some of these platforms are developed for specific applications while the others are more generic and can be used for a wider range of problems.

SmartSantander is a platform developed for research and experimentation on wireless sensor network and Internet of things in urban environment. It is a city-scale test site consisting of 20000 sensor nodes developed in 4 cities across the Europe: Belgrade, Guilford, Lübeck and Santander [15]. It is mainly viewed as a platform for experiments with the Internet of things services and infrastructures in the smart city. So far it is used for various experiments including the streaming of acoustic data [14]. GreenOrbs in China is the test site deployed using wireless sensor nodes and mainly aimed at forestry applications [1]. It has evolved into a dedicated environment monitoring system for real-time CO_2 management in the city [11].

Wisebed [10] proposes a different approach to the testbed infrastructure. It is a joint project of 9 different institutions in Europe with the testbed consisting of 750 sensor nodes (with different types of sensors, both static and mobile) organized in federation architecture. This platform also supports virtualization and co-simulation with part of the testbed. OneLab [8] is also federation of platforms but it aims to help both developers and users of the platform. It is open to third-party platforms allowing developers to promote their testbed but also users to chose among a variety of testbeds that best fits their testing platform.

Senslab is a large experimentation testbed deployed in 4 cities in France – Grenoble, Lille, Rennes and Strasbourg [7]. It consists of 1024 sensor nodes (WSN430 nodes developed specially for this project) with 2 types of radio chips and several environmental sensors. This platform's main purpose is experiments on wireless sensor networks both at high and low level of abstraction [4]. Additional tools are also developed: WSim (simulating sensor nodes) and WSNet (wireless sensor network simulator) simulators and ports to different kinds of real time operating systems (FreeRTOS, TinyOS, Contiki). Currently, there is an ongoing upgrade of the Senslab platform, called IoT Lab[3]. This platform promotes a new approach in the architecture of the testbed which is basically a service allowing users hands-on experience with real platforms. Wisebed, Smart-Santander and Senslab are parts of this platform allowing users to perform multidisciplinary research in the area of wireless networks and Internet of Things.

Emulation is widely used as a term that explains uses of both hardware and software in the process of evaluation and experimentation. The term emulation covers different approaches in testing and evaluation and it goes from pure simulation with the enhanced models [9] to the usage of smaller hardware platforms to replicate results of large scale networks [3,13].

In our laboratory we have previously developed the Airplug software distribution. This platform is the complete system that can be run several different modes including simulation, emulation and experiment mode. The user defines the parameters of Airplug usage on the start of application giving details about the scenario (type of mobility or fixed position, range, type of communication..., number of nodes as well as the mode of usage).

The solution that we are proposing in this article is enhanced comparing to the current testbed solutions in three aspects:

- Hardware and software co-development – both physical architecture and program support are developed simultaneously i.e. if we decide to use new type of wireless sensors that are measuring different set of values then we are also developing appropriate software support for that piece of hardware.
- Mobility – while the other testbed solutions also propose certain kind of mobility to the nodes (or subset of nodes) used in testbed our solution is unique in the sense that each part of the hardware platform can be easily relocated

---

[3] http://www.iotlab.eu/.

according to our needs in the experiment, also the same type of dedicated hardware (Airbox) can be used as the part of in-vehicle hardware platform making it fully mobile. Although we envisaged this testbed for vehicular network testing and development, it's concept can be used for different types of mobile agents – robots or unmanned air vehicles – and it can be easily transformed (using same software platform) in the testbed for other specific purposes (obstacle avoidance, object tracking, ...).

– Modularity – additional software elements can be easily added and they do not depend on the type of the elements used in physical implementation as long as they comply with general communication standards used in our platform. This means that if we need to test for example multi-hop routing algorithm we can develop application that implements that specific algorithm and to deploy additional number of Airbox units that will allow us to have appropriate test results.

3 Wireless Sensors Testbed

Architecture of our testbed is simple, it consists of wireless Xbee sensors and Airbox devices (dedicated embedded hardware). In the terminology of the vehicular networks Airboxes are road-side units – RSUs (we will use these two terms interchangeably throughout this paper). In the creation of the wireless sensor testbed our leading ideas were *simplicity* of usage, *modularity* and *mobility* of the given solution. Following our main concepts we have decided to use simple wireless sensors that do not communicate between themselves but only the device that is hierarchically above them in this case with the road-side units. To avoid possible data collisions we have decided to use different wireless standards for communication with sensors and for communication between RSUs. Figure 1 presents conceptual scheme of our testbed.

Wireless sensors are communicating only to the RSUs that are in their neighborhood using dedicated communication link (802.15.4 standard). All road-side units are running Airplug software distribution (more details about it in the Sect. 4) that has application dedicated to communication with Xbee sensors – given as XBE block in Fig. 1. Each RSU can serve one or more Xbee sensors. RSUs communicate between themselves using wifi communication link (802.11).

Configuration, that we are showing in Fig. 1 is fixed. This means that each Airbox used in this configuration is actually a road-side unit. Airboxes that we are using inside of the vehicle cannot communicate with wireless sensors (they are not equipped with Xbee modem), they are meant to gather data from vehicles using other interfaces, namely using CAN bus.

3.1 Wireless Sensors

We are using cheap off-the-shelf wireless Xbee sensors equipped with temperature, humidity and luminosity sensors. They are using Xbee modem based on the 802.15.4 protocol to communicate with other devices. These sensors are not

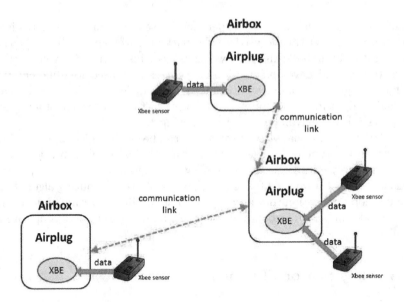

Fig. 1. Testbed with Xbee sensors, road-side units (Airboxes) and software support from Airplug software distribution

able to exchange messages between themselves, only to the device that is hierarchically above them (Fig. 2).

Usage of 802.15.4 protocol modems, so called series 1 Xbee modems, guarantee simple setup and communication between the sensors and other devices. Xbee modem is physically attached to the Airbox unit's dedicated slot. It communicates with the Airbox device using its (and device's) serial port. There are two possibilities for data gathering from Xbee sensors:

– Polling – Xbee modules are configured in such a way that they initiate periodical reading of the sensors and transfer of the retrieved data.

Fig. 2. Xbee sensor (with Xbee modem visible and enclosed)

Fig. 3. Airbox – platform based on IGEP development boards, Xbee modem with external antenna is shown attached to the dedicated slot (1 euro coin is used for the comparison of sizes)

– Upon request – RSUs are initiating reading of the sensors sending the specific read request packet to Xbee sensors; for this to be available all Xbee modems must have API mode enabled.

3.2 Road-Side Units

Road-side units are dedicated embedded hardware units, called Airbox. They are based on the IGEP platform. These units are running Linux operating system and Airplug software distribution. RSUs are equipped with Xbee modems that allow them to communicate with wireless Xbee sensors. They are also having wifi communication module that allows them to communicate between themselves and to form ad-hoc network (Fig. 3).

RSUs and Xbee modems are configured to communicate using API mode of Xbee modems. This means that they exchange dedicated packets that follow the rules imposed by their manufacturer (Digi international).[4]

3.3 Placement of the Testbed

We have chosen to deploy our testbed near to the Research Center of the Technological University in Compiégne. Three road-side units are deployed. Configuration of the RSUs is shown on the Fig. 4. RSU1 is deployed close to the garage and has continuous power supply making it available all the time. RSU2 and RSU3 are

---

[4] http://www.digi.com/support/kbase/kbaseresultdetl?id=3215.

Fig. 4. Geographical position of the road-side units, RSU3 is shown in the upper left corner encased with an energy harvesting module

deployed in the parking space, closer to the street. In this way they can exchange messages with the vehicles equipped with Airbox units. These two RSUs are using batteries as a power supply, their autonomy is roughly around 20 h.

Xbee wireless sensors (not shown on the Fig. 4) are deployed in the proximity of each road-side unit. RSU2 and RSU3 have one Xbee sensor in their proximity while RSU1 has 2 Xbee sensor, one placed inside of the garage and the other one positioned outside of the garage. In this way it is possible to obtain different readings and to use this data in other applications.

4   Program Support for Wireless Sensors Testbed

Important part of our testbed is its program support, it is based on the Airplug software platform[5]. To incorporate wireless Xbee in our testbed and in the Airplug software distribution sensors we have implemented dedicated, Airplug compatible, application that implements all the functionalities of the wireless sensors. In this section we will give brief overview of Airplug software distribution and its functionalities and the support for the Xbee sensors, more details about design philosophy and Airplug's architecture can be found out in previous publications [5].

---

[5] http://www.hds.utc.fr/airplug/.

4.1   Airplug Software Distribution

Airplug software distribution actually presents practical implementation of the Airplug framework. This framework is developed to support dynamic wireless network and it is based on the few simple guidelines:

- Independence of the programming language implementation – to follow this guideline framework is using standard input/output channels (each programming language can read/write from input/output channels).
- Portable message format – we are using ASCII text messages, binary messages, if needed, are encoded; the only limitation is that the field delimiter (for the different parts of the message) cannot be used as the part of the field.
- Simple addressing scheme – Airplug compatible applications are uniquely addressed with a pair (application_name, host_name); communication between nodes rely mainly on the broadcast in the neighborhood thus three keywords are reserved for the host: **AIR** when we broadcast to all neighboring nodes, **LCH** when we broadcast to all local applications (run by the localhost) and **ALL** includes both local and remote broadcast.

Airplug software distribution presents several implementations that are also called *modes*. These *modes* are standalone implementations and they are complementing each other, i.e. while the one is dedicated for the laboratory studies, the other is used in real experiments.

Terminal Mode is also called airplug-term, it is based on the implementation of the Airplug framework in UNIX compatible terminal. This mode is dedicated for the rapid application development and prototyping and it gives many functionalities thanks to the wide range of libraries. These libraries are dedicated to Tcl/Tk programming language but they can be also programmed in any other programming language as long as they comply with message format and addressing scheme.

Emulation Mode is also called airplug-emu, it is a *network emulator* with the upper layers (of the communication stack) same as the ones in the real experiments while lower layers (physical communication) are reproduced – simulated [2]. Emulation scenarios are described using XML files with the possibility to detail each node's mobility and the applications that are running on each of the node. This mode can be extended using *remote mode* (airplug-rmt) that allows some applications to use remote execution connecting them via sockets to a dedicated application called RMT (an application that relays messages between computers).

Live Mode also called airplug-live, is the implementation dedicated for the real experiments. It is efficient implementation written in C programming language and it manages both local and inter-node communication while running on the top of POSIX compatible operating system [6]. This implementation actually represents middleware between Airplug compatible applications and network interfaces.

4.2 Xbee Sensors Connectivity

Program support for the wireless Xbee sensors is Airplug compatible application – called XBE – written, like the most Aiplug libraries, in Tcl/Tk. This application (XBE) is compatible with all Airplug software distribution modes, meaning that it can be run both on the PC running Linux terminal, but also on the embedded devices used in experiments, running airplug-live.

XBE establishes connection between Xbee modem and serial port of the device running the XBE application, using either predefined configuration settings for the serial port or the settings that user gives upon the start of XBE application. It communicates with Xbee sensors sending the request for read packages either periodically or upon the users request (on the graphic user interface – GUI). Communication with other applications is periodic and the messages containing sensors readings, unique ID and time-stamp of the reading of each sensor are sent with the period that user defines on the start of application. Sending of message is also possible on the request but only in the terminal mode when the GUI is present. Read data can be logged into the file using Airplug saving facilities.

This application has ability to read the log files previously created with this application. We have implemented this with an intention to replicate the experiments in emulation mode using the real data previously gathered from Xbee sensors.

5   Preliminary Results and Experimentation Scenarios

In this section we will give brief overview of the usage of our testbed and explain possible scenarios of usage for the data collected using it. Testbed is used for environmental data gathering that should be used in the emulation of new protocols that we are developing. Four different scenarios of testbed and data usage are given along with an explanation on the specificity of each scenario and its possible application.

5.1   Environmental Data Gathering

As a part of our work on the distributed data fusion we were evaluating algorithms using generated environmental data. Our idea was to develop and use our own testbed equipped with sensors gathering environmental data. In the Fig. 5 we are showing the example of the data gathered by our testbed. More precisely, two wireless Xbee sensor were attached to the one road-side unit (RSU1 in the Fig. 4) one being inside of the building and the other one outside. On these graphs (Fig. 5) we are showing mean values of the data gathered (RSU has requested data from Xbee sensors each minute) in the night between 6 and 7 February.

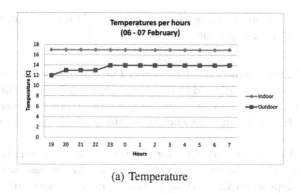

(a) Temperature

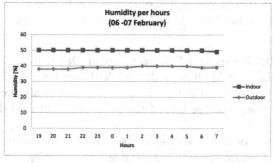

(b) Humidity

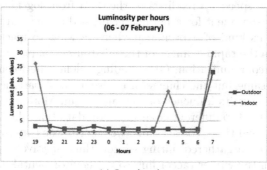

(c) Luminosity

Fig. 5. Environmental data gathered during 12 h period (from 19 h - 07 h) in the night between 6th and 7th February using Airplug software RSU unit and Xbee sensor

5.2 Different Scenarios for Testbed and Gathered Data Usage

Gathered data along with the possibilities of our testbed and different Airplug modes give us four possibilities for the experiments with environmental data. We must stress in here that only two of four possibilities are real experiments (in the sense that they are using all deployed hardware) but the idea is that they all use real data, either previously gathered or collected on the fly from the sensors running at the time of experiment.

Real Time Experiment with Collected Data. The idea is to use our test-bed hardware with data gathered in the real time. Each RSU is running Airplug-live mode with XBE application while being connected to Xbee sensor(s). Data is gathered in the real time, depending on the parameters set for each XBE application.

This kind of experiment is good for the verification of communication algorithms that depend on the environmental data but in the case when algorithm itself does not depend on the varying of environmental parameters. This is due to the change of the environmental data that is rather slow and cannot be easily observed in an experiment that lasts for a short period of time.

Experiment with Emulated Environmental Data. Real hardware is used with RSUs running Airplug-live mode and XBE application. The only difference to the previous solution is that XBE application is not gathering data in real time, it is reading the log files of the previously collected data with XBE application.

In this kind of experiment we can choose the speed of reading from the file. Knowing that we have delays between the readings in the log file we can produce the same kind of an experiment like the one using real data. In this mode we can also choose to speed up readings of the data from log file thus effectively speeding up the experiment for a parameter defined at the start of experiment.

We can use this kind of experiment when we want to track how the algorithms respond to the rapid changes of the environmental parameters. All other parameters (V2I communication, GPS readings, delays between transmissions, etc.) remain the same but the change of environmental parameters is rapid so we can better observe its influence on the algorithm. Moreover, using this scenario we can easily setup different environmental data than the ones that we are able to retrieve in real time. For example, we can run the experiment during the warm days using data gathered during winter, thus effectively we will have the experiment run in the winter (according to the environmental data).

Emulation Using Real Data. Idea behind this scenario is to use the data gathered in the real time by one or several RSUs, each of them having Xbee sensors and XBE applications running. The data is then used either in one computer running the appropriate emulation scenario with airplug-emu mode or it can be run on multiple computers using airplug-rmt mode.

For example, we can setup emulation scenario using several vehicles moving using map traces and fix several RSUs in the scenario. Some of these RSUs can be the ones that are gathering data in real time and some of them running XBE application with data gathered previously, or we can avoid using XBE applications at all on some of the RSUs (if the testing does not have use of them) and run other Airplug compatible applications that are implementing tested algorithms.

In this kind of scenario we are mixing different kind of applications and sources of data while communication parameters are emulated in the Airplug-emu mode. These kinds of scenarios are good for longer experiments in which we have repetitive change of some parameters (movements of vehicles, communication between RSUs or V2I communication) but we want to see how different environmental data may influence execution in this specific case.

Emulation Using Gathered Data. This scenario can also be called pure emulation. In this kind of experiment we do not depend on the hardware that we are using, everything can be executed using only one PC, running the airplug-emu and appropriate emulation scenario or using multiple PCs with aiplug-rmt.

This is multipurpose experiment, since we can easily change source of the data, at which rate it is being read and communication parameters. It is best suited for the rapid development of the applications and as the first step after the simulations ran on some of generic simulators (ns2, omnet++, etc.).

6   Conclusion

In this paper we have explained our solution for the usage of wireless sensors and dedicated hardware to build testbed for vehicular ad-hoc networks. While it is the truth that our current solution has only three fix nodes it can be viewed as the proof of concept of our idea of wireless sensors testbed specially suited for the urban areas and vehicular networks. Flexibility of our hardware and software platform permits us to easily deploy this kind of testbed that can contain significantly higher number of nodes (measured in tens or hundreds). This architecture can also be used with different wireless sensors, with the only limitation that they have to use same type of communication module as our dedicated Airbox units. Development and testing of different kinds of algorithms, if not already a part of the Airplug platform, is based on the development of the dedicated applications that handle data provided from other Airplug applications.

We see great potential in the future usage of our testbed architecture. We are already using it for the gathering of environmental data that we plan to use for thorough testing of previously developed algorithms. We are also planning to use this platform to solve different problems in vehicular networks (distributed data fusion using information gathered from sensors, data propagation, correlation of gathered spatial data, etc.). While primarily developed as a tool for vehicular network testing this same platform can be used for the experiments with UAVs (unmanned air vehicles) and for robot networks.

Acknowledgments. This work was carried out and funded in the framework of the Labex MS2T. It was supported by the French Government, through the program "Investments for the future?" managed by the National Agency for Research (Reference ANR-11-IDEX-0004-02).

References

1. Bo, C., Ren, D., Tang, S., Li, X.-Y., Mao, X.F., Huang, Q., Mo, L., Jiang, Z., Sun, Y., Liu, Y.: Locating sensors in the forest: a case study in greenorbs. In: INFOCOM, 2012 Proceedings IEEE, pp. 1026–1034 (2012)
2. Buisset, A., Ducourthial, B., El Ali, F., Khalfallah, S.: Vehicular networks emulation. In: ICCCN, pp. 1–7 (2010)
3. Coulson, G., Porter, B., Chatzigiannakis, I., Koninis, C., Fischer, S., Pfisterer, D., Bimschas, D., Braun, T., Hurni, P., Anwander, M., Wagenknecht, G., Fekete, S.P., Kröller, A., Baumgartner, T.: Flexible experimentation in wireless sensor networks. Commun. ACM 55(1), 82–90 (2012)
4. des Roziers, C.B., Chelius, G., Ducrocq, T., Fleury, E., Fraboulet, A., Gallais, A., Mitton, N., Noel, T., Valentin, E., Vandaele, J.: Two demos using SensLAB: very large scale open WSN testbed. In: Distributed Computing in Sensor Systems and Workshops (DCOSS), 2011 International Conference on, pp. 1–2 (2011)
5. Ducourthial, B.: Designing applications in dynamic networks: the Airplug software distribution. In: ASCoMS@SAFECOMP (2013)
6. Ducourthial, B., Khalfallah, S.: A platform for road experiments. In: VTC Spring (2009)
7. Ducrocq, T., Vandaele, J., Mitton, N., Simplot-Ryl, D.: Large scale geolocalization and routing experimentation with the SensLAB testbed. In: Mobile Adhoc and Sensor Systems (MASS), 2010 IEEE 7th International Conference on, pp. 751–753 (2010)
8. Fdida, S., Friedman, T., MacKeith, S.: OneLab: developing future Internet testbeds. In: Di Nitto, E., Yahyapour, R. (eds.) ServiceWave 2010. LNCS, vol. 6481, pp. 199–200. Springer, Heidelberg (2010)
9. Girod, L., Stathopoulos, T., Ramanathan, N., Elson, J., Estrin, D., Osterweil, E., Schoellhammer, T.: A system for simulation, emulation, and deployment of heterogeneous sensor networks. In: Proceedings of the 2nd International Conference on Embedded Networked Sensor Systems, SenSys 2004, pp. 201–213. ACM, New York (2004)
10. Hellbruck, H., Pagel, M., Kroller, A., Bimschas, D., Pfisterer, D., Fischer, S.: Using and operating wireless sensor network testbeds with wisebed. In: Ad Hoc Networking Workshop (Med-Hoc-Net), 2011 The 10th IFIP Annual Mediterranean, pp. 171–178 (2011)
11. Liu, Y., Mao, X., He, Y., Liu, K., Gong, W., Wang, J.: Citysee: not only a wireless sensor network. IEEE Netw. 27(5), 42–47 (2013)
12. Mahrenholz, D., Ivanov, S.: Real-time network emulation with NS-2. In: Distributed Simulation and Real-Time Applications, 2004, DS-RT 2004, Eighth IEEE International Symposium on, pp. 29–36, Oct 2004
13. Pavkovic, B., Radak, J., Mitton, N., Rousseau, F., Stojmenovic, I.: From real neighbors to imaginary destination: emulation of large scale wireless sensor networks. In: ADHOC-NOW, pp. 459–471 (2012)
14. Pham, C., Cousin, P.: Streaming the sound of smart cities: experimentations on the SmartSantander test-bed. In: Green Computing and Communications (GreenCom), 2013 IEEE and Internet of Things (iThings/CPSCom), IEEE International Conference on and IEEE Cyber, Physical and Social Computing, pp. 611–618 (2013)
15. Sanchez, L., Galache, J.A., Gutierrez, V., Hernandez, J.M., Bernat, J., Gluhak, A., Garcia, T.: SmartSantander: the meeting point between future internet research and experimentation and the smart cities. In: Future Network Mobile Summit (FutureNetw), 2011, pp. 1–8 (2011)

Author Index

Printed in the United States
By Bookmasters